Gene Therapy

Mauro Giacca

Gene Therapy

 Springer

Mauro Giacca
International Centre for Genetic
Engineering and Biotechnology (ICGEB)
Trieste
Italy

ISBN 978-88-470-1642-2   e-ISBN 978-88-470-1643-9

DOI 10.1007/978-88-470-1643-9

Springer Dordrecht Heidelberg London Milan New York

Library of Congress Control Number: 2010922812

9 8 7 6 5 4 3 2 1

Cover design: Ikona S.r.l., Milan, Italy

Typesetting: Ikona S.r.l., Milan, Italy
Printing and binding: Grafiche Porpora, Segrate (MI), Italy
Printed in Italy

Springer-Verlag Italia S.r.l. – Via Decembrio 28 – I-20137 Milan
Springer is a part of Springer Science+Business Media (www.springer.com)

*This book is dedicated to
Serena, Massimo and Giovanna,
for making everything possible.*

Preface

I entered the gene therapy field in the mid-1990s, being fascinated by the immense potential of genes as drugs for the treatment of human disease. Since then, I have experienced the ups and downs of this discipline, and tried to contribute with my work and that of my laboratory to the development of innovative approaches to the treatment of cardiovascular disorders. During these years, I have had several opportunities to speak on gene therapy at lectures and academic lessons, and have often noticed that the field is very attractive to scientists of all disciplines. However, as yet no comprehensive book on the subject has been published. Indeed, most books in the field are either a collection of gene transfer laboratory protocols or deal with the subject in a rather superficial manner. Hence the idea to write a gene therapy textbook that is broad and comprehensive, but at the same time provides sufficient molecular and clinical detail to be of interest to students, professors, and specialists in the various disciplines that contribute to gene therapy. I have tried to keep the language plain and, whenever possible, non-technical. Since the book is intended to be a textbook in the field of gene therapy in both the basic science and clinical areas, whenever technical descriptions are required, they are provided. For example, clinical readers might find it difficult to understand the principles of viral vector design without knowing some molecular details on viral genome organization and virus life cycle, and basic scientists might be unaware of the clinical and therapeutic advancements for the various disorders discussed in the book. In both cases, basic explanations are provided.

The book starts by providing a broad overview of the field of gene therapy (Chapter 1), and then moves on with a description of the gene therapy "tools", namely the nucleic acids with curative potential (Chapter 2) and the methods for their delivery into patients' cells (Chapter 3). Chapter 4 contains an extensive description of the clinical conditions so far addressed by gene therapy clinical trials, describing the successes and frustrations over the last 20 years of study. In this respect, I wish to point out that the purpose of the book is to be truly a gene "therapy" book, namely to only consider gene transfer approaches that have already proceeded to the clinic or are close to doing so. Basic research on gene transfer or nucleic acid modification is only discussed when close to clinical application. Finally, the social and ethical problems related to the use of genes to combat disease and to the possibility of engineering human genetic material are the subject of Chapter 5. Along with the problem of enhancing safety and minimizing health risk in gene therapy clinical applications, this chapter also deals with issues such as gene transfer into germ cells, *in utero* gene therapy, and the possibility of exploiting gene therapy for non-disease conditions, such as in the case of gene doping.

A bibliography list is provided for each section at the ends of the chapters. This list is divided into Further Reading, which cites papers specifically addressing individual topics presented in that session, and a Selected Bibliography, listing the most relevant sources of original information in the scientific literature.

The content of this book has already constituted the basis for a series of academic lessons that I have delivered in various Molecular Medicine courses held at the Scuola Normale Superiore in Pisa and the University of Trieste, Italy, over the past 5 years. The audience of these courses, and thus the potential readership of this book, included graduate students in biology, biotechnology, biochemistry, and medicine in their final years of study, PhD students in the above disciplines, and medical doctors specializing in different fields of internal and specialist medicine, as well as research scientists in various fields of current biomedical research.

Curing human disease using nucleic acids constitutes one of the most demanding challenges of science and medicine, and many hurdles remain to be overcome. However, I am deeply convinced that, once clinical success is fully attained, gene therapy will offer unprecedented possibilities for curing the vast range of human disorders that are currently beyond the scope of traditional medicine.

Trieste, March 2010

Mauro Giacca

Acknowledgments

I am indebted to Oscar Burrone, Serena Zacchigna, Lorena Zentilin, and Miguel Mano for their critical reading of various parts of the manuscript, and to Suzanne Kerbavcic for her valuable help. Many thanks to them and to all the other members of my laboratory at the International Centre for Genetic Engineering and Biotechnology in Trieste for their constant contributions to the maintenance of a scientifically and intellectually stimulating environment.

Contents

Abbreviations

αMHC	α-myosin heavy chain
AADC	aromatic L-amino acid decarboxylase
AAT	α1-antitrypsin
AAV	adeno-associated virus
ABC	ATP-binding cassette
Ab-MLV	Abelson murine leukemia virus
ACV	acyclovir
AD	Alzheimer's disease
ACE	angiotensin-converting enzyme
AC	adenylate cyclase
ACS	acute coronary syndrome
AEV	avian erythroblastosis virus
AIDS	acquired immunodeficiency syndrome
AFP	α-fetoprotein
ALS	amyotrophic lateral sclerosis
ALV	avian leukosis virus
Ang1	angiopoietin 1
AMCV-29	avian myelocytomatosis virus 29
AMD	age-related macular degeneration
AMV	avian myeloblastosis virus
ApoB	apolipoprotein B
ApoE	apolipoprotein E
APC	antigen presenting cell
APP	amyloid precursor protein
ART	artemin
ASO	antisense oligonucleotide
ASLV	avian sarcoma/leukosis virus
ASV-17	avian sarcoma virus 17
β-AR	β-adrenergic receptor
β-ARK	β-adrenergic receptor kinase

BDNF	brain-derived neurotrophic factor
BFV	bovine foamy virus
BIV	bovine immunodeficiency virus
BLV	bovine leukemia virus
BMD	Becker's muscular dystrophy
BMT	bone marrow transplantation
BNA	bridged nucleic acids
CaMKII	Ca^{2+}/calmodulin-dependent kinase II
CABG	coronary artery bypass grafting
CAEV	caprine arthritis encephalitis virus
CAD	coronary artery disease
CB1954	5-aziridine-1-il-2,4-dinitrobenzamide
CCV	clathrin coated vesicle
CDR	complementarity-determining region
CEA	carcino-embryonic antigen
CF	cystic fibrosis
CGD	chronic granulomatous disease
CLC	cardiotrophin-1-like cytokine
CLL	chronic lymphocytic leukemia
CLN2	late infantile neuronal ceroid lipofuscinosis
CMD	congenital muscular dystrophy
CT-1	cardiotrophin-1
CNTF	ciliary neurotrophic factor
CAR	coxsackie/adenovirus receptor
CD	cytosine deaminase
CDK	cyclin-dependent kinase
CML	chronic myelogenous leukemia
CMV	cytomegalovirus
CNG	cyclic nucleotide-gated
CNTF	ciliary neurotrophic factor
CNV	choroidal neovascularization
COPD	chronic obstructive pulmonary disease
cPPT/CTS	central polypurine tract/central termination sequence
CTA	cancer-testis antigen
CTE	constitutive export element
CTL	cytotoxic T-lymphocyte
DBP	deep brain stimulation
DC	dendritic cell
DD	distal muscular dystrophy
DEAE-D	diethyl-aminoethyldextran
DES	drug eluting stent
DDR	DNA damage repair
DGC	dystrophin-associated glycoprotein complex
DHFR	dihydrofolate reductase (DHFR)
DLB	dementia with Lewy bodies

DMAEMA	2-(dimethyl-amino)-ethyl methacrylate
DMD	Duchenne muscular dystrophy
EBV	Epstein-Barr virus
EcR	ecdysone receptor
EEAT2	excitatory amino acid transporter 2
EEC	excitation–contraction coupling
EF	ejection fraction
EFV	equine foamy virus
EGF	epidermal growth factor
EGFR	epidermal growth factor receptor
ER	endoplasmic reticulum
ERAD	ER-associated protein degradation
EIAV	equine infectious anemia virus
ENA	ethylene-bridged nucleic acid
EPC	endothelial progenitor cell
EPO	erythropoietin
ERV	endogenous retrovirus
ES cell	embryonic stem cell
ESE	exon sequence enhancer
ESWL	extracorporeal shock wave lithotripsy
FasL	Fas ligand
FBJ-MSV	Finkel-Biskis-Jinkins murine sarcoma virus
FH	familial hypercholesterolemia
FIV	feline immunodeficiency virus
FFV	feline foamy virus
FGF	fibroblast growth factor
FGFR	fibroblast growth factor receptor
FKBP	FK506-binding protein
FeLV	feline leukemia virus
FGF	fibroblast growth factor
FSHD	facioscapulohumeral muscular dystrophy
GA-FeSV	Gardner-Arnstein feline sarcoma virus
GAD	glutamic acid decarboxylase
GAG	glycosaminoglycans
GaLV	gibbon ape leukemia virus
GCV	gancyclovir
GGF	glial growth factor
GDNF	glial cell line-derived neurotrophic factor
GFP	green fluorescent protein
GPCR	G-protein-coupled receptor
GRK2	G-protein coupled receptor kinase 2
GST	glutathione S-transferase (GST)
GvHD	graft-versus-host disease
GvL	graft-versus-leukemia
GvT	graft-versus-tumor

Ha-MSV	Harvey murine sarcoma virus
HAART	highly active antiretroviral therapy
HAT	histone acetyltransferase
HD	Huntington's disease
HF	heart failure
HFV	human foamy virus
HGF	hepatocyte growth factor
HGFR	hepatocyte growth factor receptor
HIFU	high-intensity focused ultrasound
HIV	human immunodeficiency virus
HHV	human herpesvirus
HMWK	high molecular weight kinogen
HPV	human papillomavirus
HSC	hematopoietic stem cell
HSV	herpes simplex virus
HSPGs	heparan sulfate proteoglycans
HSV-TK	herpes simplex thymidine kinase
HTLV	human T-lympotropic virus
HVJ	hemoagglutinating virus of Japan
HZ4-FeSV	Hardy-Zuckerman 4 feline sarcoma virus
IAP	inhibitor of apoptosis
IFN	interferon
IGF	insulin-like growth factor
IL-6	interleukin-6
IN	integrase
IRES	internal ribosomal entry site
ITR	inverted terminal repeat
L-DOPA	L-3,4-dihydroxyphenylalanine
LamR	laminin receptor
LATs	latency associated transcripts
LCA	Leber's congenital amaurosis
LDL	low density lipoprotein
LGMD	limb-girdle muscular dystrophy
LICLN	late infantile neuronal ceroid lipofuscinosis
LIF	leukemia inhibitory factor
LMO2	LIM domain only 2
LNA	locked nucleic acid
LSDs	lysosomal storage disorders
LTR	long terminal repeat
M6PR/IGFIIr	mannose-6-phosphate/IGF-II receptor
MA	matrix protein
MAO	monoamino oxidase
ManR	mannose receptor
MCK	muscle creatine kinase
MD	myotonic dystrophy

MDGF	megakaryocyte growth and development factor
MeP	6-methylpurine-2'-deoxyribonucleoside
MGMT	methylguanine methyltransferase
MHC	major histocompatibility complex
MI	myocardial infarction
miRNA	microRNA
MLP	major late promoter
Mo-MLV	Moloney-murine leukemia virus
Mo-MSV	Moloney murine sarcoma virus
MLV	murine leukemia virus
MMA	methyl methacrylate
MMP	matrix metalloprotease
MMTV	mouse mammary tumor virus
MPMV	Mason-Pfizer monkey virus
MPS	mucopolysaccharidosis
MPTP	1-methyl-4-phenyl-1,2,3,6-tetrahydropyridine
MSC	mesenchymal stem cell
MVB	multivesicular body
MVM	minute virus of mice
NCL	neuronal ceroid lipofuscinosis
NCX	Na^+/Ca^{2+} exchanger
NeuGC	N-glycolyl neuraminic acid
NGF	nerve growth factor
NIPA	poly(N-isopropylacrylamide)
NLS	nuclear localization signal
NK	natural killer
NO	nitric oxide
NSCLC	non-small-cell lung carcinoma
NT	neurotrophin
NTN	neurturin
6-OHDA	6-hydroxydopamine
OIR	oxygen-induced retinopathy
OPMD	oculopharyngeal muscular dystrophy
OTC	ornithine transcarbamylase
PAD	peripheral artery disease
PAGA	poly[α-(4-aminobutyl)-L-glycolic acid
PBS	primer binding site
PCI	percutaneous coronary intervention
PD	Parkinson's disease
PDE	phosphodiesterase
PDGF	platelet-derived growth factor
PEDF	pigment epithelium-derived factor
PEG	polyethylene glycol
PEI	polyethylenimine
PIC	pre-integration complex

PLB	phospoholamban
PlGF	placental growth factor
PMO	morpholino
PNA	peptide nucleic acids
PNP	purine-deoxynucleoside phosphorylase
Pol	polymerase
POMC	pro-opiomelanocortin
PPT	polypurine tract
PSP	persefin
PTA	percutaneous transluminal angioplasty
PTCA	percutaneous transluminal coronary angioplasty
PTGS	post-transcriptional gene silencing
RBS	Rep binding site
RCA	replication-competent adenovirus
RCL	replication-competent lentivirus
RCR	replication-competent retrovirus
REV	reticuloendotheliosis virus
RISC	RNA-induced silencing complex
RNAi	RNA interference
ROP	retinopathy of prematurity
RP	retinitis pigmentosa
RPE	retinal pigmented epithelium
RRE	Rev responsive element
RSV	Rous sarcoma virus
RT	reverse transcriptase
RXR	retinoic acid receptor
RXRE	Rex responsive elemet
SA	splice acceptor (3' splice site)
SAP	sphingolipid activator protein
scAAV	single-chain AAV
SCF	stem cell factor
scFv	single chain variable fragment
SCID	severe combined immunodeficiency syndrome
SCT	stem cell transplantation
SD	splice donor (5' splice site)
SF	scatter factor
SIV	simian immunodeficiency virus
SFV	simian foamy virus
shRNA	short hairpin RNA
siRNA	short interfering RNA
SIN	self inactivating
SM-FeSV	Susan McDonough feline sarcoma virus
SMA	spinal muscular atrophy
SMC	smooth muscle cell
SOD	superoxide dismutase

SSV	simian sarcoma virus
ST-FeSV	Snyder-Theilen feline sarcoma virus
STLV	simian T-lymphotropic virus
TAA	tumor-associated antigen
TAR	transactivation response element
TCR	T-cell receptor
TetR	tetracycline repressor
TFO	triple-helix-forming oligodeoxynucleotide
TGF-β	transforming growth factor-β
TGN	trans-Golgi network
TK	thymidine kinase
TIL	tumor infiltrating lymphocyte
TLR	Toll-like receptor
tPA	tissue plasminogen activator
TPO	thrombopoietin
TSA	tumor-specific antigen
TNF	tumor necrosis factor
TRS	terminal resolution site
USP	ultraspiracle
UTR	untranslated region
XGPRT	xanthine-guanine phosphoribosyltransferase
Y73SV	Y73 sarcoma virus
VEGF	vascular endothelial growth factor
VEGFR	vascular endothelial growth factor receptor
VSV	vesicular stomatitis virus
VSV-G	vesicular stomatitis virus G protein
VZV	varicella-zoster virus
vWF	von Willebrandt factor
XIAP	X-linked inhibitor of apoptosis
WMSV	woolly monkey sarcoma virus
WDSV	Walleye dermal sarcoma virus

Introduction to Gene Therapy

1

1.1
Genes as Drugs

The idea to use "genes" as "drugs" for human therapy was originally conceived in the United States around the 1970s. It was the logical consequence of at least two major advancements that were occurring in those years, namely the exponentially growing knowledge of human gene function and the impact of their mutations, and the development of progressively more effective technologies for the delivery of DNA into mammalian cells.

The first foreseen applications of gene therapy were for the treatment of patients with inherited monogenic disorders with recessive inheritance. Examples of these disorders are cystic fibrosis, muscular dystrophies, lysosomal storage disorders, hemophilia, and several thousand other different conditions. These diseases are due to single gene defects and the pathologic phenotype becomes evident only when both alleles are defective. Most importantly, the relatives of patients carrying one normal copy of the gene that is mutated in the patients are perfectly healthy; thus, should we be able to transfer one normal copy of the disease gene in the affected individuals, we would have found a cure for the disease.

These initial objectives of gene therapy immediately highlight the main characteristic of this therapeutic modality, which is based on the delivery of additional copies of a gene instead of aiming at the direct correction of the endogenous genetic defects. The latter goal, indeed, would require the development of technologies for homologous recombination-based gene targeting, based on the substitution of a genomic DNA tract carrying a mutation with one carrying the normal sequence, administered exogenously. Homologous recombination can indeed occur in mammalian cells and is currently used in embryonic stem cells to generate genetically modified animals. However, our knowledge of the mechanism regulating this process is still scant, and the probability that a DNA fragment inserted into the nucleus might recombine with its endogenous homologous sequence is about one thousand fold lower than the probability that it integrates randomly, which, in itself, is a rare event. Thus, homologous recombination is now limited to cultured cells, in which a desired recombination event can be selected; therapeutic application to patients is still remote.

Gene Therapy. Mauro Giacca
© Springer-Verlag Italia 2010

In 1980 an American hematologist and geneticist, Martin Cline, surreptitiously conducted the first gene therapy clinical experimentation on two patients with thalassemia, with no therapeutic effect. This attempt was vigorously criticized both scientifically and ethically, since it was not based on any solid experimental ground. In addition, lacking permission from the United States' authorities, the treatment was performed, in one patient, in Israel and, in the other, in Italy, two countries in which, at that time, specific legislation in this field did not exist.

The first gene therapy trial was official approved in the United States in October 1988 and started in the first months of 1989 at the National Institutes of Health (NIH) in Bethesda, Maryland. This was a gene-marking study, lacking a therapeutic goal. The principal investigator was the oncologist Steven Rosenberg, who wanted to genetically mark the cells obtained from a group of cancer patients to later follow their fate once reinjected in the same patients. More specifically, these were T lymphocytes recovered from the tumors (tumor infiltrating lymphocytes, TILs) of five patients with terminal melanoma. Cells were transduced *ex vivo* with a retroviral vector able to transfer a maker gene, which rendered them traceable after reinfusion. Provirus-containing TILs were indeed found, in all the treated patients, in both peripheral blood and in tumor biopsies at least 2 months after infusion, thus proving the property of TILs to home to the tumor after *ex vivo* expansion and manipulation. This initial experimentation was the foundation of several current trials of cancer immunotherapy.

The first therapeutic clinical trial was conducted in 1990, again at the NIH in Bethesda, by Michael Blease and French Anderson in two patients with adenosine deaminase (ADA) deficiency, a hereditary condition causing a serious form of primary immunodeficiency. In this trial, gene transfer was also performed *ex vivo*, in T lymphocytes obtained from the peripheral blood of patients and expanded in the laboratory, before reinfusion of the treated cells back into the patients. The gene coding for a normal ADA enzyme was conveyed into the cells using a retroviral vector. Despite the apparent improvement of the clinical conditions of the patients, the real efficacy of this first trial has remained controversial, since the treated patients continued to be treated with the ADA enzyme in the form of a recombinant protein.

Already during these first phases of gene therapy development, the scientific community started to perceive that the same technologies allowing the delivery of replacement genes might indeed also be used to transfer other kinds of genes, for example those controlling cell survival or modifying various aspects of cell behavior. Furthermore, the genes to be transferred did not necessarily have to code for a protein, but might consist in any DNA or RNA fragment exerting any regulatory function. These concepts enormously broadened the perspectives of gene therapy and contributed essentially to its development. In this respect, it should be recalled that diseases with monogenic inheritance and recessive phenotype, of which several thousands are known, nonetheless constitute less than 30% of all hereditary disorders and, most importantly, the monogenic disorders afflict less than 2% of the general population. Rather, the major causes of morbidity and mortality in the western population are cardiovascular disorders, cancer, and the degenerative disorders of the central nervous system, all conditions for which there is absolute urgency to develop novel therapies. The perception that gene therapy might play a fundamental role in finding innovative approaches for these conditions was at the basis of the enormous development that this discipline witnessed in the early 1990s.

At the end of 1990s, however, it started to become apparent that a series of technical and conceptual problems were significantly lowering the initial expectation of a rapid success of gene therapy. The initial enthusiasm for this therapeutic modality was reduced, at first, by the observed difficulty in transferring genes into primary human cells with adequate efficiency. Later, a general skepticism pervaded the scientific community, mainly because of the serious adverse events that had occurred in some of the clinical trials. In particular, in 1999 a patient with an inherited disorder of metabolism, a deficiency of ornithine transcarbamylase, died in a clinical trial in Philadelphia after administration of an adenoviral vector to the liver. Later, first in Paris and then in London, a few patients with severe combined immunodeficiency due to mutations of the interleukin receptor gamma chain gene, which had been successfully treated by retrovirus-mediated gene transfer into hematopoietic stem cells, developed leukemia due to mutagenic retrovirus insertion. These and other episodes clearly highlighted that several unknowns still hampered broad clinical application of gene transfer.

Notwithstanding these difficulties, gene therapy has recently made significant advances in other fields. In particular, over the last five years, progressively more attention has been focused on viral vectors based on the adeno-associated virus (AAV), which are now generating very encouraging results in the treatment of Leber's congenital amaurosis, an inherited form of blindness, together with Alzheimer's and Parkinson's diseases. When the overall results obtained by gene therapy from 1989 to date are evaluated objectively, it might appear that progress has been very slow. Indeed, still today gene therapy remains a young discipline, based on unconventional tools, which therefore require further time to fully develop, certainly longer than conventional medicines. However, for several disorders, gene therapy continues to represent the only hope for cure.

It might seem somehow suprising that, several years after the identification and cloning of the genes responsible for different human diseases (for example, the cystic fibrosis and the Duchenne muscular dystrophy genes, which were cloned in 1987), this information has still not led to any significant molecular therapy for most of these disorders. The history of medicine shows that innovative therapies require a long incubation prior to success, since this often depends on several small, incremental steps. Antineoplastic therapy, for example, was originally proposed in children with acute lymphatic leukemia in the 1960s, however it took a further 40 years for therapeutic success to rise from the 5–10% of the early days to the current 80–90%. Similarly, the first bone marrow transplantation for the treatment of hematological disorders was conducted in 1957, while its first success was only reported in 1970, and several additional years were subsequently required to broaden its fields of application to a vast series of non-neoplastic disorders such as aplasia and immunodeficiency.

1.2
Gene Therapy: An Overview

The development of any gene therapy approach requires an accurate evaluation of a series of parameters, including assessment of the likelihood that a given disease might realistically be treated by gene therapy, followed by the choice of the correct therapeutic gene, regulatory signals, should this gene be expressed in the cells, delivery system, and administration route.

1

Choice of Therapeutic Gene. A vast spectrum of nucleic acids with potential therapeutic function is now available to gene therapy. This includes protein-coding genes (or, more often, cDNAs) and a broad series of small, non-coding nucleic acids, among which oligonucleotides, usually used as antisense to a target mRNA, ribozymes, siRNAs, RNA and DNA decoys, and aptamers. The properties of these different classes of therapeutic nucleic acids are the subject of the chapter on 'Therapeutic Nucleic Acids'.

Administration Route. Gene therapy can follow one of two general routes of administration, based on either the isolation of the patient's cells followed by gene transfer in the laboratory (*ex vivo* gene therapy) or on the direct delivery of the therapeutic gene into the patients (*in vivo* gene therapy).

In *ex vivo* gene therapy, cells are recovered from the patients and cultured. During this period, the therapeutic gene is transferred, and the cells are eventually reinfused back into the same patients from which they have been collected. Among the advantages of this approach are the possibility to expand different cell populations *ex vivo* (for example, T lymphocytes), to select the cells in which gene transfer has occurred, and to avoid the possibility of immune response against the vector that might neutralize gene transfer. However, the procedure is significantly more cumbersome and expensive than *in vivo* gene transfer, and needs to be personalized for each patient. The list of the possible cell types that can be either maintained or expanded *ex vivo* has grown in recent years, and now includes, besides lymphocytes and hematopoietic stem cells, stem cells of various derivation, keratinocytes, satellite cells, and hepatocytes. Of note, the possibility to introduce genes into these various cell types *ex vivo* offers important therapeutic possibilities by which gene therapy becomes a support to cell therapy. A similar condition applies to the treatment of both hereditary disorders (for example, muscular dystrophy or some hereditary diseases of the liver) and degenerative disorders of the adult (for example, Parkinson's disease or myocardial infarction).

In *in vivo* gene therapy, the therapeutic gene is directly administered to the patient. In principle, this approach appears simpler than *ex vivo* gene transfer and, once optimized, can be applied to a vast series of patients with the same disease. However, some limitations exist. First, several tissues are difficult to reach or to transduce at a significantly level *in vivo*. For example, it is difficult to obtain broad gene transfer to the brain, cartilage, connective tissue, or bone; in these cases, gene therapy is necessarily limited to the administration of the therapeutic gene to specific districts (for example, to the joints or to specific brain regions). Second, once administered *in vivo*, the gene might enter cells other than the desired targets, thus causing unwanted effects (for example, injection of a gene in the heart might also lead to its diffusion to other organs). Third, should a vector be used for gene delivery, it might be subject to inactivation (by complement, as in the case of gammaretroviral vectors, or by specific neutralizing antibodies) or in any case elicit an immune response. Fourth, most cells in our body, including cardiomyocytes, neurons, vascular endothelial cells, and hepatocytes are post-mitotic or, at any rate, resting cells. This essentially limits application of some vectors, such as those based on gammaretroviruses, which require the target cells to be in active replication.

Gene Delivery System. Probably the most crucial factor conditioning the success of any gene therapy approach is the efficiency with which therapeutic nucleic acids enter the target cells. The hydrophobic plasma membrane of mammalian cells represents a formidable barrier for large polyanions such as DNA or RNA. With very few exceptions, naked nucleic acids are, therefore, very poorly internalized by the cells. Gene transfer must be facilitated using physical (for example, electroporation or high pressure injection), chemical (cationic lipids or polymers), or biological (viral vectors) tools. In the case of naked DNA, gene transfer into mammalian cells is known as *transfection*; when a viral vector is used instead, delivery of a gene is called *transduction*. At least four families of viruses are currently considered in clinical experimentations, those based on retroviruses (gammaretroviruses and lentiviruses), adenoviruses, adeno-associated viruses (AAVs), and herpesviruses. The main methods for gene delivery are described in the chapter on 'Methods for Gene Delivery'.

Cell Targeting. In *in vivo* gene therapy, it would be important that the therapeutic gene might be internalized exclusively by the desired cell type. As a matter of fact, this requirement recapitulates the notion originally put forward by the scientist and physician Paul Ehrlich, winner of the Nobel prize for Physiology and Medicine in 1908, who introduced the concept of the *magic bullet*, namely an ideal drug that, upon systemic injection, could selectively target a specific cellular target (which, in the concept of Ehrlich, was mainly a pathogenic microorganism).

We are far from having met Ehlrich's ideal objective, although some significant steps toward this goal have been made. Protein ligands or monoclonal antibodies (the latter molecules are in fact closest to the ideal concept of a *magic bullet*) that recognize specific cellular receptors, for example the c-ErbB2 receptor or the asialoglycoprotein receptor for breast cancer and liver gene therapy, can be used. The incorporation of such specific ligands in cationic lipid/DNA complexes (where non-viral transfection is used) or their conjugation on the surface of viral vectors (for transduction) allows targeting of the therapeutic gene towards a specific cell type. Alternatively, the viral vector capsid itself can be modified by the insertion of specific ligands, an objective that can be met using both adenoviral and AAV vectors. In this respect, it is however important to highlight that the efficiency of these targeting strategies is still modest, mainly because the targeting procedures entailing modification of the viral capsid usually determine a remarkable loss of overall viral infectivity.

Another approach to obtain cell targeting is to restrict transgene expression in a defined cell type at the transcriptional level, by placing the transgene itself under the control of a tissue-specific promoter. In this manner, regardless of the number of cell types in which gene transfer will occur, the transgene will only be expressed in the cells in which the promoter driving its expression is active.

Persistence of the Therapeutic Gene. Another highly important parameter to consider in gene therapy is the persistence of gene transfer. This is essentially linked to the characteristics of the nucleic acid and the delivery system used. When the therapeutic gene is a small regulatory RNA or RNA (antisense oligonucleotide, aptamers, ribozymes, siRNAs) injected systemically, disappearance from circulation and elimination via the liver, kidney, and reticulo-endothelial system is usually very rapid. This poor pharmacokinetics can, however, be counteracted by various means, including chemical modification of the phosphate backbone or ribose sugars of these molecules, conjugation with specific ligands such as polyethylene glycol (PEG) or cholesterol, or association with cationic lipid or polymer carriers. Repeated or continuous administration of these molecules is, in any case, required.

In the case of viral vectors, those based on gammaretroviruses and lentiviruses integrate their genome into that of the infected cells, thus determining permanent transduction. These are therefore the vectors of choice for the therapy of monogenic hereditary disorders or for gene transfer in stem cell populations, which undergo subsequent rounds of division either *ex vivo* or *in vivo*. In different cell types, however, gene expression from retroviral constructs undergoes silencing over time. For example, in the case of myeloid cells, wherein retroviral gene expression after gene transfer into the $CD34^+$ hematopoietic stem cell is progressively switched off along myeloid differentiation, unless driven by promoters specifically active in mature granulocytes and macrophages.

AAV vectors also determine permanent transduction since, albeit not integrating their genome into the host cell DNA, they target post-mitotic, long-surviving cells, such as neurons, cardiomyocytes, muscle fibers, and retinal cells, in which their genomes persist indefinitely in episomal form.

In contrast, transduction with adenoviral vectors is transient. Not only do these vectors not integrate into the host cell genome, they also elicit a strong immune response, which determines destruction of the transduced cell a couple of weeks from injection, at least in the case of first-generation vectors. These vectors, therefore, can only be considered for applications in which high-level but transient gene expression is desirable, such as genetic vaccination or gene therapy of cancer.

Expression of the Therapeutic Gene. When the therapeutic gene needs to be expressed in the target cells, as is the case for genes coding for proteins or shRNAs, an essential issue is to obtain high, persistent, and, possibly, tissue-specific levels of expression. In some situations, it is instead mandatory that expression of the therapeutic gene be more physiological. An example of this requisite is gene therapy of thalassemias, in which synthesis of the globin chains needs to occur in the erythroblast in a balanced manner, to avoid precipitation of the proteins in excess inside the cells, determining their damage and death. Since the sequences of promoters driving expression of endogenous, regulated genes are often poorly known or, in any case, too large to be used, it is essential to identify new tissue-specific promoters that are sufficiently short to be accommodated into viral vectors.

In the case of gene therapy of cancer, the situation is completely different. Should a vigorous immune response against cancer cells be sought through gene therapy, it is neither relevant that all cells are reached by the treatment not that gene transfer is permanent. In this case, adenoviral vectors can find good application, since these vectors can express large amounts of immunogenic and immunostimulatory proteins. The approaches that can be followed include direct administration of the vectors into the tumors or gene transfer into cultured tumor cells (or allogenic cell tumor cell lines) followed by their irradiation before injection.

Immune Response to Gene Therapy. One of the problems encountered in gene therapy, and which often threatens to invalidate its efficacy, is the immune response elicited by either the gene delivery method or the transgene itself. For example, in the case of adenoviral and AAV vectors when administered *in vivo*: the presence of pre-existing immunity against the virus, a common condition given the diffusion of both viruses in the general population, can block transduction due to the presence of neutralizing antibodies. Alternatively, should the patient not have pre-existing immunity against these viruses, the viral capsid can stimulate an immune response at the moment of injection, which will prevent the possibility of further injections of the same vectors. In addition, should the vector continue to express viral proteins after *in vivo* injection, as is the case with first-generation adenoviral vectors, the immune system will recognize the transduced cells and activate cytotoxic T lymphocytes to eliminate these.

In other conditions it is the therapeutic gene rather than the vector that becomes a target for immune attack. In particular, this can occur in hereditary disorders in which the genetic defect determines the complete absence of a protein: once production of the missing protein is obtained upon gene therapy, this is recognized as a foreign antigen by the immune system, with the consequent effect of both antibodies neutralizing its function and cytotoxic T lymphocytes destroying the producing cells.

Further Reading

Couzin-Frankel J (2009) Genetics. The promise of a cure: 20 years and counting. Science 324:1504–1507

Ledley FD (1995) Nonviral gene therapy: the promise of genes as pharmaceutical products. Hum Gene Ther 6:1129–1144

Miller DA (1992) Human gene therapy comes of age. Nature 357:455–460

Pearson H (2009) Human genetics: one gene, twenty years. Nature 460:164–169

Ross G, Erickson R, Knorr D et al (1996) Gene therapy in the United States: a five-year status report. Hum Gene Ther 7:1781–1790

Thomas CE, Ehrhardt A, Kay MA (2003) Progress and problems with the use of viral vectors for gene therapy. Nat Rev Genet 4:346–358

Zaldumbide A, Hoeben RC (2007) How not to be seen: immune-evasion strategies in gene therapy. Gene Ther 15:239–246

Therapeutic Nucleic Acids

2

As introduced in the previous chapter, the term "gene therapy" refers to a vast series of applications, both *in vivo* and *ex vivo*, based on the utilization of nucleic acids for therapeutic purposes.

This chapter describes in an analytical manner the various types of molecules that are part of gene therapy. These can be grouped into one of two classes: (i) DNA sequences coding for proteins having various cellular functions; (ii) nucleic acids (DNAs or RNAs) with regulatory function, either synthesized as small synthetic molecules or, in the case of RNAs, expressed inside the cells after transfer of the corresponding genes. Table 2.1 reports a list of the various types of nucleic acids with a possible therapeutic function.

2.1
Protein-Coding Genes

The concept of gene therapy was originally developed with the idea of supplying a missing cellular function by transferring a normal copy of the altered gene into the relevant cells. In reality, in human cells the average size of protein-coding genes is ~27 kb, by far longer than the maximum length fitting the most common gene delivery systems. For this reason, gene therapy is most commonly based on the transfer of cDNAs (i.e., the double-stranded DNA copies derived from the gene mRNAs; average length: ~2.5 kb) or of their protein-coding portion (average length: ~1.5 kb, corresponding to about 500 codons).

From the molecular point of view, the transfer of a gene, or its cDNA or its cDNA coding region has essentially different properties. Both cDNAs and their coding portions need to be transcribed from promoters that are usually different from the natural ones, which are commonly too large to be used. In addition, the cDNA coding portions alone lack the regulatory elements controlling gene expression at the post-transcriptional level, which are usually contained in the introns or in the untranslated regions (UTRs) at the 3' and 5' ends of the cDNAs. These regions are commonly involved in the regulation of cDNA stability, transport, subcellular localization, and translation of the cellular mRNAs. On the

Gene Therapy. Mauro Giacca
© Springer-Verlag Italia 2010

Table 2.1 Therapeutic nucleic acids

Protein-coding DNA sequences	Proteins substituting missing or mutated cellular proteins	
	Proteins modulating cellular functions	
	Secreted growth factors and cytokines	
	Proteins regulating cell survival and apoptosis	
	Antigens for vaccination	
	Antibodies and intracellular antibodies	
	T-cell receptor (TCR) subunits	
Non-coding nucleic acids	Oligonucleotides and modified oligonucleotides	Phosphorothioate oligonucleotides
		2'-Ribose modified oligonucleotides
		Locked nucleic acids (LNA) and ethylene-bridged nucleic acids (ENA)
		Morpholinos (PMO)
		Peptide nucleic acids (PNA)
	Catalytic RNAs and DNAs	Ribozymes and DNAzymes
	Small regulatory RNAs	siRNAs and shRNAs, microRNAs
	Long antisense RNAs	
	Decoys	
	Aptamers	

other hand, however, in several instances the levels at which proteins are produced are not very important, and thus a tight translational or post-translational regulation of gene expression is not required. For example, this is the case of proteins replacing missing cellular functions in the hereditary disorders of metabolism, or antigens for anti-cancer vaccination, or secreted antibodies. In these cases, transfer of the protein-coding region under the control of a strong promoter, such as the promoter for the cytomegalovirus (CMV) immediate-early (IE) genes, is adequate. In several circumstances, inclusion of a small intron upstream of the cDNA coding sequence, or of a 3' UTR downstream of it, are known to facilitate expression of the protein of interest.

The proteins encoded by the therapeutic genes can have very different functions, ranging from the substitution of a missing cellular protein to the modulation of the immune system, as summarized in Table 2.1 and discussed in detail in the various sections of the chapter on 'Clinical Applications of Gene Therapy'.

2.1.1
Proteins Substituting Missing or Mutated Cellular Proteins

As already reported above, gene therapy was conceived with the purpose to express proteins that are missing or defective, thus curing autosomal recessive and X-linked disorders. These replacement proteins can exert their functions inside the cells (for example, in the case of gene therapy for muscular dystrophies) or on the cell membrane (for example, the CFTR gene in cystic fibrosis), or be secreted into the extracellular environment or the blood stream (as is the case of coagulation factors in the hemophilias).

2.1.2
Proteins Modulating Cellular Functions

The objective of several gene therapy applications is not to supply a missing cellular function, but to express proteins able to modulate cell behavior. The approaches utilized are very variegate. For example, in the cancer gene therapy field several experimentations take advantage of the possibility to induce arrest of cell proliferation using cyclin-dependent kinase (CDK) inhibitors such as p27 or p21, or checkpoint proteins such as p53. Again in the cancer gene therapy field, another interesting class of proteins is those that increase the therapeutic index of chemotherapy. On several occasions, myeloid toxicity limits the dose of antineoplastic drugs that can be administered to a patient with a solid cancer. In these cases, it is possible to confer resistance to $CD34^+$ hematopoietic precursors by transferring into these cells a gene coding for a membrane transporter, such as the *mdr* gene, able to prevent the intracellular accumulation of a large series of anticancer drugs.

Other examples of proteins modulating cellular functions are those used for gene therapy of viral infections, which block activity of some viral proteins and thus impair viral infection. Among these proteins, RevM10 is a mutated form of the HIV-1 Rev protein, acting as a transdominant negative mutant: when this protein is present in $CD4^+$ T lymphocytes, it blocks wild-type Rev function and thus inhibits viral replication.

Finally, the immune response against cancer cells can be increased by expressing, into these cells, the genes coding for co-stimulatory proteins, such as B7, ICAM-1, or LFA-3, which are commonly downregulated in tumors as a mechanism of immune escape and are however necessary for proper antigenic presentation to cytotoxic T lymphocytes (CTLs).

2.1.3
Secreted Growth Factors and Cytokines

A number of gene therapy applications are based on the delivery of genes coding for secreted factors and cytokines, having various activities. For example, in cancer gene therapy, several clinical approaches take advantage of the genes coding for interleukin-2 (IL-2), IL-12, IL-7, IL-4, GM-CSF, and other cytokines to increase antigen stimulation or modulate the immune response against the transformed cells.

In the field of cardiovascular disorders, vascular endothelial growth factor (VEGF) gene transfer is used by a number of applications aimed at the induction of therapeutic angiogenesis in patients with cardiac or peripheral ischemia, due to the powerful activity of this factor in inducing new blood vessel formation.

Finally, gene therapy for a number of neurodegenerative disorders is based on the delivery of genes coding for neurotrophic factors, including nerve growth factor (NGF) in Alzheimer's disease, neurturin (NTN) in Parkinson's disease and ciliary neurotrophic factor (CNTF) in Huntington's disease.

2

2.1.4
Proteins Regulating Cell Survival and Apoptosis

Several pathologic conditions are caused by the death of some cell types or, on the contrary, by inappropriate survival of others. Examples of the former condition are neurodegenerative disorders, which are caused by accelerated apoptotic death of some neuronal populations; in contrast, tumors represent a paradigmatic example of a condition in which lack of cell apoptosis essentially contributes to disease development. Since apoptosis ensues as the net result of the function of various proteins favoring or contrasting this process, transfer of the genes coding for these proteins might have a therapeutic role. For example, one of the gene therapy approaches for amyotrophic lateral sclerosis (ALS), a motoneuron disorder, is based on the transfer of the *bcl-2* gene, coding for an anti-apoptotic protein. In contrast, apoptosis in tumors can be induced by oligonucleotides targeting a few antiapoptotic proteins, including Bcl-2 itself, Survivin, or the X-linked inhibitor of apoptosis (XIAP).

Another way to interfere with cell survival is the so-called "suicide gene" approach. Cells are transduced with a vector expressing the thymidine kinase gene of the herpes simplex virus type 1 (HSV-TK), an enzyme that is innocuous *per se* but becomes toxic in the presence of the drug gancyclovir. The result of this activation is a block in DNA synthesis and consequent cell death by apoptosis.

2.1.5
Antigens for Vaccination

A vast range of gene therapy applications have the immune system as their target. A couple of these applications have already been mentioned above. These consist in the utilization of gene transfer to activate the immune system, by either the delivery of genes coding for immunomodulatory cytokines or by transferring, into cancer cells, genes coding for the co-immunostimulatory proteins necessary for antigenic presentation.

A growing field of gene therapy applications in this area is that of DNA-based vaccination (genetic vaccination). This is based on the *in vivo* delivery of genes usually coding for viral or tumor cell antigens, to seek activation of the immune system against the encoded proteins. The antigen gene can be transferred using naked plasmid DNA ("DNA vaccination" tout court), sometimes using physical methods such as gene transfer facilitators (for example, the "gene gun" approach), or in the context of one of the several available viral (adenovirus, vacciniavirus) or bacterial (attenuated *Salmonella* strains, BCG, or attenuated *Mycobacterium bovis*) strains.

Genetic vaccination offers several advantages over the common vaccination strategies based on the administration of either attenuated or inactivated microorganisms, or on protein antigens. These include the possibility to evoke a cytotoxic immune response, since the processed exogenous proteins are directly expressed in the context of MHC Class I, together with the relative safety, reduced cost, and possibility of prolonged storage.

Cancer and HIV-1 infection are among the disorders in which genetic vaccination is most commonly considered.

2.1.6
Antibodies and Intracellular Antibodies

A peculiar class of therapeutic genes are those coding for antibodies. Natural antibodies have a typical Y-shaped structure, composed of 4 polypeptide chains: two identical heavy (H) chains (~440 amino acids each) and two identical light (L) chains (~220 amino acids). Both the H and L chains contain a variable region V (V_H and V_L), which together recognize the antigen. The H chains then contain three constant (C) regions (C_H1, C_H2, and C_H3) and a flexible "hinge" (h) region between C_H1 and C_H2, while the L chains only have one constant region (C_L) (Fig. 2.1A).

The H and L chains are synthesized separately in the B-cell endoplasmic reticulum and assemble thanks to the formation of both inter- and intra-chain disulfide bonds. Digestion of an antibody molecule with the proteolytic enzyme papain generates three fragments, which can be separated by chromatography. Two of these are identical and correspond to the antigen-binding portion of the antibody (antigen-binding fragment, Fab), composed of the whole L chain (V_L+C_L) and the N-terminal portion of the H chain, including the variable region (V_H) and the C_H1 constant region. The third fragment (crystallizable fragment, Fc) is instead only composed of the C-terminal region of the H chain, which is able to bind

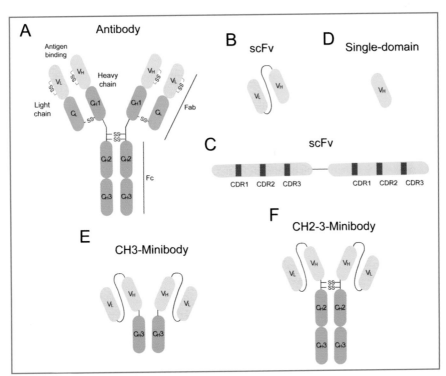

Fig. 2.1 Antibodies and antibody derivatives. **A** Schematic representation of natural antibodies. **B** and **C** scFvs, with the indication of the complementarity-determining regions, CDRs. **D** Single-domain antibodies. **D** and **F** Minibodies. See text for description

complement and different receptors, including the Fc receptor, on various cell types, thus mediating part of the immune response against the antigen recognized by the antibody.

The fact that an antibody consists of four polypeptide chains poses obvious problems for its expression by gene transfer. However, it is possible to obtain synthetic antibodies consisting of a single polypeptide chain formed by the V portions of the H and L chains (V_H and V_L) of the natural antibodies, separated by a flexible amino acid linker (Fig. 2.1B). This construct, named single-chain Fv (scFv), contains all the structural determinants allowing specific binding of the molecule to its target antigen. While the mass of a natural IgG antibody is about 150 kDa, that of a scFv is 29 kDa, corresponding to about 250 amino acids.

The V_H and V_L regions contain three hypervariable regions that essentially participate in antigen recognition (complementarity-determining regions, CDRs) (Fig. 2.1C). An even simpler form of antibody only contains the CDR regions of one chain. These are named single-domain antibodies, with analogy to the antibodies produced by some camelids (camel and llama), in which the antigen-binding site is composed of a single chain, analogous to the mammalian H chain (referred to as VHH; Fig. 2.1D). Neither scFv nor single-domain antibodies exert any effector activity (complement fixation, binding to cellular receptors), since these are commonly associated with the Fc portion of natural antibodies.

The simplest manner to obtain an scFv antibody is to clone the V_H and V_L region of a monoclonal antibody (mAb) having high affinity for the antigen of interest. Alternatively, the scFv can be obtained by the direct screening of an scFv library, for example a phage display library, in which the antibody is displayed on the phage surface. The scFv-coding gene can thus be cloned into a plasmid or a viral vector and transferred into the cell. When the scFv gene coding sequence is preceded by a secretion signal, the antibody is secreted outside the cells.

In order to stabilize an scFv antibody and increase its avidity – one of the major limitations of scFvs is indeed their lower avidity compared to natural antibodies, mainly due to their monovalent rather divalent nature – the V_L–V_H region can be followed by a portion of the antibody Fc fragment, for example, the C_H3 or both the C_H2 and C_H3 regions. This can induce dimerization of two scFvs and, when the hinge/C_H2 regions are included, formation of disulfide bonds. These types of engineered antibodies are known as minibodies (Figs. 2.1E and 2.1F).

Natural antibodies are secreted in serum or expressed on the surface of B lymphocytes and recognize their target antigens only when these are present in the extracellular environment or expressed on the cell surface. In the absence of a secretion signal, the scFv antibodies can instead be translated in the cytosol. In this reducing compartment, disulfide bonds cannot be formed, however some scFvs can still fold properly in the absence of these bonds and retain their binding specificity. These intracellular antibodies, also named intrabodies, thus have the important potential to target intracellular antigens.

Both scFv and single-domain intracellular antibodies can be used for a vast series of therapeutic applications. They can block interaction between two intracellular proteins, or protein binding to DNA, or inhibit function of an enzyme. Alternatively, they can re-localize a protein to a cellular compartment different from its normal, thus blocking its activity. For example, intracellular antibodies having a nuclear localization signal can bind a cytosolic antigen and relocalize it into the nucleus.

A vast series of pre-clinical studies have shown the potential efficacy of intracellular antibodies for gene therapy of various disorders. In the field of gene therapy for viral infections, the cell can be rendered resistant to infection by intracellular antibodies blocking function of the proteins, of either viral or cellular origin, that are essential for viral replication. In gene therapy of cancer, intracellular antibodies are used to modify the intracellular localization, and thus inhibit the function, of oncogenes such as ErbB-2 in breast and ovary carcinoma, of the IL-2 receptor in some leukemias, and of the EGFR in glioblastoma. In the field of gene therapy for neurodegenerative disorders, the use of intracellular antibodies can block the accumulation of pathological proteins, such as Tau in Alzheimer's disease or the PrPsc protein in prion disease.

2.1.7
T-Cell Receptor (TCR) Subunits

An interesting class of protein-coding genes is that including variants of the T-cell receptor (TCR), used to modify the target specificity of T-lymphocytes. In both viral infection and cancer a potentially powerful therapeutic approach would consist in the *ex vivo* activation of the patients' CD8+ CTLs reacting against specific viral or tumor antigens, followed by their reinfusion *in vivo*, a procedure known as adoptive immunotherapy. More efficacious than recovering the endogenous antigen-specific CTLs, which are usually limited in number, this can be achieved by modifying the specificity of any CTL by transducing these cells with the genes coding for a TCR of choice.

The TCR consists of a complex of at least 6 different proteins (Fig. 2.2A). Antigen recognition specificity is conferred by a heterodimer of α/β (in most cases) or γ/δ chains. These associate with the invariant subunits of the CD3 complex, composed of the heterodimers CD3γ/CD3ε and CD3δ/CD3ε which, together with the invariant homodimer

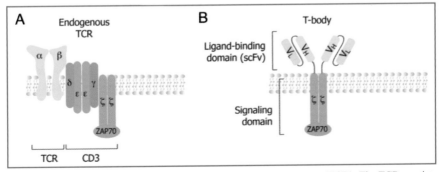

Fig. 2.2 Structure of the T-cell receptor and of a T-body. **A** T-cell receptor (TCR). The TCR consists of a protein complex formed by the α/β (or γ/δ) heterodimer, which confers antigen specificity, together with the CD3 γ/ε and δ/ε heterodimers that, together with the ζ/ζ homodimer and the ζ-chain-associated protein kinase 70 (Zap70), mediate signal transduction. **B** T-body. T-bodies consist of a fusion between a single chain antibody recognizing the antigen of interest and the TCR CD3 ζ chain

zeta/seta (ζ/ζ), couple antigen recognition with intracellular signal transduction. With the exception of the ζ chain, the other 5 components of the TCR are members of the immunoglobulin superfamily of proteins. In particular, the α and β chains consist of an N-terminal variable domain and a constant C-terminal portion. The variable domains of these chains contain three hypervariable regions, which are therefore equivalent to the complementarity-determining regions (CDR) of the immunoglobulin L and H chains (cf. above).

The simplest manner to confer TCR specificity to T lymphocytes is to transfer, into these cells, the genes coding for the α and β chains specific for an antigen of interest, obtained from a previously selected natural cell clone. Gene transfer is commonly achieved using a retroviral vector in which the genes coding for the two chains are separated by an IRES (see section on 'Vectors Based on Gammaretroviruses').

A potential limitation of this approach, however, is that the exogenously introduced TCR molecules might complex with those endogenously expressed by the T cells, thus generating TCRs with different, and potentially undesired, target specificity.

An effective manner to overcome these problems is to exploit alternative ways to modify TCR specificity. One of these consists in the use of single-chain TCRs, in which mispairing between α and β chains is prevented. Another possibility is offered by the so called *T-bodies* (Fig. 2.2B). These consist of a genetic fusion between a single-chain antibody recognizing the antigen of interest and the TCR CD3 ζ chain, able to activate the signal transduction cascade leading to cell activation. This approach both renders the engineered CTL independent from MHC restriction and prevents mispairing between the exogenous and endogenous TCR α and β chains.

2.1.8
Control of Therapeutic Gene Expression

As already mentioned above, several of the current gene therapy applications based on the intracellular expression of protein-coding genes take advantage of strong constitutive promoters, such as the CMV IE gene promoter. In several situations, however, expression of the therapeutic protein (or regulatory RNA, see below) must be limited to one specific cell type, or at least controlled in terms of levels, or restricted to a defined temporal period. For example, the genes coding for the different hemoglobin chains (α- and β-globin in the adult) must be transcribed in a balanced manner, to avoid monomer aggregation and precipitation in the erythroblasts, causing apoptosis of these cells: in gene therapy of diabetes, insulin must be precisely produced in response to blood glucose concentration; expression of some pro-apoptotic genes can be beneficial only if limited to cancer cells; finally, over-expression of some growth factors, including those inducing new blood vessel formation, might be detrimental if expressed for excessively long periods; and so on.

In several of these conditions, an optimal solution would be to use the natural promoter of the therapeutic gene to direct its expression *in vivo*. In most cases, however, this is not possible, since natural promoters are usually very large and thus cannot be accommodated in most viral vectors.

2.1.8.1
Tissue-Specific Promoters

The first strategy to tackle the issue of regulated transgene expression is to limit its transcription to the tissue of interest, using a promoter specific for this tissue and sufficiently short to be cloned in a gene transfer vector. Examples of such transcriptional elements are the muscle creatine-kinase (MCK) enhancer or the β-actin gene promoter for the skeletal muscle, the α-myosin heavy chain gene promoter (α-MHC) for the heart, the insulin gene promoter for the pancreas, the albumin gene promoter for the liver, the transthyretin gene promoter for the retina, and so on.

An elegant manner to obtain the opposite effect, namely block transgene expression in a given tissue, is to act post-transcriptionally and exploit the presence, in several cell types, of natural microRNAs (miRNAs) targeting various cellular genes (see section on 'Small Regulatory RNAs'). If the transgene mRNA sequence is the target of a miRNA – which can be obtained by cloning the miRNA recognition sequence downstream of the transgene coding region and before the polyadenylation site – protein expression will be selectively inhibited in the cell types expressing this miRNA while remaining unaffected in other cells. This property can be used, for example, to block expression of a protein of interest in antigen-presenting cells (APCs), thus blocking its presentation to the cells of the immune system.

2.1.8.2
Inducible Promoters

Besides the use of tissue-specific promoters, an alternative manner to regulate therapeutic gene expression is by using inducible promoters, in which transcription can be selectively activated. Some of these promoters are naturally present in the genome and direct gene expression under specific physiologic conditions. Among these promoters are those of genes expressed in response to heavy metals – for example the methallothionein gene, heat – heat shock proteins 70 and 90 (Hsp70 and Hsp90), hormones – the mouse mammary tumor virus (MMTV) long terminal repeat (LTR) promoter, sensitive to dexamethazone induction, and hypoxia – for example the VEGF angiogenic protein. Most of these natural promoters, however, cannot be easily used since transcriptional control is not very stringent, the levels of induced expression are too low, or they are activated in conditions that cannot be easily reproduced in a therapeutic setting, due to the pleiotropic undesired effects elicited by the stimulating agent itself (for example, high temperature or hormone administration).

A very interesting category of promoters, in contrast, consists of a few synthetic promoters, the activity of which can be controlled pharmacologically, i.e., by the administration of simple drugs. These promoters respond to specific transcriptional activators that can be assigned to one of at least three classes: (i) transcriptional activators regulated by small molecules; (ii) intracellular steroid hormone receptors; or (iii) synthetic transcription factors in which dimerization is controlled by the antibiotic rapamycin.

(i) *Transcriptional activators controlled by small molecules.* This class of inducible regulators is based on the use of transcription factors that change their conformation (and are thus either activated or deactivated) upon binding one small chemical molecule. The prototype of this class of regulators is the Tet repressor (TetR), which, in *E. coli*, regulates expression of the *Tn10* operon genes. This operon encodes a system of transporters that determine exit of tetracycline from the bacterial cell, thus conferring resistance to this antibiotic. In the absence of tetracycline, TetR binds a DNA sequence (the Tet operator, *tetO*) positioned upstream of the Tet operon and suppresses transcription. When tetracycline is instead present, this binds TetR and causes a conformational change that blocks its interaction with *tetO*: in this manner, transcription of the *Tn10* operon is de-repressed and the bacterium becomes resistant to the antibiotic. This transcriptional control strategy was adapted to mammalian cells by generating a two-component system: on one hand, the gene of interest is placed under the transcriptional control of a minimal promoter – which dictates the localization of the transcription start site – with a series of *TetO* repetitions positioned upstream; on the other hand, another construct codes for a fusion protein between TetR and an RNA polymerase II transcriptional activator, such as the carboxy-terminal domain of the herpes simplex virus type 1 (HSV-1) protein VP16 (the fusion protein between TetR and VP16 is known as tTA). In the absence of tetracycline, tTA binds in multiple copies to the promoter and activates transcription (Fig. 2.3A). When tetracycline or, better, its derivative doxycycline is present, transcription is instead switched off since the antibiotic binds the TetR moiety of tTA and the allosteric change determined by this interaction causes detachment of the hybrid transcription factor from the promoter. This system of transcriptional regulation is known as *Tet-Off*, since the antibiotic blocks transcription. A variant of this system was subsequently developed, based on the use of a four-amino-acid mutant of the TetR protein, known as reverse TetR (rTetR), which is unable to bind the promoter unless the antibiotic is present. When this protein is fused to VP16 to generate the reverse tTA (rtTA) regulator, transcription is switched on by the addition of the antibiotic (*Tet-On* system; Fig. 2.3B).

Conceptually analogous to the TetR model, other inducible transcription systems have also been developed. A couple of these are derived from *E. coli* and exploit the antibiotic-dependent binding of other activators/repressors to their cognate DNA sites; these include the PIP transactivator binding the PIR binding site, which is regulated by pristinamycin, and by the E transactivator binding the ETR binding site, regulated by erythromycin. Alternatively, a system regulated by gaseous acetaldehyde was also developed, based on the AlcR transactivator derived from the fungus *Aspergillus nidulans*.

(ii) *Intracellular receptors for steroid hormones.* Intracellular receptors are proteins that, once complexed with their ligands, directly activate target gene expression in the nucleus. The prototypes of this class of molecules are the steroid hormone receptors. In particular, the estrogen and progesterone receptors are sequestered in the cell cytosol by the Hsp90 protein, which masks their nuclear localization signal; binding to the respective hormones induces a conformational change that releases the receptors, allows their nuclear import, and thus activates target gene transcription. To exploit this system for the inducible expression of a desired gene, the natural receptor protein was engineered by fusing it with the DNA binding domain of the yeast transcription factor Gal4, in order to target this fusion protein towards synthetic promoters containing Gal4 binding sites, and by mutating its amino acid sequence to lower its affinity for the natural hormones and increase that for synthetic analogues, including tamoxifene (for the estrogen receptor) and mifepristone (RU486, for the proges-

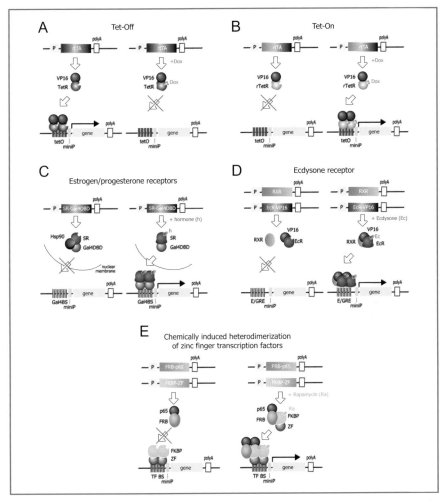

Fig. 2.3 Promoters for inducible gene expression. **A** Tet-Off system. A fusion factor between the bacterial tetracycline repressor (TetR) and the viral VP16 transactivator only binds the *tet* operator (*tetO*) and activates transcription in the absence of doxycycline. **B** Tet-On system. The reverse-tetracycline repressor (rTetR) fused to VP16 binds *tetO* only in the presence of doxycycline. **C** Estrogen/progesterone receptors. The steroid receptor (SR) fused to the Gal4 DNA binding domain (Gal4DBD) are sequestered in the cytosol by Hsp90; when the hormone (h) is administered, the factor is released and binds a cognate Gal4 binding site (Gal4BS) in the promoter region of the gene, driving its transcription. **D** Ecdysone receptor. Only in the presence of ecdysone (Ec), a fusion protein between the ecdysone receptor (EcR) and the VP16 transactivator heterodimerizes with the retinoic acid receptor (RXR) and thus activates transcription of genes containing a hybrid binding site between the ecdysone receptor and glucocorticoid response element (E/GRE). **E** Ligand-induced transcription factor dimerization. The FRB domain of the mTOR protein is fused to a transcriptional activation domain, for example the p65 protein of NF-kB, while FKBP is fused to the zinc finger (ZF) domain of a transcription factor (TF). In the presence of rapamycin, the two fusion proteins heterodimerize, and thus the ZF domain tethers the transcriptional activation domain to a gene containing the TF ZF binding site (BS) in its promoter

terone receptor). When cells are transfected with a construct containing the gene of interest under the control of the promoter containing the Gal4 binding elements and another construct expressing the engineered receptor, transcription of the gene of interest can be selectively controlled by administering the synthetic hormone (Fig. 2.3C).

A peculiar type of steroid hormone receptor is the *D. melanogaster* ecdysone receptor (EcR). In insects, the steroid hormone ecdysone has an essential role in stimulating metamorphosis. The hormone acts by entering the cells and binding a receptor composed of a heterodimer of two proteins, the EcR itself and the *ultraspiracle* (USP) gene product. This inducible regulatory system can also be reproduced in mammals by transferring two genes, one coding for a modified form of EcR fused to the transcriptional activation domain of VP16 and the other one corresponding to the human homologue of USP, which is the retinoic acid receptor RXR; the gene of interest is placed under the control of a promoter binding the heterodimer of the two proteins (Fig. 2.3D). In this manner, expression of the gene of interest only occurs after administration of ecdysone or its synthetic analogues muristeron A or ponasteron A. In mammals, the pharmacokinetics of these hormones is very favorable, since they neither accumulate nor are eliminated too rapidly, are not toxic and are inactive in the absence of their cognate receptors. Furthermore, the lipophilic nature of these compounds allows them to freely cross the plasma membrane and the blood–brain barrier.

(iii) *Transcriptional control by ligand-induced transcription factor dimerization.* A third system to achieve pharmacologically induced transcription is based on the property of the drug sirolimus (rapamycin, a macrolide antibiotic originally extracted from a *Streptomyces* found in the Easter Island, known as Rapa Nui in the indigenous language). This drug has been used for over a decade as an immunosuppressant in transplanted patients and is now also being tested for cancer therapy. Inside the cells, rapamycin interacts with a member of the FKBP (FK506-binding protein) family, the immunophilin FKBP12. The rapamycin/FKBP12 complex then binds a specific domain (the FRB domain) of mTOR (mammalian target of rapamycin, a protein of the PI3K family) protein kinase and blocks its action. Since this kinase is essential for signal transduction in T cells, rapamycin causes cell cycle arrest in the G1 phase of the cell cycle; lack of T-cell proliferation explains the immunosuppressive effect of the drug. Rapamycin is thus the prototype of a drug selectively inducing dimerization of two proteins, FKBP12 and mTOR. When FKBP is fused to the zinc finger (ZF) domain of a transcription factor and its FRB partner to a transcriptional activation domain, the presence of rapamycin induces the formation of a heterodimeric transcription factor; this synthetic factor binds a specific target sequence thanks to the ZF domain and activates transcription through its activation domain (Fig. 2.3E). To extensively utilize this system, however, it is essential to identify rapamycin analogues retaining high-affinity binding to the engineered FKBP variants but not interfering with the functions of the endogenous FKBP12 protein.

The various inducible transcription systems considered above appear, in general, to be ingenious and effective; in some instances, transcription can be activated over 10,000-fold in the induced condition compared to the basal level. However, they require that, together with the therapeutic gene, the gene coding for the inducible regulator is also transferred into the same cell. In addition, the regulator often consists of two different proteins, as in the case of the ecdysone- or rapamycin-regulated systems. This poses important limitations to the clinical experimentation, since it obliges the experimenter to accommodate several transcriptional cassettes within the same vector, and to use vectors allowing cloning of large DNA inserts. In addition, the regulatory proteins must be expressed in a constitutive manner; however, these proteins derive from different organisms (for example, the tetracycline repressor from *E. coli*, the VP16 transactivator from HSV-1, the EcR from *D. melanogaster*, and so on) and are thus potentially immunogenic.

Finally, the above-described inducible systems can efficiently regulate RNA polymerase II transcription, while they are difficult to adapt to RNA polymerase III regulation, which, by its own nature, tends to be constitutively active. Thus, the inducible systems are effective in regulating protein- and microRNA-coding genes, but much less the genes coding for shRNAs or other small regulatory RNAs (cf. also below).

2.2
Non-Coding Nucleic Acids

Besides protein-coding nucleic acids, the spectrum of gene therapy applications is enormously increased by the possibility of using short nucleic acids (DNAs or RNAs) with regulatory function. These molecules belong to one of at least six possible classes:
(i) oligonucleotides and modified oligonucleotides;
(ii) small catalytic RNAs and DNAs (ribozymes and DNAzymes respectively);
(iii) small regulatory RNAs (siRNAs and microRNAs);
(iv) long antisense RNAs;
(v) decoy RNAs and DNAs;
(vi) RNAs binding to other molecules thanks to their tridimensional structure (aptamers).

DNA oligonucleotides and modified oligonucleotides must be administered to the cells from the outside. In contrast, the RNA molecules can also be synthesized inside the cells by transferring their coding DNA sequences.

Fig. 2.4 schematically shows the sites of action of these different classes of molecules inside the cell. Virtually all steps regulating gene expression can be controlled by small regulatory molecules, including transcription and splicing (oligonucleotides), mRNA decay (oligonucleotides, ribozymes, DNAzymes, siRNAs and long antisense RNAs), protein synthesis (siRNAs), and protein function (aptamers and decoys).

2.2.1
Oligonucleotides

The simplest form of nucleic acids with potential therapeutic function consist in short, chemically synthesized, single-stranded DNA molecules, usually 15–100 nt long. Use of these oligodeoxynucleotides is based on the intrinsic property of a DNA strand to pair with its complementary sequence thanks to the formation of hydrogen bonds between the nucleotide bases (the so-called Watson and Crick bonds). Although each of these bonds is non-covalent and thus weak, their overall number renders the overall affinity between two complementary DNA strands very high: the K_d of double-stranded DNA is in the order of 1×10^{-15} mol/l, far higher than that, for example, of an antibody for its ligand (K_d between

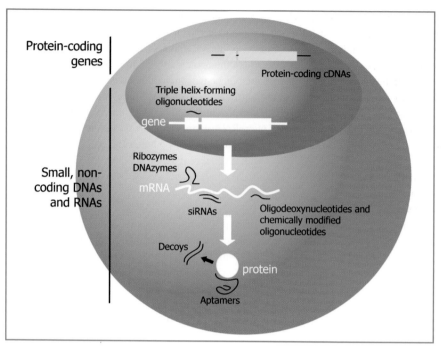

Fig. 2.4 Therapeutic nucleic acids. The site of action of the various nucleic acid-based therapeutics is shown along the pathway of gene expression

1×10^{-6} and 1×10^{-10}) or a growth factor for its receptor (K_d between 1×10^{-8} and 5×10^{-10}). Furthermore, base pairing is extremely specific in terms of target recognition, since a 17-nt DNA stretch is statistically present only once in the entire human genome.

Thanks to their properties, synthetic oligodeoxynucleotides can be used in at least four different types of applications: the first two targeting cellular mRNA and the second two genomic DNA; namely:

(i) to block gene expression, exploiting base paring between the oligodeoxynucleotides and a target RNA sequence;

(ii) to modulate pre-mRNA splicing, thus favoring or impeding inclusion of one exon in the mature mRNA;

(iii) to block transcription, by promoting formation of triple helix DNA structures, usually in the promoter region of a target gene;

(iv) to promote gene correction, exploiting oligodeoxynucleotide pairing with a homologous genomic DNA segment.

(i) *Antisense oligodeoxynucleotides (ASOs) blocking gene expression.* To selectively inhibit expression of a cellular or viral gene, 17–22-nt oligodeoxynucleotides can be used, having a sequence complementary to that of the target mRNA. Once introduced into the cell, pairing of the ASO with its target blocks ribosomal translation and stimulates degradation of the RNA:DNA hybrid by cellular RNase H enzymes.

(ii) *Antisense oligonucleotides (ASOs) that modulate splicing.* Thanks to its complementarity to target mRNA, an ASO can also be used to induce exclusion of an exon during pre-mRNA maturation. In some diseases, presence of a point mutation in an exon can generate a Stop codon, leading to the synthesis of a prematurely truncated protein, or shifting the open reading frame downstream of the mutation. In some situations, such as Duchenne muscular dystrophy (DMD), the disease caused by such mutations is much worse than that eventually induced by the absence of the entire exon containing the mutation (cf. section on 'Gene Therapy of Muscular Dystrophies'). One strategy to induce exon skipping from an mRNA is to treat cells with an ASO complementary to the signals that regulate pre-mRNA splicing. These ASO typically target the 5' and 3' splice sites or the sequences internal to the exon that determine its retention in the mature mRNA (exon sequence enhancers, ESEs).

(iii) *Triple-helix-forming oligodeoxynucleotides (TFOs).* TFOs are single-stranded oligodeoxynucleotides binding the DNA major groove in a sequence-specific manner. In particular, they form very stable bonds (Hoogsteen bonds; Fig. 2.5) with a DNA duplex of

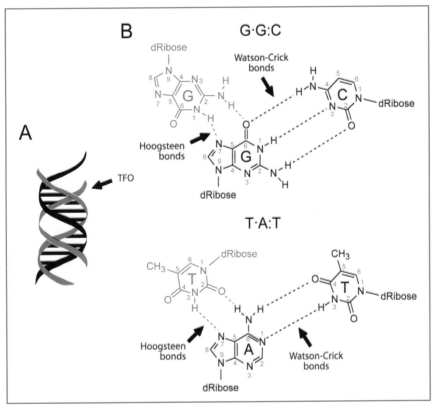

Fig. 2.5 Triple helix forming oligonucleotides (TFOs). **A** Schematic representation of an oligodeoxynucleotide (*red*) forming a triple helix with a B-type DNA target. **B** Hydrogen bonds formed by guanine and cytosine (*upper*) or adenine and thymine (*lower*) in B-type, natural DNA on antiparallel chains ("Watson-Crick" bonds, *black*) and hydrogen bonds formed by an additional guanine (*upper*) or thymine (*lower*; Hoogsteen bonds, *red*)

10–30 bp, containing a stretch of purines and a complementary stretch of pyrimidines. The third helix is formed by the TFO, which, according to the target sequence, can be made of either purines or pyrimidines. The TFO binds the target duplex through the formation of two hydrogen bonds between each of its bases and the purine-rich strand of the target DNA. Formation of a triple helix is thus an intrinsic property of DNA structure and does not require any cellular protein. When the TFO target is in the promoter of a gene, or immediately downstream of the transcription start site, the triple helix structure impairs transcription factor binding or duplex DNA unwinding, thus blocking gene expression. In some instances, formation of a triple helix was also shown, although at low efficiency, to promote DNA repair, thus inducing the correction of mutations, by recruiting the cellular machineries responsible for excision repair or homologous recombination.

(iv) *Oligonucleotides inducing correction of point mutations.* One of the most ambitious goals of gene therapy is to directly modify the genetic information to obtain the correction of a pathological mutation. This objective can be met by introducing, into the cells, a DNA stretch having a sequence identical to the region to be corrected, however without the mutation, and then exploiting the cellular machinery involved in DNA repair for the substitution of the genomic DNA sequence with that administered exogenously. This gene conversion process is usually based on the cellular machinery responsible for homologous recombination, and uses, as substrates, long, single-stranded DNA stretches, composed of several thousand nucleotides. In some cases, however, short, single-stranded oligonucleotides, composed of 20–110 nt and having mismatched nucleotides in their central position, can stimulate gene correction by favoring the recruitment of cellular DNA mismatch repair protein. Although this approach is experimentally interesting, the frequency at which gene correction can be obtained is usually very low (usually less than 0.1% of the treated cells) and still restricted to cultured cells.

2.2.2
Modified Oligonucleotides

Among the essential problems limiting application of oligodeoxynucleotides *in vivo* is their limited tissue distribution, cytotoxicity, and, mostly, low stability. Both in the extracellular environment and inside the cell, short natural DNA molecules, in which nucleotides are connected by phosphodiester bonds, are rapidly degraded by endo- and 3'-exonucleases (DNases).

To overcome this problem and thus increase bioavailability of these molecules *in vivo*, the oligodeoxynucleotides' structure can be altered by introducing various chemical modifications (Fig. 2.6).

(i) In first-generation modified oligonucleotides, one non-bonding oxygen atom in the phosphate group is substituted by a sulfur atom (*phosphorothioate oligodeoxynucleotides*). This modification confers higher stability to the molecules and thus increases their half-life. *In vitro*, phosphorothioate oligodeoxynucleotides can be efficiently delivered to cells using lipofection (cf. chapter on 'Methods for Gene Delivery'). *In vivo*, they can be used in the form of naked molecules, however their half-life is very short (less than two hours in serum and four hours in tissues) and therefore they need to be administered

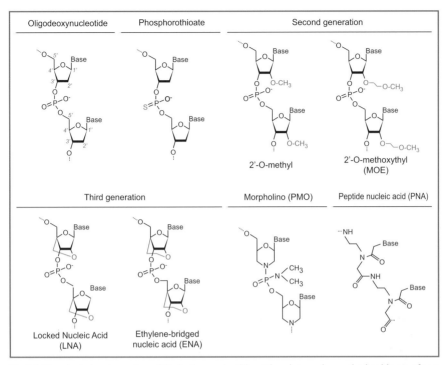

Fig. 2.6 Chemical structure of modified oligonucleotides. The picture shows the backbone of two adjacent nucleotides in natural DNA (*upper left* structure) and in the various types of modified oligonucleotides. For phosphorothioates, second- and third-generation modified oligonucleotides, the relevant chemical modifications over natural DNA are indicated in *red*

by continuous intravenous infusion. Phosphorothioate oligodeoxynucleotides are relatively well tolerated, with minimal side effects probably due to the interaction of the polyanion backbone with some serum proteins.

(ii) A second generation of modified oligonucleotides was obtained by the introduction of an alkyl group in the 2' position of the ribose molecule of nucleotides. The generated molecules include *2'-O-methyl-* and *2'-O-methoxyethyl-RNA*. These molecular species are less toxic than phosphorothioate oligodeoxynucleotides. The conformation changes induced by these modifications and, in particular, the presence of ribose, improve target binding, however they also abrogate the oligonucleotide:target mRNA duplex ability to activate RNase H, a crucial aspect of ASO activity. Thus, the inhibitory effect of these molecules is only exerted through the inhibition of mRNA translation and is thus lower than that of phosphorothioates. This drawback can be avoided by generating chimeric oligonucleotides with 2'-modified nucleotides placed only at the ends of the oligonucleotide, thereby leaving a central RNase-compatible DNA gap. These hybrid oligonucleotides are known as *gapmers*.

(iii) An even more drastic chemical modification characterizes the so-called bridged nucleic acids (BNAs), better known as *locked nucleic acids* (LNAs). In these molecules, the ribose contains an extra bond connecting the 2' and 4' carbons (2'-O, 4'-C methylene bridge). This bond blocks ribose in the C3'-endo structural configuration, which is often present in the DNA or RNA "A" configurations. In the variant *ethylene-bridged nucleosides* (ENAs), the 2' and 4' carbons are linked by a 2'-O, 4'-C ethylene bond. Within an oligonucleotide, LNA nucleotides can be interposed at will between natural nucleotides or deoxynucleotides; therefore, these modified nucleotides can also be placed in ribozymes or siRNAs. The presence of LNA nucleotides confers, to the oligonucleotides, higher hybridization affinity for their complementary strand compared to natural oligodeoxynucleotides. Therefore, these molecules also find utilization as *in vitro* substrates for microarray hybridization, or as probes for fluorescent *in situ* hybridization (FISH) or real-time PCR.

Modification of the oligonucleotide backbone can be so drastic as to completely alter the chemical structure of the nucleic acid, thus generating molecules that are still able to base pair with their targets but have completely different properties compared to natural nucleic acids. This third generation of modified nucleic acids includes morpholinos and peptide nucleic acids (PNAs).

(iv) *Morpholinos*, also known as PMO (*phosphorodiamidate morpholino oligomers*), are synthetic molecules, usually composed of 25 nucleotides, differing from natural nucleic acids in that their bases are connected to the morpholino ring rather than to deoxyribose, and a phoshorodiamidic bond substitutes the natural phosphodiester bond. These molecules inhibit gene expression by binding to a target mRNA and blocking its processing or translation in a steric manner, independent from RNase H activity. Morpholinos are extensively used in the laboratory to study animal development, especially in *Xenopus* and *Zebrafish*, in which they are microinjected in embryos to stably and specifically knock out expression of a gene of interest. In mammalian cells, entry of these molecules does not occur spontaneously and must be facilitated by conjugating them, either covalently or non-covalently, with peptides able to transport them through the plasma or endosomal membranes. Examples of such peptides are those derived from the *Drosophila* homeotic protein Antennapedia or the HIV-1 Tat protein. Morpholinos have already entered clinical experimentation to block replication of the hepatitis C and the West Nile viruses, and to prevent vascular restenosis after angioplasty using a drug-eluting stent to release an anti-cMyc PMO. Another interesting application is to induce exon skipping for gene therapy of DMD.

(v) *Peptide nucleic acids* (PNAs) are synthetic polymers in which the bases are connected by a backbone consisting of a repetition of the modified amino acid N-(2-aminoethyl)-glycine, connected by peptide bonds. These bonds are electrically neutral compared to the phosphate bonds, which are negatively charged. Therefore, PNAs form very stable bonds when paired to their complementary DNAs or RNAs, or form a triple helix when pairing, through Hoogsteen base-pairing, to the poly-purine or poly-pyrimidine tracts. Furthermore, their non-natural chemical structure renders PNAs very resistant to degradation. Despite these very appealing characteristics for gene therapy, the clinical application of PNAs have so far been hampered by their modest bioavailability once administered *in vivo*.

2.2.3
Clinical Applications of Oligonucleotides

Notwithstanding the huge amount of experimental work performed in the oligonucleotide field over the last almost 30 years, the global efficiency of these molecules as a gene therapy tool still remains suboptimal, as also demonstrated by the relatively modest success of the vast number of clinical trials performed over the last few years. The first approved oligonucleotide clinical trial in the USA and Europe was for the treatment of CMV retinitis in patients with HIV-1 injection. This trial was based on the intraocular administration of a phosphorothioate ASO, named fornivirsen, having the CMV immediate early gene mRNA as a target. This drug, however, despite having shown efficacy, was recently withdrawn from the market in countries of the European Community for purely marketing considerations, since CMV retinitis in Western countries has become very rare in HIV-1 patients, due to the success of the current combined antiretroviral therapies.

In contrast, treatment of other viral infections, such as those by HSV-1 and HIV-1, in which ASOs were administered systemically, was much less satisfactory. To obtain a therapeutic effect in these conditions, it is mandatory that the oligonucleotide concentration is maintained permanently high without causing important side effects, an objective that is not reasonably attainable with the currently available delivery methods.

There are numerous clinical trials based on ASOs for cancer therapy. In this case, the oligonucleotide targets are various genes essential for cell replication or regulating apoptosis. The purpose of these trials is, on one hand, to block tumor cell growth while, on the other hand, to promote their apoptosis. Table 2.2 reports some of the cancer gene therapy clinical applications based on ASO administration, usually administered by continuous intravenous infusion either as naked nucleic acids or complexed with lipids, most commonly in combination with conventional chemotherapy. The overall results of these trials have been modest, especially as far as the use of first-generation, phosphorothioate ASOs is concerned. Some of the second- (phosphorothioates with 2'-O-methoxyethyl ribonucleotides and LNAs) and third- (PNAs and PMOs) generation molecules have proven higher efficiency in the animal models and are currently under clinical experimentation.

Finally, two additional therapeutic applications of oligonucleotides deserve mention, due to the promising preclinical results they have shown. In the first applications, ASOs were aimed at lowering the levels of apolipoprotein B-100, an important structural protein present on the surface of atherogenic lipoproteins, such as VLDL and LDL, which facilitates cellular internalization of cholesterol by binding the LDL receptor. Systemic administration of a modified anti-ApoB-100 ASO was found to significantly lower the plasma LDL-cholesterol levels and delay the development of atherosclerotic plaques in the vessel wall. The second promising application, which was already mentioned above, is based on PMO injection into the skeletal muscle to induce exon skipping in DMD patients.

Currently, it is still unclear and debatable whether the discovery of the phenomenon of RNA interference (cf. below) will progressively lead to the use of siRNAs rather than ASOs in the applications aimed at inducing silencing of gene expression.

Table 2.2 Examples of oligonucleotides used in cancer gene therapy clinical trials

Target gene	Gene function	Drug name	Structure of oligonucleotide	Cancer type
Bcl-2	Apoptosis inhibitor	G3139 (Oblimersen)	Phosphorothioate	Melanoma, chronic lymphatic leukemia, multiple myeloma, non-small-cell lung carcinoma
Clusterin	Protein chaperone	OGX-011	Phosphorothioate with 2'-*O*-methoxyethyl ribonucleotides (gap-mer)	Prostate carcinoma, breast carcinoma, non-small-cell lung carcinoma
Protein-kinase Cα (PKCα)	Signal transducer	ISIS 3621	Phosphorothioate	Non-small-cell lung carcinoma
Survivin	Apoptosis inhibitor	LY2181308	Phosphorothioate with 2'-*O*-methoxyethyl ribonucleotides	Solid cancers
Myb	Oncogene, transcription factor	LR3001	Phosphorothioate with 2'-*O*-methoxyethyl ribonucleotides	Chronic myelogenous leukemia (bone marrow purging)
XIAP (X-linked inhibitor of apoptosis)	Apoptosis inhibitor	AEG35156	Phosphorothioate with 2'-*O*-methoxyethyl ribonucleotides	Chronic myelogenous leukemia
HSP27	Heat shock protein, apoptosis inhibitor	OGX-427	Phosphorothioate with 2'-*O*-methoxyethyl ribonucleotides	Prostate carcinoma
STAT-3	Signal transducer and transcription factor	ISIS 345794	Phosphorothioate with 2'-*O*-methoxyethyl ribonucleotides	Different cancers

2.2.4
Catalytic Nucleic Acids

In the 1970s and 1980s it was originally discovered that the RNAs of some organisms, including the ciliated protozoon *Tetrahymena thermophila*, possess enzymatic activity, since they catalyze cleavage of the phosphodiester bonds present on their own or on other RNA molecules. The RNAs having this enzymatic activity are called *ribozymes*.

At least seven different classes of natural ribozymes are now known. These include:

(i and ii) Group I and II introns, which undergo splicing through an autocatalytic process.

(iii) The RNA subunit of *E. coli* ribonuclease P (RNase P), which is responsible for the maturation of the tRNA 5' ends. In bacteria, this enzyme consists of an RNA subunit (M1 RNA), with catalytic activity, and a protein subunit, having structural function (in humans, RNase P is composed of an RNA subunit, the H1 RNA, whose enzymatic activity is only apparent under specific circumstances, and 10 protein subunits).

(iv) Hammerhead ribozymes, present in the RNA genome of different plant viroids and virusoids, where they are essential for rolling circle RNA replication.

(v) Hairpin ribozymes, also naturally present in the satellite RNAs of some plant viruses, where they participate in viral genome RNA replication.

(vi) The hepatitis δ virus (HDV) pseudoknot ribozyme.

(vii) The *Neurospora* VS satellite RNA ribozyme.

While group I and II introns and RNase P are several hundred nucleotides long and depend, for their enzymatic function, on the formation of complex secondary RNA structures, the ribozymes belonging to one of the last four classes (in particular, hammerhead and hairpin ribozymes) are shorter (50–150 nt) and determine cleavage of the target RNA phosphodiester bond after simple base pairing. It is thus possible to incorporate the ribozyme catalytic core within an antisense DNA molecule, with the purpose of directing its enzymatic activity towards a specific cellular or viral mRNA, thus determining its cleavage. In contrast to ASOs, ribozymes have enzymatic activity: after digestion of an RNA molecule, they detach from this and recycle to attack and cleave other targets.

Hammerhead ribozymes are typically ~40 nt long and have a secondary structure formed by three helical domains (domains I, II, and III), surrounding a junction containing the catalytic core, defined by the presence of specific nucleotides (Fig. 2.7A). While

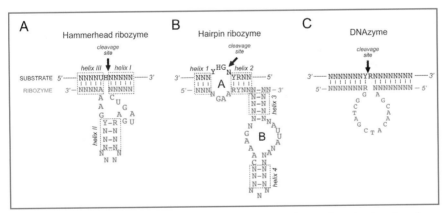

Fig. 2.7 Small catalytic nucleic acids. **A** Consensus sequence and secondary structure for a hammerhead ribozyme (*red*) base-paired to its substrate (*black*). N represents any nucleotide, R purines, Y pyrimidines, and H any nucleotide except G. The hammerhead ribozyme consists of three helices and 11 non-helical residues located in the highly conserved central region. In commonly used ribozymes, the intramolecular helix II is formed by four base pairs joined by a loop. The site of substrate cleavage is indicated. **B** Minimal consensus sequence for a hairpin ribozyme paired to its substrate. **C** Sequence and secondary structure of a DNAzyme

helix II is formed intramolecularly by the ribozyme itself, helices I and III depend on pairing of the single-stranded ribozyme sequence with its complementary sequence in the substrate, and can thus be chosen at will to match the sequence of the target RNA.

Hairpin ribozymes consist of 4 paired helices (helices H1–H4) and two internal loops (A and B; Fig. 2.7B). Enzymatic cleavage occurs at the level of a bond in loop A. In the plant satellite RNAs, cleavage takes place within the RNA molecule itself; however the sequences of helices H1 and H2 can be modified in order to render them complementary to a desired nucleotide sequence of the target RNA. In contrast to hammerhead ribozymes, hairpin ribozymes do not require the presence of a metal ion in the catalytic core.

A non-natural variant of ribozymes are the so-called deoxyribozymes or DNAzymes, in which the nucleic acid is DNA instead of RNA (Fig. 2.7C). Similar to ribozymes, the molecules also consist of a catalytic domain flanked by sequences pairing with the target RNA. The advantage of DNAzymes compared to ribozymes is the increased stability of DNA compared to RNA.

The fields of application of ribozymes (and DNAzymes) essentially overlap with those of ASOs, since both classes of molecules aim at targeting mRNA. They include gene therapy of genetic disorders with dominant inheritance, in which destruction of the disease allele is sought, gene therapy of viral diseases, in which viral RNAs are targeted, and, finally, gene therapy of cancer. In the last case, the objective is to destroy inappropriately expressed mRNAs contributing to the transformed phenotype, such as those from oncogenes, or cellular mRNAs coding for antiapoptotic proteins.

A specific application of ribozymes in the field of cancer gene therapy is for the selective destruction of pathologic cellular mRNAs ensuing in the process of malignant transformation and contributing to tumor cell proliferation. An example of this condition is chronic myelogenous leukemia (CML), in which a reciprocal translocation between chromosomes 9 and 22 generates the *bcr-abl* fusion gene; the transcript of this gene has been used in a number of pre-clinical studies as a target for ribozymes recognizing the junction between the two sequences (see section on 'Gene Therapy of Cancer').

The enzymatic activity of ribozymes, allowing them to recycle on multiple substrate molecules, renders them more effective than ASOs, which bind their targets in a stoichiometric manner. As in the case of ASOs, however, the main limitations of ribozymes are their low stability, limited tissue distribution, and difficulty entering cells when administered systemically. These characteristics have so far limited the clinical application of these molecules. A ribozyme targeting expression of vascular endothelial growth factor receptor-1 (VEGFR-1), administered systemically as a synthetic molecule, has been used to inhibit tumor angiogenesis in various types of cancers, with limited success. Another synthetic ribozyme targeting the proliferating cell nuclear antigen (PCNA), a co-factor of DNA polymerase δ that is essential to maintain cell proliferation, was injected into the vitreus to inhibit endothelial cell proliferation in proliferative diabetic retinopathy. Finally, ribozymes were expressed from gammaretroviral vectors in different trials for gene therapy of HIV-1 infection. Further details on these applications are presented in the respective sections.

Similar to ASOs, it is unclear at the moment which clinical applications ribozymes will find used in the future, and in which siRNAs will be more successful.

2.2.5
Small Regulatory RNAs (siRNAs, microRNAs)

At the end of the 1990s, an unexpected molecular phenomenon was identified in the worm *Caenorhabditis elegans*, by which double-stranded RNA molecules in which one of the two strands was complementary to a target mRNA were able to induce silencing of expression of that mRNA, much more effectively than single-stranded DNA, both sense and antisense. Now we know that this phenomenon is part of a broader mechanism of regulation, collectively known as *RNA interference* (RNAi), which was originally discovered in plants in the 1980s and operates in both invertebrates and vertebrates. RNAi regulates endogenous gene expression post-transcriptionally, by selectively blocking mRNA translation or inducing mRNA degradation.

A synthetic cartoon summarizing RNAi is shown in Fig. 2.8. The process is triggered by the presence, in the cytoplasm, of small, double-stranded RNA (dsRNA) molecules of 21–26 bp, with two unpaired bases at the 3' ends and phosphorylated 5' ends. One strand of these dsRNA molecules (the *guide* strand) is assembled into a ribonucleoprotein complex called RISC (*RNA-induced silencing complex*), which selectively binds a target

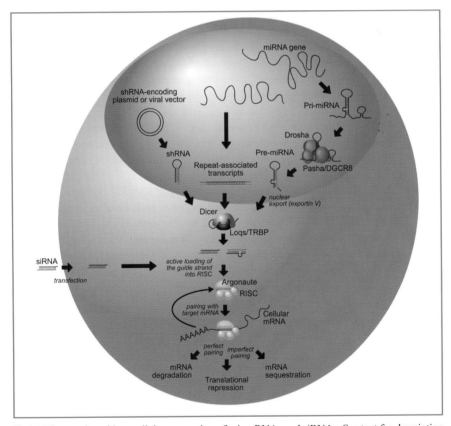

Fig. 2.8 Biogenesis and intracellular processing of microRNAs and siRNAs. See text for description

RNA, inducing its cleavage or blocking its translation. By pairing with the target mRNA, the guide strand confers binding specificity to RISC.

The dsRNA molecules that activate RNAi can be generated by one of at least two pathways: the microRNA (miRNA) and the short interfering RNA (siRNA) pathways. MicroRNAs are processed from stem-loop structures in primary transcripts, while siRNAs arise by cleavage of long double-stranded precursors or by chemical synthesis.

2.2.5.1
MicroRNAs

MicroRNAs are generated by the cells from specific RNA transcripts forming stem-loop structures (i.e., palindromic regions interrupted by a loop of non-complementary nucleotides). The genes coding for these transcripts are localized within introns of protein-coding genes (40% of cases) or in intergenic regions (60% of cases); in the latter situation, the same genomic region often contains a cluster of different miRNA genes, one close to the other. Most miRNA genes are transcribed by RNA polymerase II as long (in some cases, over 1000 nt) RNA transcripts; transcription can be regulated by the promoter of the host gene (for intronic miRNAs) or, most commonly, by a specific promoter. These transcripts, known as primary miRNA (pri-miRNA), contain a CAP and a poly-A at their 5' and 3' ends respectively. They are processed in the nucleus by a complex consisting of at least two components: the RNase III enzyme Drosha and a dsRNA-binding protein called Pasha in *Drosophila* or DGCR8 (DiGeorge critical region 8) in mammals. This complex, also known as the Microprocessor, processes the pri-miRNA to generate the pre-miRNAs, which consists of a short stem-loop structure of 65–75 nt. These are transported from the nucleus to the cytoplasm by exportin-5, and here are further processed by a complex of proteins that includes the RNase III Dicer and the protein *Loquacious* (Loqs) in *Drosophila* or TRBP in humans to generate the final miRNAs, presenting two characteristic overhang nucleotides at their 3' ends. Processing by Dicer is coupled with loading of the guide strand into RISC, which is the final effector of RNAi.

Post-translational regulation of gene expression by miRNAs plays a fundamental role in the regulation of a number of processes, including embryonic development, differentiation, apoptosis, and cell transformation. At least 800 human miRNAs have been identified so far and their number is certainly destined to increase. Since perfect pairing with a target mRNA is not required for miRNAs to induce silencing (cf. below), it is estimated that each miRNA controls expression of about 200 different cellular genes. Bioinformatics estimations indicate that expression of at least 30% of human genes is regulated post-transcriptionally by miRNA-induced RNAi.

2.2.5.2
Short Interfering RNAs (siRNAs)

Unlike miRNAs, siRNAs are generated from longer precursor dsRNAs, which can be endogenously introduced into the cells or produced endogenously. Similar to miRNAs, these dsRNA molecules are also processed by Dicer and the final, RISC-mediated RNAi pathway is in common with that of miRNAs. With analogy to the function of Loqs/TRPB, the dsRNA-binding protein R2D2 binds Dicer and favors RNA assembly into RISC.

The siRNAs were originally identified as mediators of RNAi after the introduction of exogenous dsRNA molecules into the cells; however, intracellular sources of siRNA production were later discovered, most of which derive from the transcription of long repeated sequences present in the genome. These are also known as *repeat-associated siRNAs* (rasiRNAs), a class of *PIWI-interacting RNAs* (piRNAs), since they selectively bind the protein PIWI, a member of the Argonaute family, within the RISC complex (cf. below). In some organisms, these rasiRNAs derive from the heterochromatic regions present at the chromosome centromeres and telomeres. Here they act by activating a complex that represses transcription of their own coding sequences, called the RITS (*RNA-induced transcriptional silencing*) complex. In mammals, piRNAs are almost exclusively expressed in the germ cells of ovary and testis. Here, these molecules would have the function to induce silencing of the vast fraction of "parasitic" genomic DNA elements, with particular reference to transposons. Thus far, 140 rat and mouse piRNA clusters have been described, each of which is most likely transcribed as a unit and subsequently processed into mature piRNAs. In *C. elegans*, plants, and fungi, production of long dsRNAs, acting as precursors of endogenous siRNAs, also occurs thanks to a specific RNA-dependent RNA polymerase, able to generate dsRNA molecules starting from a transcript. Such a polymerase, however, has not been found in *Drosophila* or mammals.

2.2.5.3
Mechanisms of Gene Silencing

RNAi leads to post-translational silencing of gene expression by multiple mechanisms, which include cleavage of target RNA, inhibition of translation, and sequestration into specific cytoplasmic compartments.

The final effector of RNAi, from either the miRNAs or the siRNA pathway, is the ribonucleoproteins complex RISC, the catalytic component of which includes various members of the Argonaute (Ago) family of proteins. These proteins contain a PAZ domain, which binds miRNA/siRNA, and a PIWI domain, similar to that of RNases H, which is the effector of target RNA cleavage in correspondence of a specific nucleotide, corresponding to nucleotide at position 10 of the guide filament. This endonucleolytic activity (also known as *Slicer* activity, which is apparently exerted by Ago2 but not by other members of the Argonaute family within RISC) requires perfect pairing between the guide filament in RISC and the target RNA, a condition that is usually sought when siRNAs are administered exogenously to the cell. In this condition, RISC functions as an enzyme, progressively destroying one molecule of substrate after the other. In the case of most endogenously produced miRNAs, in contrast, pairing usually occurs with the 3' UTR of the target mRNA and is not perfect, only requiring 7 nt at the 5' end of the guide filament (from nucleotide in position 2 to nucleotide in position 8); these 7 nucleotides form the so-called *seed* sequence. In this case, silencing of gene expression is mainly due to the inhibition of translation of the target mRNA. Finally, a third mechanism of gene silencing is due to the ability of the RISC-associated Argonaute proteins to sequester the target mRNAs within specific cytoplasmic compartments, called *P bodies*, in which RNA degradation is believed to occur.

The above-described gene expression silencing mechanisms have, as a target, the transcribed mRNA and are thus collectively known as post-transcriptional gene silencing (PTGS). However, in plants, yeasts, and *Drosophila*, siRNAs can also switch off gene expression at the transcriptional levels (transcriptional gene silencing, TGS). In this case, the siRNAs act by inducing a conformational change of chromatin at the levels of the genes coding for the target mRNAs and by promoting DNA methylation at these regions (deacetylation and compaction of chromatin and cytosine DNA methylation are the essential epigenetic modifications associated with silencing of gene expression). Recent experimental information indicates that analogous TGS mechanisms might also operate in mammalian cells, for example, in the above-mentioned case of the rasiRNAs.

2.2.5.4
Therapeutic Use of RNAi

The introduction into the cells of long dsRNA molecules, composed of several hundred base pairs, is a very efficient way to induce RNAi in yeasts, plants, *Drosophila*, and *Caenorhabditis*. However, in mammalian cells, the use of exogenous siRNAs was initially discouraged by the observation that, in these cells, dsRNA molecules longer than 30 bp activate the production of interferon. Indeed, this cytokine is normally secreted by several cell types upon infection with RNA viruses as a defense mechanism (single-stranded RNA viruses also generate double-stranded RNA forms as replication intermediates). However, it was subsequently observed that dsRNA molecules of 21–22 nt with two overhang nucleotides at their 3' ends are incapable of activating interferon production, thus paving the way for the experimental and later clinical utilization of RNAi.

A few examples of the potential utilization of RNAi for human gene therapy are shown in Table 2.3. At the pre-clinical level, different studies have shown that these molecules are well tolerated and show an overall efficacy and specificity definitely higher than ASOs or ribozymes. However, siRNAs are less amenable than ASOs to the introduction of chemical modifications that might improve their stability and *in vivo* biodistribution, with the exception of the possibility to introduce a few phosphorothioate links and ribose 2'-*O*-alkyl groups in some of the ribonucleotides.

The first filing to the United States FDA for an siRNA-based therapeutic trial dates back to 2004. Ongoing or planned siRNA applications include treatment of syncytial respiratory virus (RSV), hepatitis C virus (HSV) and human immunodeficiency virus-1 (HIV-1) infections and of various neurodegenerative disorders, including Huntington's disease. The rationale behind the use of siRNAs in these applications is discussed in the chapter on 'Clinical Applications of Gene Therapy'.

Phase I and II gene therapy clinical trials have already been concluded, with three different siRNAs for age-related macular degeneration (AMD), an important disorder of the eyes leading to blindness (cf. section on 'Gene Therapy of Eye Diseases'). These trials have taken advantage of the possibility to directly administer the siRNAs in a relatively restricted anatomical space, such as the vitreus. One of the pathogenetic mechanisms of wet AMD is retinal neoangiogenesis, by which the hypoxic environment stimulates new blood vessel formation, a process largely sustained by the overproduction of VEGF. One of the siRNAs used in these trials had the VEGF mRNA as a target, another the VEGF-R1, and the third the hypoxia-induced gene RTP801. The results obtained were largely reassuring in terms of safety and encouraging in terms of efficacy.

One of the most relevant problems of the siRNAs is the possibility that they might silence in an inappropriate manner cellular genes different from their specific target (the so-called *off-target effect*). Since the presence of a complementarity sequence of only 7 nucleotides (the *seed* sequence) is sufficient to inhibit translation, it is almost inevitable that each siRNA might silence different targets in addition to its specific one. Experimentally, this drawback can be modulated by controlling the intracellular concentration of siRNAs or, better, by using more than one siRNA to silence a single gene, each one targeting a different mRNA sequence and administered at low doses. It is possible that the same strategy might also be used in the future *in vivo* to modulate the specificity of siRNAs at the clinical level.

Finally, it is interesting to observe that the mechanism of RNAi can also be exploited to silence expression of endogenous miRNAs by using short antisense RNAs complementary to specific miRNA sequences. These molecules, which can be chemically stabilized or be composed of LNAs, are known as *antagomirs*. Since they inhibit endogenous miRNAs, they consequently upregulate expression of the genes that miRNAs negatively control. Antagomirs are often conjugated to cholesterol to facilitate their internalization by the cells upon *in vivo* delivery.

Table 2.3 Potential siRNA-based gene therapy approaches

	Disease	Target gene
Monogenic or multifactorial diseases	Familial hypercholesterolemia	Apolipoprotein B
	Age-related macular degeneration (AMD)	VEGF, VEGFR1, RTP801
	Amyotrophic lateral sclerosis (ALS)	SOD1
	Spinocerebellar ataxia type 1	Ataxin 1
	Alzheimer's disease	Tau, APP
	Huntington's disease	Mutated huntingtin allele
	Parkinson's disease	α-Synuclein
Cancer	Different tumors	Bcl-2
	Acute myeloid leukemia (AML)	AML1/MTG8
	Chronic myelogenous leukemia (CML)	Bcr-Abl
	Glioblastoma	MMP-9, uPAR
Infectious diseases	Hepatitis B	HBsAg
	Hepatitis C	NS3, NS5B, E2
	Influenza	Nucleoprotein, polymerase
	HIV-1 infection	Viral or cellular genes required for viral replication
	HSV-1 infection	Glycoprotein E
	Syncytial respiratory virus (RSV)	P, N, L genes

2.2.6
Decoys

The term "decoy" refers to small DNA or RNA molecules containing the same binding site of a DNA or RNA binding protein. As a consequence, these molecules are able to compete, once present in high concentration inside the cells, with the endogenous nucleic acids for binding to this protein, thus diverting it from its natural target.

An applicative example of a decoy is a double-stranded oligodeoxynucleotide containing the recognition sequence of transcription factor E2F-1, a protein essential for cell replication since it controls transcription of a vast series of genes required for the G1-S transition of the cell cycle. When administered to the cell, this oligodeoxynucleotide competes with the natural promoters binding E2F-1 and thus blocks expression of the genes controlled by this factor. This decoy molecule was used in a few trials aimed at inhibiting the vascular stenosis occurring after coronary artery bypass grafting (CABG) when using a vein as the grafted vessel. In these conditions, the vein wall often undergoes a process of hyperplasia based on the proliferation of smooth muscle cells of the tunica media, which eventually leads to stenosis of the bypass, by a mechanism similar to post-angioplasty restenosis (see section on 'Gene Therapy of Cardiovascular Disorders'). Proliferation of the smooth muscle cells can be diminished by blocking activity of E2F-1. The decoy-based gene therapy trials were based on the *ex vivo* treatment of the vein to be grafted with a high concentration of an E2F-1 oligonucleotide decoy. Unfortunately, this approach did not prove successful in preventing stenosis after implantation of the treated veins, probably due to the very short duration of the decoy effect.

Other types of decoys find application in gene therapy for infectious disorders. For example, replication of HIV-1 is highly dependent on binding of two of its proteins, the Tat transactivator and the regulatory Rev protein, to specific cis-acting sequences on viral RNA, the TAR (trans-acting response) and RRE (Rev-responsive element) respectively (see section on 'Gene Therapy of HIV Infection'). In the absence of Tat, transcription of proviral DNA does not occur, while the absence of Rev blocks export of the viral unspliced mRNAs into the cytoplasm. The intracellular expression of high amounts of short RNA sequences corresponding to either TAR (TAR decoy) or RRE (RRE decoy) diverts the respective binding proteins from their natural targets and thus inhibits viral replication. One of these decoy sequences (multimeric TAR) is present in one of the lentiviral vectors that have recently entered a clinical trial.

The success of the decoy-based therapeutic approaches is usually limited by the high concentration at which these molecules must be continuously present in the cells to act in a competitive manner. Such a concentration is usually difficult to obtain by extracellular administration. At least for the RNA decoys, this problem can be overcome by the intracellular expression of these molecules upon delivery of the respective genes. In addition, the DNA decoys are usually less efficient than the natural promoter DNA in binding a transcription factor, since, in the latter case, factor binding is stabilized by chromatin structure and interaction with the other proteins involved in transcription.

2.2.7
Aptamers

In the 1990s, a few laboratories developed a method, called SELEX, for the selection of RNA molecules able to specifically bind proteins or small organic molecules, starting from a large library of oligoribonucleotides with random sequence. The selected molecules, known as aptamers, exploit the secondary structure of nucleic acids (i.e., their spatial conformation) rather than the complementarity of their nucleotide sequence to bind their specific target. These are usually short (25–40 nt) RNA segments binding their target molecules at relatively high affinity (K_d varying between 1×10^{-9} and 1×10^{-12} mol/l) and blocking their function, either inside the cells or in the extracellular environment. These molecules can be used for either developing diagnostic assays or for therapeutic purposes; therefore, the applications of aptamers are conceptually similar to those of antibodies. Different from antibodies, however, aptamers are easier to obtain, are not immunogenic, and are much more stable when their backbone is modified, for example by including modified pyrimidines with a fluoride atom in the ribose 2' position, or 2'-*O* alkyl groups, or by modifying their 3' end. These modifications have the overall purpose of rendering these molecules resistant to nucleases and preventing their rapid elimination once administered systemically.

The first of these molecules to reach clinical experimentation was a 28-mer, modified RNA aptamer, conjugated with two polyethylene glycol molecules (cf. below), called pegaptanib, able to bind the angiogenic factor VEGF with high affinity. As discussed above, the uncontrolled release of this factor is one of the main factors involved in the development of pathologic retinal neovascularization in the course of wet AMD. Several thousand patients have so far received pegaptanib in the vitreus in different Phase I, II, and III trials; the drug has so far shown very good results overall.

2.2.8
Modes of Delivery or Intracellular Synthesis of Small Regulatory RNAs

The different classes of small, non-coding nucleic acids can be either administered to the cells *in vivo* as synthetic molecules or, in the case of the RNAs, directly expressed inside the cells after delivery of the corresponding genes, under the control of appropriate promoters.

2.2.8.1
Exogenous Administration

In case nucleic acids are administered exogenously, a great variability exists in the efficiency with which different cell types *in vitro* and different tissues *in vivo* internalize these molecules. The pharmacokinetic properties of small DNAs and RNAs can be variously modified by the introduction of different chemical modifications in the phosphate backbone or the sugar moieties of nucleotides, as described in the section on 'Modified

2

Oligonucleotides'. In some instances, these modifications allow acceptable tissue distribution and target cell internalization; in other cases, it is possible to promote internalization of small RNAs using cationic lipids, liposomes, or cationic polymers, as detailed in the chapter on 'Methods for Gene Delivery'.

When these molecules are administered *in vivo*, unless injection is performed in an anatomically constrained area such as the posterior chamber of the eye, small nucleic acids are eliminated from the circulation with excessively fast kinetics. To circumvent this problem, one of the most used technologies is to conjugate the nucleic acids with molecules such as polyethylene glycol (PEG), which increases the half-life of the compounds in the blood stream. This process, known as *PEGylation*, is analogous to that used by the pharmaceutical industry to improve the pharmacokinetics of various recombinant proteins used for human therapy (for example, erythropoietin, interferon, and several others).

Finally, a very interesting strategy to selectively favor entry of small nucleic acids (oligonucleotides, siRNAs, antagomirs, ribozymes) into the cells and, in particular, into hepatocytes in the liver, is to covalently conjugate these molecules with cholesterol. In this manner, the complex cholesterol-nucleic acid is internalized upon its interaction with the low-density lipoprotein (LDL) receptor, which binds the cholesterol moiety and internalizes the complex through an active endocytosis process.

2.2.8.2
Intracellular Synthesis of Regulatory RNAs

Thanks to their specificity, different classes of RNA (in particular, long antisense RNAs, ribozymes, siRNA and miRNAs, RNA decoys, and aptamers) have an enormous therapeutic potential. However, several of the conditions in which application of these molecules would be desirable, including therapy of viral infections by targeting viral RNAs, therapy of cancer by blocking essential cancer cell regulators, or therapy of autosomal dominant disorders by targeting the disease allele, require the prolonged or continuous presence of the therapeutic RNA. This condition is difficult to achieve by exogenous administration. In these cases, a much more effective approach is the delivery, into the relevant cells, of a gene coding for the RNA of interest, thus exploiting the process of transcription for its intracellular production.

Vast experience of the intracellular expression of small RNAs has been achieved using ribozymes. The intracellular production of these molecules can be obtained by inserting their coding DNA sequences within the UTR of genes transcribed by cellular RNA polymerase II (Pol II), such as the actin gene. Alternatively, ribozymes can be transcribed by inserting their coding sequences within those of tRNAs (transcribed by Pol III) or, more efficiently, of the small nuclear RNAs (snRNAs) U1 (Pol II) or U6 (Pol III), involved in the splicing process, or the small adenoviral RNA VA-I (Pol III). The ribozyme usually replaces a portion of these small RNAs and is thus transcribed in the context of these molecules, using, as a promoter, the regulatory regions endowed in the small RNAs themselves. This strategy allows the ribozymes to be expressed at relatively high levels in these cells, since these small RNA transcripts are usually very abundant. Additionally, fusion with a cellular RNA confers the ribozymes structural stability and, in some instances, results in specific subcellular localization, usually the same as the target RNA molecules. For example, the adenovirus VA-I RNA is exclusively found in the cytosol, the same location of cellular mRNAs. Fig. 2.9 shows, as an example, the strategy for cloning a ribozyme targeting the bcr-abl fusion transcript into the U1 and VA-I RNAs.

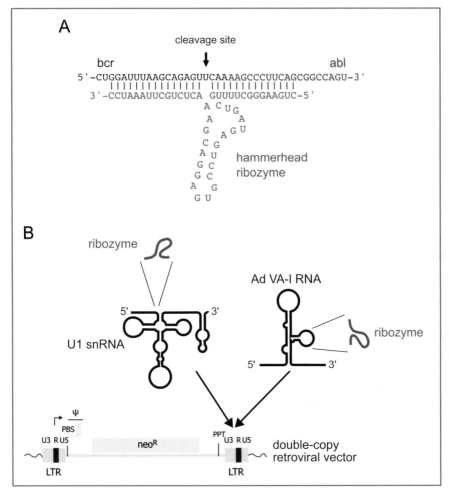

Fig. 2.9 Anti-bcr-abl ribozyme expressed in the context of U1 or VA-I RNAs and conveyed by a retroviral vector. **A** Sequence of an anti bcr-abl hammerhead ribozyme and of its target DNA sequence. The target sequence corresponds to the junction between the cellular bcr (*black*) and abl (*blue*) genes. The sequence of the hammerhead ribozyme is shown in *red*. The cleavage site is indicated by an arrow. **B** Schematic representation of the secondary structure of the U1 snRNA (*left*) and of the adenovirus VA-I RNA (Ad VA-I RNA; *right*) and of the sites of cloning of the ribozyme. The RNA-ribozyme cassettes can be inserted into the 3' U3 region of a gammaretroviral vector, containing, in the example, the selectable neomycin-resistance gene (*neo*^{R}) under the transcriptional control of the viral LTR. Upon reverse transcription in the transduced cells, the RNA-ribozyme cassette is also copied to the 5' LTR (*double-copy vector*)

The transcriptional cassettes consisting of the ribozyme-containing tRNA, snRNA, or VA-I RNA can be cloned into retroviral, adenoviral, or AAV vectors. In case retroviral vectors are used, an interesting option is to clone the transcriptional cassette in the U3 region of the vector 3' LTR, in order to exploit the process of reverse transcription, occurring in the transduced cells, to also duplicate this sequence in the 5' LTR (double-copy vector, cf. section on 'Vectors Based on Gammaretroviruses'). Besides duplicating the number of genes coding for the small RNA-ribozyme, this strategy also allows these molecules to be expressed independently from the viral mRNA, thus minimizing the problem of transcriptional interference between promoters.

Analogous strategies can also be applied to obtain the intracellular expression of RNA decoys, aptamers, or short antisense sequences, the DNA of which has been efficiently inserted into short Pol II (snRNA U1) or Pol III (U6 and U7 snRNA, VA-I RNA, tRNA, rRNA 5S) transcripts, followed by the insertion of these transcriptional cassettes into various vectors. In the case of antisense transcripts, these can also be relatively long and thus be produced as canonical Pol II mRNAs, with the addition of a CAP and a poly-A tail. An example of such a situation can be derived from a clinical trial for HIV-1 gene therapy, entailing the delivery of a lentiviral vector expressing a long antisense RNA sequence complementary to the HIV-1 *rev* gene (cf. section on 'Gene Therapy of HIV Infection').

RNA Pol III promoters can also be exploited for the intracellular expression of siRNAs, since Pol III transcripts are abundant in human cells and have defined 3' and 5' ends, lacking the 5'-cap and poly-A tail proper of mRNAs.

Two kinds of Pol III promoters, from either the U6 snRNA or the H1 RNase P gene, have been used to generate siRNAs *in vivo*. These promoters are classified as Type III promoters, based on their simple and compact structure and absence of regulatory signals downstream of the transcription start site, with the exception of the U6 promoter, which requires the presence of a G at position +1. Other U6 Pol III constructs used for the expression of siRNAs also contain the first 27 nt of the transcribed sequence, which direct methylation of the 5'-γ-phosphate, a modification that stabilizes the transcript. Transcription terminates when it encounters stretches of 4–6 Us yielding small RNA transcripts with two 3' overhangs, which are essential for Dicer processing.

To express siRNAs intracellularly, at least three strategies can be followed. The first is based on the expression of two short complementary RNAs, of 21–23 nt each, from two different transcriptional cassettes (Fig. 2.10A). The second, which is by far the most utilized, entails the expression of a single DNA construct consisting of a siRNA sequence of 21–29 nt, a short 3–9 nt loop region, and the reverse complement of the siRNA sequence and a short terminator (5–6 T residues). Once transcribed, this sequence folds back to form a short hairpin RNA (shRNA). These shRNAs are directly exported from the nucleus by exportin-5 and then, analogous to endogenous miRNAs and other dsRNA precursors, are processed by Dicer to generate 21–23-nt long siRNAs (Fig. 2.10B).

Contrary to Pol III promoters, RNA Pol II transcription units, responsible for transcription of protein-encoding cellular genes, are usually less suitable for the expression of fully active siRNA molecules. Indeed, long dsRNA transcripts from Pol II promoters are rapidly translocated to the cytoplasm where they induce the generalized interferon response. Moreover, functional siRNAs would require minimal or no terminal overhangs; the presence of the 5'-cap and the long 3' poly-A terminal modifications, which characterize Pol II

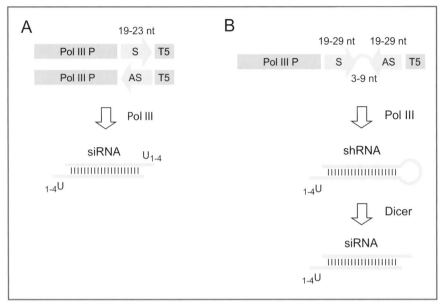

Fig. 2.10 Intracellular synthesis of siRNAs. **A** Generation of *siRNAs* from two different transcriptional cassettes. Two transcriptional cassettes express complementary short RNAs from RNA Pol III promoters, contained in two distinct vectors or positioned in tandem on the same expression unit. The sense (*S*) and antisense (*AS*) RNA strands anneal *in vivo* and generate a double-stranded siRNA molecule. Stretches of 1–4 Uridines (*U*) are present at the 3' ends of each strand. *T5*: transcriptional termination signal consisting of 5 Ts. **B** Expression of short hairpin RNAs (shRNAs). Inverted repeated sequences of appropriate length, separated by a loop spacer of typically 3–9 nucleotides, are transcribed from a Pol III promoter. The resulting RNA fold back into a hairpin structure, which is recognized and processed by the cellular endonuclease Dicer into a siRNA molecule

transcripts, do not adapt well to the structural requirements of siRNA molecules. However, Pol II promoters can be successfully considered when an shRNA-containing cassette is cloned within one of the loops of a natural miRNA, for example by substituting the sequence of the natural loop of miR-30 that is recognized and processed by Drosha (Fig. 2.11A). This arrangement not only allows transcription by RNA Pol II, but also confers stability to the shRNA. A construct obtained in this manner, defined shRNAmir, undergoes the same maturation process as endogenous miRNAs, including cleavage by the RNase III Drosha, followed by nuclear export and subsequent processing by Dicer (Fig. 2.11B). The use of shRNAmir constructs is very interesting, also because the genes coding for these molecules can be placed under the control of tissue-specific or inducible RNA Pol II promoters, similar to the protein-coding genes (cf. above). In contrast to RNA Pol II, Pol III promoters are indeed much more difficult to modify to render them inducible.

Analogous to ribozymes, transcriptional cassettes containing a Pol III or Pol II gene encompassing an shRNA or shRNAmir, respectively, can be vectored into the cells using plasmids or, more efficiently, retroviral, adenoviral, or AAV vectors. In the case of retroviral vectors, the Pol III cassettes are usually inserted into the viral LTR (double-copy vec-

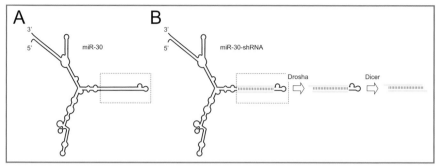

Fig. 2.11 Expression of shRNAs in the context of microRNAs (shRNAmir). **A** Structure of miR-30. The boxed area shows the microRNAs sequence that is substituted with the shRNA. **B** Mir-30-shRNA. The shRNA sequence undergoes the same maturation process as endogenous miRNAs, including cleavage by the RNase III Drosha and subsequent processing by Dicer

tors), while those transcribed by Pol II can be either inserted into this position or, alternatively, between the LTRs, thus becoming the only proviral transcript.

Finally, it should be highlighted that the intracellular expression of shRNAs is fraught with potential toxicity and should be examined with much scrutiny. Similar to externally introduced synthetic siRNAs, endogenous expression of siRNAs can also cause side effects such as activation of innate immunity via induction of the interferon response as well as off-target gene silencing. In addition, because shRNA and shRNAmir constructs exploit the miRNA machinery for their generation and export, they also pose the risk of competing with cellular miRNAs. As an example, in mice, high-level intracellular expression of a shRNA in the liver under the control of the U6 Pol III promoter, delivered using an AAV vector, was recently found to cause dose-dependent liver injury and led to mortality at the highest doses. This toxic effect was attributed to the saturation of exportin-5, the protein that is essential for export of endogenous miRNAs from the nucleus.

2.1 Protein-Coding Genes

Further Reading

Baron U, Bujard H (2000) Tet repressor-based system for regulated gene expression in eukaryotic cells: principles and advances. Methods Enzymol 327:401–421

Brown BD, Naldini L (2009) Exploiting and antagonizing microRNA regulation for therapeutic and experimental applications. Nat Rev Genet 10:578–585

Clackson T (1997) Controlling mammalian gene expression with small molecules. Curr Opin Chem Biol 1:210–218

Duca M, Vekhoff P, Oussedik K et al (2008) The triple helix: 50 years later, the outcome. Nucleic Acids Res 36:5123–5138

Guo ZS, Li Q, Bartlett DL, Yang JY, Fang B (2008) Gene transfer: the challenge of regulated gene expression. Trends Mol Med 14:410–418

Harvey DM, Caskey CT (1998) Inducible control of gene expression: prospects for gene therapy. Curr Opin Chem Biol 2:512–518

Lobato MN, Rabbitts TH (2003) Intracellular antibodies and challenges facing their use as therapeutic agents. Trends Mol Med 9:390–396

Magnenat L, Schwimmer LJ, Barbas CF (2008) Drug-inducible and simultaneous regulation of endogenous genes by single-chain nuclear receptor-based zinc-finger transcription factor gene switches. Gene Ther 15:1223–1232

Muyldermans S, Cambillau C, Wyns L (2001) Recognition of antigens by single-domain antibody fragments: the superfluous luxury of paired domains. Trends Biochem Sci 26:230–235

Toniatti C, Bujard H, Cortese R, Ciliberto G (2004) Gene therapy progress and prospects: transcription regulatory systems. Gene Ther 11:649–657

Selected Bibliography

Baum C, Margison GP, Eckert H-G et al (1996) Gene transfer to augment the therapeutic index of anticancer chemotherapy. Gene Ther 3:1–3

Brown BD, Gentner B, Cantore A et al (2007) Endogenous microRNA can be broadly exploited to regulate transgene expression according to tissue, lineage and differentiation state. Nat Biotechnol 25:1457–1467

Gossen M, Bujard H (1992) Tight control of gene expression in mammalian cells by tetracycline-responsive promoters. Proc Natl Acad Sci USA 89:5547–5551

Lobato MN, Rabbitts TH (2003) Intracellular antibodies and challenges facing their use as therapeutic agents. Trends Mol Med 9:390–396

Meyer-Ficca ML, Meyer RG, Kaiser H et al (2004) Comparative analysis of inducible expression systems in transient transfection studies. Anal Biochem 334:9–19

No D, Yao TP, Evans RM (1996) Ecdysone-inducible gene expression in mammalian cells and transgenic mice. Proc Natl Acad Sci USA 93:3346–3351

Weber W, Fussenegger M (2006) Pharmacologic transgene control systems for gene therapy. J Gene Med 8:535–556

2.2 Non-Coding Nucleic Acids

Further Reading

Brown BD, Naldini L (2009) Exploiting and antagonizing microRNA regulation for therapeutic and experimental applications. Nat Rev Genet 10:578–585

Carthew RW, Sontheimer EJ (2009) Origins and mechanisms of miRNAs and siRNAs. Cell 136:642–655

Castanotto D, Rossi JJ (2009) The promises and pitfalls of RNA-interference-based therapeutics. Nature 457:426–433

Duca M, Vekhoff P, Oussedik K et al (2008) The triple helix: 50 years later, the outcome. Nucleic Acids Res 36:5123–5138

Dykxhoorn DM, Palliser D, Lieberman J (2006) The silent treatment: siRNAs as small molecule drugs. Gene Ther 13:541–552

Fichou Y, Férec C (2006) The potential of oligonucleotides for therapeutic applications. Trends Biotechnol 24:563–570

Hannon GJ (2002) RNA interference. Nature 418:244–251

Opalinska JB, Gewirtz AM (2002) Nucleic-acid therapeutics: basic principles and recent applications. Nat Rev Drug Discov 1:503–514

Que-Gewirth NS, Sullenger BA (2007) Gene therapy progress and prospects: RNA aptamers. Gene Ther 14:283–291

Rao DD, Vorhies JS, Senzer N, Nemunaitis J (2009) siRNA vs. shRNA: similarities and differences. Adv Drug Deliv Rev 61:746–759

Ryther RC, Flynt AS, Phillips JA 3rd, Patton JG (2005) siRNA therapeutics: big potential from small RNAs. Gene Ther 12:5–11

Shi Y (2003) Mammalian RNAi for the masses. Trends Genet 19:9–12

Sioud M, Iversen PO (2005) Ribozymes, DNAzymes and small interfering RNAs as therapeutics. Curr Drug Targets 6:647–653

Stevenson M (2004) Therapeutic potential of RNA interference. N Engl J Med 351:1772–1777

Tuschl T (2002) Expanding small RNA interference. Nat Biotechnol 20:446–448

Wagner RW (1994) Gene inhibition using antisense oligodeoxynucleotides. Nature 372:333–335

Wall NR, Shi Y (2003) Small RNA: can RNA interference be exploited for therapy? Lancet 362:1401–1403

Zentilin L, Giacca M (2004) In vivo transfer and expression of genes coding for short interfering RNAs. Curr Pharm Biotechnol 5:341–347

Selected Bibliography

Bertrand E, Castanotto D, Zhou C et al (1997) The expression cassette determines the functional activity of ribozymes in mammalian cells by controlling their intracellular localization. RNA 3:75–88

Brummelkamp TR, Bernards R, Agami R (2002) A system for stable expression of short interfering RNAs in mammalian cells. Science 296:550–553

de Fougerolles AR (2008) Delivery vehicles for small interfering RNA in vivo. Hum Gene Ther 19:125–132

Dollins CM, Nair S, Sullenger BA (2008) Aptamers in immunotherapy. Hum Gene Ther 19:443–450

Fedor MJ (2000) Structure and function of the hairpin ribozyme. J Mol Biol 297:269–291

Gleave ME, Monia BP (2005) Antisense therapy for cancer. Nat Rev Cancer 5:468–479

Good PD, Krikos AJ, Li SX et al (1997) Expression of small, therapeutic RNAs in human cell nuclei. Gene Ther 4:45–54

Elmen J, Lindow M, Schutz S et al (2008) LNA-mediated microRNA silencing in non-human primates. Nature 452:896–899

James W, al-Shamkhani A (1995) RNA enzymes as tools for gene ablation. Curr Opin Biotechnol 6:44–49

Krutzfeldt J, Rajewsky N, Braich R et al (2005) Silencing of microRNAs in vivo with 'antagomirs'. Nature 438:685–689

Lim LP, Glasner ME, Yekta S et al (2003) Vertebrate microRNA genes. Science 299:1540

Matteucci MD, Wagner RW (1996) In pursuit of antisense. Nature 384:20–22

Mendoza-Maldonado R, Zentilin L, Giacca M (2001) Purging of chronic myelogenous leukemia cells by retrovirally expressed anti-bcr/abl ribozymes with specific celluar compartmentalization. Cancer Gene Ther 9:71–86

Mishra PK, Tyagi N, Kumar M, Tyagi SC (2009) MicroRNAs as a therapeutic target for cardiovascular disease. J Cell Mol Med 13:778–789

Paddison PJ, Caudy AA, Hannon GJ (2002) Stable suppression of gene expression by RNAi in mammalian cells. Proc Natl Acad Sci U S A 99:1443–1448

Prislei S, Buonomo SB, Michienzi A, Bozzoni I (1997) Use of adenoviral VAI small RNA as a carrier for cytoplasmic delivery of ribozymes. RNA 3:677–687

Reynolds A, Leake D, Boese Q et al (2004) Rational siRNA design for RNA interference. Nat Biotechnol 22:326–330

Rossi JJ (2008) Expression strategies for short hairpin RNA interference triggers. Hum Gene Ther 19:313–317

Scherer LJ, Rossi JJ (2003) Approaches for the sequence-specific knockdown of mRNA. Nat Biotechnol 21:1457–1465

Seidman MM, Glazer PM (2003) The potential for gene repair via triple helix formation. J Clin Invest 112:487–494

Soutschek J, Akinc A, Bramlage B et al (2004) Therapeutic silencing of an endogenous gene by systemic administration of modified siRNAs. Nature 432:173–178

Stein CA, Chen YC (1993) Antisense oligonucleotides as therapeutic agents: Is the bullet really magical? Science 261:1004–1012

Suarez Y, Sessa WC (2009) MicroRNAs as novel regulators of angiogenesis. Circ Res 104:442–454

Thompson JD, Macejak D, Couture L, Stinchcomb DT (1995) Ribozymes in gene therapy. Nat Med 1:277–278

Weng DE, Masci PA, Radka SF et al (2005) A phase I clinical trial of a ribozyme-based angiogenesis inhibitor targeting vascular endothelial growth factor receptor-1 for patients with refractory solid tumors. Mol Cancer Ther 4:948–955

Xia H, Mao Q, Paulson HL, Davidson BL (2002) siRNA-mediated gene silencing in vitro and in vivo. Nat Biotechnol 20:1006–1010

Methods for Gene Delivery

3

The success of any gene transfer procedure, either through *in vivo* inoculation of the genetic material or after gene transfer into the patient's cells *ex vivo*, strictly depends upon the efficiency of nucleic acid internalization by the target cells. As a matter of fact, making gene transfer more efficient continues to represent the most relevant challenge to the clinical success of gene therapy.

3.1
Cellular Barriers to Gene Delivery

The plasma membrane lipid bilayer, which is apolar and hydrophobic, constitutes an impermeable barrier to large and charged macromolecules such as DNA and RNA, since, at physiological pH, phosphates in the nucleic acid backbone are deprotonated and thus negatively charged. Therefore, entry of these polyanions into the cells needs to be facilitated, usually by exploiting the same cellular mechanisms that normally allow macromolecule internalization. Alternatively, nucleic acids can be vectored into the cells within biological particles that are naturally capable of crossing biological membranes, such as viruses.

3.1.1
Endocytosis

In physiological conditions, entry of large, polar macromolecules into the cells occurs through a mechanism involving formation of membrane vesicles at the cell surface, followed by their internalization and intracellular trafficking. This process is collectively known as "endocytosis". Over the last several years, a number of different endocytosis mechanisms have been discovered, which are distinguished by the size of the vesicle formed and the molecular machinery involved. The four best understood and most relevant endocytosis pathways are depicted in Figure 3.1.

Gene Therapy. Mauro Giacca
© Springer-Verlag Italia 2010

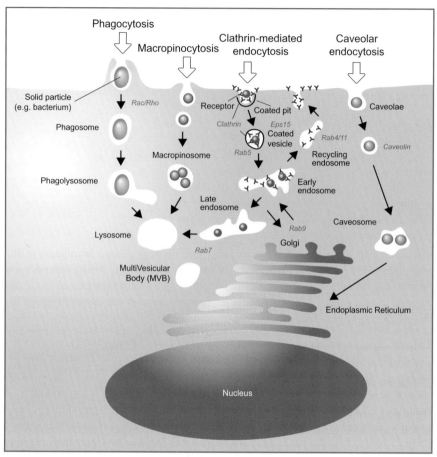

Fig. 3.1 Endocytosis. The four major mechanisms of endocytosis are schematically shown. See text for discussion

Phagocytosis (literally: *cell-eating*) is the process by which specialized eukaryotic cells (in mammals, typically neutrophils and macrophages) internalize large particles (>500 nm in diameter, including cells that have undergone apoptosis or bacteria). Cell surface receptors bind the extracellular particles, initiate local intracellular signals, and reorganize the actin cytoskeleton to induce the change in cell shape needed to engulf the particle, eventually resulting in the formation of a large vacuole known as a *phagosome*. A similar process also occurs in *macro-pinocytosis* (literally, *cell-drinking*), where large endocytic vacuoles (typically >500 nm in diameter) continuously form, resulting in the engulfment of large amounts of both extracellular medium and plasma membrane, containing solutes and single molecules such as proteins, which are thus internalized in a non-specific manner. Both these types of endocytosis eventually end up in the fusion of the vesicles with lysosomes, which represent the major hydrolytic compartment of the cell. These are large (1–2 μm in diameter), acidic (~pH 4.5) vacuoles containing a variety of acid hydrolases (lipase, carbohydrases, nucleases, proteases), targeted to these organelles from the Golgi apparatus through the addition of a mannose-6-phosphate tag.

Receptor-mediated or *clathrin-mediated endocytosis* is a more specific, active event where the plasma membrane folds inward to form pits coated with the cytosolic protein clathrin. These regions of the plasma membrane usually contain specific protein receptors, including the receptors for transferrin, low-density lipoproteins, growth factors, antibodies and several others. Once these receptors bind their specific ligands, an active process is triggered by which small (~100 nm in diameter) vesicles form, having a morphologically characteristic crystalline coat made up of a complex of proteins associated with clathrin (*clathrin-coated vesicles*, CCVs). These vesicles progressively undergo maturation to first become early endosomes, which show a tubulo-vesicular morphology (vesicles up to 1 μm in diameter connected by tubules of ~50 nm in diameter) and mildly acidic pH. These are principally sorting organelles, where many ligands dissociate from their receptors due to the acidic pH and from which many of the receptors recycle to the cell surface. Most of the early endosomes mature into late endosomes, which essentially receive internalized material en route to lysosomes.

A fourth macromolecule internalization pathway is *caveolar endocytosis*. Caveolae are non-clathrin-coated, plasma membrane flask-shaped invaginations (~50 nm in diameter), which exist on the surface of several cell types, including adipocytes, endothelial cells, smooth muscle, and fibroblasts. These microdomains are often associated with the protein caveolin – which, however, is not essential *per se* to the process of clathrin-independent endocytosis – and mostly correspond to the regions of the plasma membrane in which the lipid bilayer is enriched in cholesterol and sphingolipids and is characterized by decreased fluidity; caveolae thus represent one category of the detergent-resistant microdomains of the cell membrane collectively known as *lipid rafts*. Caveolae bud from the plasma membrane and lead to the formation of caveolin-containing endosomes, named *caveosomes*. These are pH-neutral, long-lived compartments, which are known to eventually fuse with the endoplasmic reticulum (ER) or the Golgi apparatus, thus delivering their contents into these compartments. Similar to clathrin-mediated endocytosis, and unlike the other types of endocytosis, caveolar endocytosis requires the GTPase enzyme dynamin. Of note, all types of endocytosis except caveolar endocytosis end up with the delivery of the internalized material into the lysosome compartment, where it is destined for degradation. This has obvious relevance for both gene and drug delivery, since escape from degradation is an essential requisite for efficient treatment.

3.1.2
Escape from the Intracellular Vesicle Compartment

The material internalized into the cells within endosomal vesicles is outside of the cytoplasm, thus still virtually resident in the extracellular environment. Therefore, efficient gene delivery requires the nucleic acids to pass through a biological membrane bilayer, either by destruction of the membrane itself or by physical passage. In this respect, nature has evolved different mechanisms allowing entry into the cytoplasm of macromolecules contained inside the various vesicle compartments, which are variously exploited by a number of cell pathogens to enter the cytoplasm and the nucleus. In particular, viruses and toxins essentially use two main pathways to gain access to the cytosol: they can either be transferred from early or late endosomes into the cytosol in response to low pH, or enter the Golgi and then the ER and be transferred to the cytosol from this destination. An example of the first mechanism is offered by the diphtheria toxin, for which the relatively low pH found in endosomes triggers a conformational change in the toxin that drives formation of a membrane pore allowing its direct entry into the cytosol. Other molecules instead reach the Golgi apparatus and the ER following a route in reverse of the classic secretory pathway. These molecules include plant toxins, such as ricin, and bacterial toxins, such as Shiga toxin, part of cholera toxin and *Pseudomonas* exotoxin A; these often consist of two chains or domains, one responsible for cellular internalization and trafficking, and the other one exerting the biological effect proper of the toxin. Classical secretion is characterized by transport of newly synthesized proteins from the ER to the Golgi, followed by budding of vesiculated cargos from the trans-Golgi network (TGN), vesicle sorting in the cytoplasm, and eventual vesicle fusion with the plasma membrane. Toxin cell entry is enabled by binding to a cell surface molecule, followed by endocytosis. Once in the early or late endosomes, toxins escape lysosomal degradation by re-routing the vesicles toward the TGN and, from this, to the ER, either directly or through the Golgi apparatus. Once toxins have reached the ER, the cytoplasm is accessed by taking advantage of the ER protein auditing system known as ERAD (ER-associated protein degradation). This mechanism eliminates misfolded proteins from the ER by discarding them into the cytoplasm through a pore known as the Sec61 translocon, which is actually also used by several of the toxins. Finally, certain mammalian viruses, such as polyomaviruses, influenza viruses, coronaviruses, and echoviruses, and some toxins, such as cholera toxin, use caveolae-mediated transport from the cell surface to reach the ER. This has been particularly studied for the simian virus 40 (SV40), which is internalized from lipid rafts and, through caveosomes, is eventually released into the ER, from which it escapes to the cytoplasm and gains access to the nucleus via the nuclear pore complex.

In the case of viruses of interest to the gene therapy field, access to the cytoplasm is either attained by direct fusion of the viral envelope with the cell plasma membrane (for retroviruses) or by escape from the endocytic degradation route through the endosomolytic activity of the viral capsid (for adenoviruses and AAVs).

3.1.3
Nuclear Targeting

Finally, a therapeutic nucleic acid, once escaped from the intracellular vesicles, must find its way to the relevant subcellular compartment where its function is exerted, which is usually either the cytosol or the nucleus. Short regulatory RNAs are often active in the former compartment, while coding genes must access the latter to be transcribed. The final destination of the nucleic acid is commonly dictated by the proteins to which it binds once in the cytosol. For example, siRNAs are loaded onto the RNA-induced silencing complex (RISC) and remain in the cytosol (see section on 'Small Regulatory RNAs'). In contrast, coding nucleic acids are bound by various DNA binding proteins, including transcription factors, which direct them to the nucleus thanks to their nuclear localization signal (NLS), or gain access to the nucleus during mitosis. Nuclear targeting can be enhanced through the delivery of nucleic acids complexed with short peptides binding to proteins of the importin/karyopherin family, a set of proteins that actively transport macromolecules into the nucleus.

3.1.4
Methods for Gene Delivery: An Overview

The gene transfer methodologies that are clinically exploitable by gene therapy can be divided into four categories.

(i) Simple utilization of naked plasmids (circular, covalently closed DNA molecules) or short regulatory nucleic acids (oligonucleotides, siRNAs, and others), not complexed with other molecules and simply injected *in vivo* or added to the extracellular milieu of cultured cells.

(ii) Facilitation of nucleic acid entry into the cells by physical methods.

(iii) Transport of nucleic acids into the cells by lipofection.

(iv) Embedding of nucleic acid sequences within viral genomes, then exploiting the natural property of viruses to enter target cells at high efficiency.

Table 3.1 reports a concise view of the main advantages and disadvantages of these methodologies. The production of synthetic DNAs and RNAs, or the use of plasmid DNA produced in bacteria, which can be obtained in large quantities, offer important advantages in terms of simplicity and safety of use compared to viral vectors. Indeed, the efficiency of viral vectors is strictly related to the biological characteristics of the parental virus they derive from, which, in several cases, are still not very well understood. Furthermore, production of viral vectors requires the development of complex procedures based on cell culture and infection in order to obtain packaging of the therapeutic nucleic acids inside the virions. These procedures pose important safety problems, due to the possibility that either the packaging cells contain other infectious agents or the viral vector itself might recombine to generate a replication-competent virus. Finally, some viral vectors induce a powerful inflammatory and immune response once injected into the patients. Despite these problems, the efficiency of gene transfer that can be attained by viral gene transfer both *in vivo* and *ex vivo* is by far superior to that of any non-viral method. In addition, only viral vectors allow persistent, often permanent, expression of the therapeutic gene in their target tissues *in vivo*.

Once entered into the cells, the fate of the delivered nucleic acid strictly depends on its internalization route and chemical structure: plasmids, oligonucleotides, and small RNAs are usually degraded and lost with a kinetics ranging from a few hours (for small RNAs) to several days (for plasmids). However, synthetic nucleic acids, including oligonucleotides, siRNAs, and aptamers, can be chemically modified in order to escape degradation by cellular nucleases; in this way, their persistence inside the cells is significantly increased. When therapeutic nucleic acids are carried into the cells by viral vectors, their destiny depends on the biological characteristics of the vector that is used. Vectors based on adenoviruses persist for prolonged periods in an episomal, non-integrated form inside the nucleus of the transduced cells; however, the cells themselves are usually rec-

Table 3.1 Pros and cons of the major gene transfer procedures for gene therapy

Strategy	Method	Pros	Cons
Naked DNA or RNA	Direct injection *in vivo*	Simplicity of production and use Potential use as genetic vaccines	Low efficiency Transitory effect Internalization only in skeletal and cardiac myocytes and in antigen presenting cells
Physical methods	Electroporation	Relatively easy to set up for skeletal muscle and skin; invasive for other organs	Low efficiency Transitory effect Limited spectrum of applications
	Increase of hydrodynamic pressure	Usually invasive	Low efficiency Transitory effect
	Ultrasounds (sonoporation)	Relatively easy to set up	
	Bombardment with DNA-coated gold particles (gene gun)	Relatively easy to set up Stimulation of an effective immune response	Limited to gene transfer to the skin
	Jet injection		
Chemical methods	Liposomes Cationic lipids Cationic polymers Proteins	Relatively easy to set up and use	Relatively low efficiency Transitory effect
Viral vectors	Vectors based on: gammaretroviruses, lentiviruses, adenoviruses, adeno-associated viruses (AAVs), herpesviruses	High efficiency of gene transfer both *in vivo* and *ex vivo* For some vectors, persistence of therapeutic gene expression *in vivo*	Possible induction of immune and/or inflammatory response Limited cloning capacity Complexity of production For some viruses, tropism limited to specific cell types Insertional mutagenesis (for gammaretroviruses) In most cases, incomplete knowledge of the molecular mechanisms governing viral replication

ognized and destroyed by the immune system in a 1–2-week period. The same immune response prevents any possibility to re-inoculate a vector displaying the same serotype. In contrast, vectors based on the adeno-associated virus (AAV) persist in episomal form in non-replicating cells for month- or year-long periods, while retroviral vectors become integrated into the host genome. Both these vectors are thus useful for applications in which prolonged or permanent gene expression is desirable.

Given the broad spectrum of biological properties displayed by both non-viral and viral methods for gene transfer, it is evident that no perfect universal system exists. Thus, the choice of the proper gene transfer methodology strictly depends on the characteristics of the disease for which gene therapy is developed and the attainable modality of gene transfer.

The main gene transfer methodologies that are currently available for gene therapy are described and discussed in the following sections.

3.2
Direct Inoculation of DNAs and RNAs

As outlined above, the chemical and physical characteristics of the plasma membrane prevent the direct passage of large and charged macromolecules, such as plasmid DNA. Different cell types, however, have the capacity to internalize small nucleic acids, including oligodeoxynucleotides, RNA decoys, or siRNAs through an active endocytic process, usually exploiting the clathrin-mediated endocytosis pathway. As discussed above, most of the content of the endocytic vesicles formed along this pathway is destined to lysosomal degradation. However, a tiny fraction can escape the early or late endosomes, cross the endosomal membrane, gain access to the cytosol, and, from this compartment, be transported to the nucleus. This process, although highly inefficient, forms the basis of a few clinical trials taking advantage of chemically modified oligonucleotides or siRNAs, administered to the patients intravenously or injected in anatomically defined compartments such as the eye's posterior chamber in the form of naked nucleic acids (cf. sections on 'Gene Therapy of Cancer' and 'Gene Therapy of Eye Diseases' respectively).

In some cell types, the process of internalization of extracellular nucleic acids and their release into the cytosol is relatively more efficient. In particular, this is the case of striated muscle fibers and cardiomyocytes, which are also able to internalize naked plasmids simply injected into the skeletal muscles or the heart *in vivo*. This property, although modest, is exploited by a few clinical trials, especially aimed at inducing therapeutic angiogenesis in the ischemic tissues (cf. section on 'Gene Therapy of Cardiovascular Disorders').

Another cell type showing striking capacity to internalize plasmid DNA present in the extracellular milieu is the professional antigen-presenting cell (APC). These cells, which comprise dendritic cells – including Langerhans cells of the skin – macrophages, and B-lymphocytes, are very efficient at internalizing foreign antigens, by either phagocytosis or receptor-mediated endocytosis, followed by their processing and presentation to T cells via both MHC class II and MHC class I molecules. Although the process of internalization of plasmid DNA by these cells has limited efficiency, this is still sufficient to induce the intracellular synthesis of the proteins encoded by the plasmids, followed by their presentation as antigens to induce an immune response. This process is exploited by the strategy of genetic vaccination (or DNA vaccination) and will be discussed in the section on 'Gene Therapy of Cancer'.

In all other cases, entry of both short nucleic acids and large plasmids into the cells needs to be facilitated by physical or chemical treatments, or by using viral vectors.

3.3
Physical Methods

Over the last several years, relevant progress has been made in the utilization of physical methods to facilitate entry of plasmid DNA or short regulatory DNAs or RNAs into the cells. These methods are essentially aimed at bringing the nucleic acids in strict contact with the plasma membrane and/or determining the temporary localized disassembly of the membrane itself.

3.3.1
Electroporation

Electroporation (also termed *electropermeabilization* or *electrotransfer*) was originally developed as a means to deliver genes into cultured cells. Subsequently, it has also been applied to *in vivo* gene transfer to the skin, muscle, liver, and, more recently, to a variety of other organs, including the kidney, lung, heart, and retina. The method consists in the application of a series of electric pulses (typically, in the order of ~200 V/cm for tens of milliseconds, or higher voltages for microseconds) in order to induce a transient increase in membrane permeability and thus allow entry of large, charged macromolecules, including plasmid DNA. To discharge the pulse, two electrodes of various shapes are positioned flanking the site of inoculation of a solution containing the nucleic acids. The electric pulse induces the formation of hydrophilic pores in the cell membrane and the subsequent passive passage of DNA through these pores thanks to a local electrophoretic effect. At the end of the stimulus, the membrane acquires its normal properties again.

One of the most important problems of electroporation is the induction of tissue damage due to the electric pulses, which essentially limits the application of this technology. In addition, the expression of the internalized plasmid DNA is often transitory and usually lost within a few days. In the skeletal muscle, which represents one of the most interesting tissues for electroporation, the efficiency of gene transfer can be increased by the administration, prior to electric discharge, of the enzyme hyaluronidase, which degrades hyaluronic acid in the extracellular matrix and thus increases gene transfer efficiency by favoring diffusion of the nucleic acids.

3.3.2
Hydrodynamic Intravascular Injection

Transient local increase of hydrostatic pressure significantly augments cellular internalization of nucleic acids circulating in the blood. This strategy, named *hydrodynamic gene transfer*, first allows DNA or RNA to cross endothelial cell junctions, by inducing their separation, and later determines the transient formation of pores or microdefects in the plasma membranes of the target cells underlying the endothelium, a process named *hydroporation*.

Hydrodynamic gene transfer can be applied to different organs *in vivo*, including liver, skeletal muscle, and heart. An increase of hydrostatic pressure can be generated by injecting a solution containing the plasmid, oligonucleotide, or siRNA of interest at high pressure into the relevant area (for example, into the femoral artery to achieve diffuse transfection of the lower limb skeletal muscles), or by transiently occluding the veins draining from the area (for example, the superior vena cava for the diaphragm or the coronary sinus for the heart), in order to selectively increase blood pressure in the region where the therapeutic gene is injected.

3.3.3
Sonoporation

Both electroporation and hydrodynamic gene transfer are quite invasive, and thus difficult to apply for gene transfer to most organs. In contrast, ultrasound waves are used clinically for a variety of diagnostic and therapeutic applications. Low-intensity ultrasound permits a number of non-invasive diagnostic examinations (echography), while high-intensity ultrasound is used for the non-invasive treatment of urinary calculosis (extracorporeal shock wave lithotripsy, ESWL) and high-intensity focused ultrasound (HIFU) finds application for the thermal destruction of tumors. All these different modalities of ultrasound application can facilitate the transfer of plasmids and other small nucleic acids into the cells. This methodology is also collectively known as *sonoporation*.

The facilitation of gene transfer by sonoporation is due to the capacity of ultrasound to generate acoustic cavitation, which ultimately determines formation of micropores in the plasma membrane. Cavitation is increased by agents causing nucleation, such as echographic contrast agents based on gas microbubbles (typically, microbubbles filled with perfluoropropane with an albumin shell). In this case, rupture of the microbubbles caused by ultrasound increases permeability of the membrane and thus facilitates gene transfer.

Sonoporation can be achieved by injecting a plasmid or an oligonucleotide in the blood and focusing the ultrasound beam on a specific body region, typically the vascular wall, the heart, or the skeletal muscle; entry of nucleic acids into the endothelial cells, cardiomyocytes, or skeletal muscle fibers, respectively, is induced by the local and transitory increase of membrane permeability.

Notwithstanding the relative ease of assembly and the non-invasiveness of this procedure, the extent of gene transfer using sonoporation is still difficult to standardize and very variable according to the different experimental conditions.

3.3.4
Bombardment with DNA-Coated Microparticles ("Gene Gun")

Among the physical methods for gene transfer, one very interesting approach is the possibility of delivering DNA into the cells by bombarding them with micron-sized beads carrying plasmid DNA adsorbed onto their surface. The most utilized version exploits a special type of gun ("gene gun") shooting gold or tungsten particles at very high velocity into the tissues. These particles can easily cross the cell and nuclear membranes and release the DNA adsorbed on their surface into the nucleus. This method, also named *ballistic* or *biolistic transfection*, is derived from experience gained from gene transfer in plants, where it was originally invented as a way to cross the rigid plant cell wall.

Biolistic transfection now finds application for gene therapy of accessible tissues, such as skin, with the main purpose of delivering genes coding for antigenic protein in the context of DNA vaccination against tumor or viral antigens. Once in the dermis, microparticles are taken up by APCs, which can thus process the encoded antigens and present them to T lymphocytes for immune stimulation (see section on 'Gene Therapy of Cancer'). Another tissue in which biolistic bombardment was successfully performed in animal experimental models is the cornea.

3.3.5
Injection of DNA using High-Pressure Jets ("Jet Injection")

One possibility to facilitate entry of naked DNA or RNA into the cells is so-called *jet injection*. In this technology, a solution containing the nucleic acid is applied to the skin as a jet of high velocity with the force to penetrate the skin and the underlying tissues, thus determining spread transfection of the areas of interest. Jet injection has deeper penetration capacity compared to ballistic bombardment (down to 1 cm in depth) and, besides the skin and the underlying tissues, can also be applied to other accessible tissues, including solid tumors. A Phase I clinical study was performed in which jet injection was used for gene therapy of skin metastasis in patients with breast cancer and melanoma to assess the extent of gene transfer.

3.4
Chemical Methods

The purpose of physical methods for gene transfer is to facilitate entry of nucleic acids into the cells essentially by modifying the properties of biological membranes using physical forces such as pressure or electricity. Instead, chemical methods are aimed at modifying the properties of nucleic acids themselves, by promoting their association with molecules able to reduce their hydrophilicity and neutralize their charge, ultimately leading to an increased cellular uptake.

The molecules used to facilitate gene transfer can be classified into one of three categories: lipids (liposomes and cationic lipids), proteins, and cationic polymers.

3.4.1
Liposomes and Cationic Lipids (Lipofection)

Liposomes are closed vesicles formed by one or more lipid bilayers surrounding a core aqueous compartment; a variant of liposomes are micelles, consisting of lipid spheres

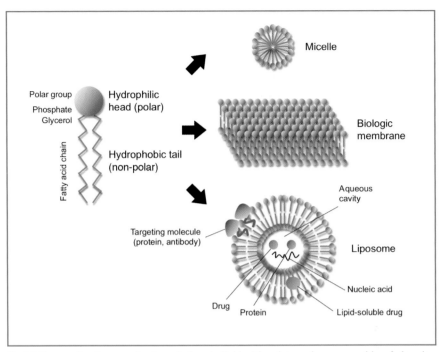

Fig. 3.2 Supramolecular organization of phospholipids. The picture shows assembly of phospho-lipids (*left*) into micelles (*upper right*), biological membranes (*middle right*), or liposomes (*lower right*). Macromolecules, such as chemical drugs, nucleic acids, or proteins, can be transported in the liposome core or, if hydrophobic, within the lipid bilayer. Proteins mediating specific cellular targeting can be embedded on the liposome surface

lacking the inner aqueous compartment (Figure 3.2). Liposomes were originally developed in the 1960s and are now extensively used to convey different molecules in a variety of applications, ranging from chemotherapy (for example, transport of antiblastic or antifungal drugs to prevent unspecific toxicity), diagnostic imaging, and cosmetic applications. The first developed liposomes were based on the same phospholipids forming biological membranes, having a polar head and a lipophilic tail formed by fatty acids. These molecules have amphipathic (or amphiphilic) characteristics: once dispersed in an aqueous solution, they tend to spontaneously assemble into a bilayer, first forming a sheet and then closing up into a vesicular structure with a central aqueous core. When liposome formation occurs in a solution containing a drug or a nucleic acid, these are eventually found in the aqueous core of the liposome and can thus be transported by it. Once in contact with a cell, liposomes can directly fuse with the plasma membrane, thus liberating their content into the cytosol, or, more frequently, be actively endocytosed.

The biological properties of liposomes derive from those of the amphiphilic lipids they are composed of; according to the characteristics of the polar head groups, they can be classified into anionic, cationic, zwitterionic, and non-ionic liposomes. Conventional liposomes, non-ionic or neutral, interact inefficiently with a large polyanion such as DNA. In contrast, DNA binding is much more effective using cationic lipids. Figure 3.3 shows the structure of two of the most used cationic lipids, DOTMA (the first one to be utilized, in 1987) and DOTAP. Both consist of two fatty acyl chains joined to a positively charged propylammonium group through an etheric and esteric bond, respectively. The positive moiety of the cationic lipid binds negatively charged DNA very efficiently and induces its condensation. Furthermore, the DNA/lipid complex maintains a positive net charge and is thus capable of electrostatically interacting with the negatively charged cell surface. In addition, the complex displays fusogenic properties, thus promoting fusion of the liposome with the cell or the endosome membrane and favoring release of DNA into the cytosol. The DNA–lipid complex is named *lipoplex*.

Over the last several years, a variety of other cationic lipids have been produced, differing in hydrophobic moiety, number of positive charges, or presence of other chemical groups mediating interaction between the polar and hydrophobic portions. One of the developed lipids is DC-Chol (Figure 3.3), in which the hydrophobic moiety consists of a sterol skeleton. This lipid is currently used in different gene therapy applications including a few for cystic fibrosis. Some of the cationic lipids, once mixed to DNA, form micellar rather than vesicular structures, and are thus more efficient at inducing DNA condensation.

In general, the maximum efficiency of gene transfer is achieved when a cationic lipid is mixed with cholesterol or with a zwitterionic lipid, which displays an overall neutral charge despite carrying both negative and positive charges on different atoms; an extensively used zwitterionic lipid is DOPE (Figure 3.3). These co-lipids, once taking part in the formation

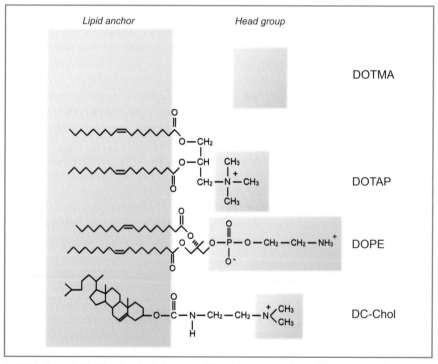

Fig. 3.3 Cationic lipids. The chemical structure of four lipids commonly used for gene transfer is shown

of the cationic lipid-DNA complex, facilitate fusion or destabilization of the cellular membranes, thus favoring transfection. As a matter of fact, the mixture DOTMA/DOPE is currently one of the most successful commercial lipid formulations for gene transfer into cultured cells.

A lipoplex is typically obtained in the laboratory by mixing a cationic lipid, a co-lipid, and DNA in appropriate concentrations, followed by sonication of the mixture in order to reduce the size of the complexes that are spontaneously formed. The final structures of the different lipoplexes are still not completely defined and probably vary from preparation to preparation according to the chemical composition and the molar ratio of the two or three lipoplex components. Molecular aggregation can lead to the formation of simple structures, consisting in DNA covered by a cationic lipid bilayer, or in aggregates of cationic liposomes surrounding DNA such as a pearl chain, or to the generation of more complex arrangements, such as multilamellar structures or hexagonal structures formed by DNA covered by monolayers of cationic lipids assembled in a bi-dimensional hexagonal lattice (Figure 3.4). Accordingly, the final lipoplex preparations can have very different sizes, with a diameter varying from a few nanometers (unilamellar liposomes) to several hundred nanometers (multilamellar and complex lipoplexes).

From the gene therapy point of view, the ideal lipoplex should provide protection from nuclease degradation of DNA, mediate very efficient cellular internalization, and exert minimal cell toxicity. At the same time, it should also display a neutral or negatively charged surface in order to escape unspecific interaction with blood components. None of the currently available lipoplex formulations completely satisfies all these requisites. As a matter of fact, positively charged lipoplexes induce DNA condensation more efficiently, leading to higher *in vitro* and *in vivo* levels of transfection. For reasons still not completely explained, lipoplexes having a size >200 nm are more efficient than smaller ones (50–100 nm).

Despite the extensive use of cationic liposomes for *in vivo* and *in vitro* gene transfer, the mechanism by which they release their DNA into the cells is still unclear, and probably varies according to the chemical and structural properties of the various lipoplexes. Most studies indicate that, following interaction with the cell membrane, the lipoplex is internal-

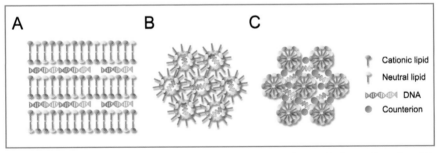

Fig. 3.4 Lipoplexes. Schematic representation of structures formed upon the interaction of nucleic acids with polar lipids. **A** Lamellar structure, where DNA molecules are sandwiched between lipid bilayers formed by an alternation of cationic and neutral lipids. **B** Inverted hexagonal structure, where DNA molecules are coated with a lipid monolayer arranged on a hexagonal lattice. **C** Intercalated hexagonal structure, where DNA molecules are interspersed with lipid micelles arranged on a hexagonal lattice

ized through a pathway of clathrin-mediated endocytosis rather than directly fusing at the plasma membrane level. In this respect, and notwithstanding the continuous progress in the generation of more efficient lipoplexes, it should be remarked that the number of DNA molecules effectively reaching the nucleus only represents a tiny fraction of those having entered the transfected cells (in the order of 1 out of 1×10^4–10^5). In fact, most of the lipoplexes remain trapped in the endosomes and are eventually degraded by lysosomes. Exit from the endosomes before reaching the lysosomes is favored by the inclusion, in the lipoplexes, of zwitterionic lipids such as DOPE, since the progressive acidification of the endosomal compartment favors the membrane-destabilizing properties of this type of lipid.

3.4.2
Cationic Polymers

Another interesting class of molecules with the property of binding DNA and favoring its transfer into the cells consists of cationic polymers. These molecules include poly-(L-lysine), poly-(L-ornithine), linear or branched polyethylenimine (PEI), diethyl-aminoethyldextran (DEAE-D), poly-(amido amine) dendrimers, and poly-[2-(dimethyl-amino)-ethyl methacrylate (poly-(DMAEMA)). These polymers, which can have linear, branched, or dendrimeric structures (Figure 3.5), usually carry a protonable amine group, which, thanks to its positive charge, binds DNA and induces its condensation. Similar to lipoplexes, the DNA/polymer complexes, named *polyplexes*, enter the cells through an active endocytosis process. Once in the endocytic vesicles, the positively charged amine groups of the polymers are believed to exert a so-called "proton sponge" effect, according to which the low endosomal pH determines entry, into the endocytic vesicles, of chloride ions, followed by osmotic rupture of endosomes and release of DNA into the cytosol.

One of the major problems related to the use of polyplexes for gene transfer is their toxicity, due to the positive charge of the polymers and the large size of the polymer–DNA complexes that are eventually formed. For this reason, several laboratories are currently investigating the possibility of improving polymer architecture and biophysical properties. One class of very interesting polymers in this respect is the amphipathic block co-polymers, consisting of alternating blocks of simple hydrophobic homopolymers and simple hydrophilic homopolymers. Such block co-polymers, which are significantly less toxic than conventional cationic polymers, display the property of interacting, at the same time, with DNA through their hydrophilic moieties and with the plasma or endosomal membranes through their hydrophobic moieties.

Another interesting class of polymers showing low toxicity and high biocompatibility are biodegradable polymers. An example of such polymers is the degradable polyester poly[α-(4-aminobutyl)-L-glycolic acid] (PAGA), a derivative of poly-[L-lysine], which binds DNA and subsequently releases it once the polymer is degraded.

Finally, an additional family of polymers with attractive properties for gene transfer includes the so-called "intelligent polymers". These polymers can undergo ample and often discontinuous variations in their chemical and physical characteristics in response to environmental changes, such as pH, temperature, ionic force, or presence of electric or magnetic fields. The polymer modification consists in variations of the size, three-dimen-

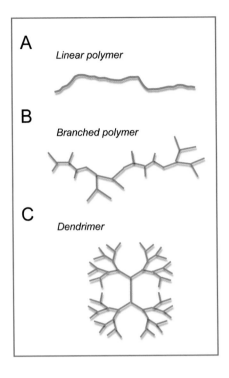

Fig. 3.5 Cationic polymers. **A** Linear polymer;
B branched polymer; **C** dendrimer

sional structure, or reactivity towards other molecules. Typical examples of intelligent polymers consist of block co-polymers formed by methyl methacrylate (MMA) and 2-(dimethylamino)-ethyl methacrylate (DMAEMA). MMA is hydrophobic, while DMAEMA is hydrophilic; the block co-polymer is thus formed by alternating hydrophobic and hydrophilic blocks, with a prevailing overall hydrophilic characteristic. DMAEMA, however, becomes more hydrophilic at lower pH and more hydrophobic at higher pH, thus determining precipitation of the co-polymer. Varying the ratio between the two co-polymers improves transfection efficiency and minimizes cytotoxicity.

Intelligent polymers sensitive to temperature behave according to a similar principle. For example, poly-(N-isopropylacrylamide) (NIPA) is soluble in water below 32°C and becomes insoluble above this temperature. In principle, polymers containing this molecule should permit *in vitro* assembly of polyplexes at low temperature in the laboratory; once injected *in vivo*, precipitation of the co-polymer induces the formation of a gel, which should allow progressive release of DNA over prolonged periods of time.

Finally, a peculiar class of polymers is dendrimers (from the Greek "dendron", tree). These consist of a central molecule, acting as a root for the progressive synthesis of a vast number of branches, which are structured in an ordered and symmetric manner (Figure 3.5C). Similar to cationic polymers, dendrimers have the capacity to efficiently complex with DNA and mediate its cellular internalization by endocytosis, followed by endosomal release by osmotic swelling. Dendrimers can be potentially used for the delivery of long stretches of DNA (several tens of megabases) and display, in a few experimental systems, higher efficiency of gene transfer than linear polymers.

3.4.3
Proteins

The efficiency of viruses to convey their nucleic acids into the cells is due to the presence, within the viral particles, of specific proteins. These mediate a series of essential functions, such as condensation of viral nucleic acids, protection against extracellular nucleases, binding to cell surface receptors, fusion of viral envelope with cell membranes or endosome disruption, and, finally, transport of viral DNA or RNA to the nucleus. Some of these processes can be mimicked by using specific proteins or protein domains in the context of non-viral gene delivery.

Some basic proteins, such as the polycationic polypeptide protamine or histones, bind negatively charged DNA with high affinity, promoting its condensation and preventing its degradation. Once in the extracellular environment, these molecules are also able to bind heparan sulfate proteoglycans (HSPGs), a family of negatively charged glycosylated proteins expressed on the cell surface and released into the extracellular matrix. HSPGs exposed on the cell surface continuously undergo a process of endocytosis; thus, molecules interacting with HSPGs outside the cells are also internalized within endocytic vesicles. Since the efficiency of this process is relatively modest and the endocytosed molecules still need to cross the endosomal membrane, the simultaneous utilization of liposomes or cationic lipids together with basic proteins favors DNA transfection.

A specific protein able to favor the internalization of other covalently linked macromolecules is the Tat protein encoded by HIV-1. This factor is a powerful transactivator of viral gene expression and is thus essential for viral replication and infectivity. For completely unknown reasons, Tat is released by HIV-1-infected cells through a non-canonical, Golgi-independent pathway of secretion and, once in the extracellular milieu, is taken up by cells upon the interaction of a highly basic, 9-amino acid domain of the protein with cell surface HSPGs, which mediate its endocytosis through the caveolar pathway. When this small amino acid domain of Tat is fused to heterologous proteins, synthetic nanoparticles, liposomes, or small nucleic acids (for example, siRNAs), Tat mediates their cellular internalization. Since this process occurs through caveolar endocytosis, a significant part of the Tat-fusion cargo escapes lysosomal degradation, finds access to the cytosol, and, from here, is transported to the nucleus.

As discussed above, internalization of lipoplexes and polyplexes occurs through an active endocytic process, mainly mediated by clathrin-coated vesicles. It is thus possible to associate, to the lipoplexes or polyplexes, proteins of various derivation, able to recognize specific cellular receptors involved in endocytosis. The ultimate purpose of this approach is twofold, namely on one hand to increase the overall efficiency of the gene transfer process while, on the other hand, targeting transfection towards specific cell types or tissues expressing the receptors of interest. The proteins that have been used for this purpose include various lectins (proteins that are very diffuse in nature, having the capacity to bind the glycidic moiety of various glycoproteins and glycolipids); the asialoglycoproteins (i.e., glycoproteins devoid of sialic acid, which specifically bind a receptor expressed on the surface of hepatocytes; this receptor, named ASGP-R, recognizes glycoproteins carrying a galactose at their extremity and removes them from the circulation by endocytosis followed by lysosomal degradation); integrin ligands (for example, peptides carrying the amino acid sequence

Arg-Gly-Asp, RGD); peptides derived from apolipoprotein E (which bind the low-density lipoprotein receptor, LDL-R, expressed by hepatocytes); and transferrin (which binds the transferrin receptor, expressed by a variety of cell types and overexpressed in tumor cells). Finally, lipoplexes and polyplexes can also be targeted towards specific cellular receptors through their association with monoclonal antibodies or single-chain antibodies (scFvs; cf. section on 'Antibodies and Intracellular Antibodies'). Both scFvs and peptides binding a specific receptor of interest can be selected thanks to the phage display technology.

Collectively, however, it should be remembered that binding a specific cellular receptor does not necessarily translate into a parallel increase of transfection efficiency. Indeed, this parameter strictly depends not only on the efficiency of internalization but also, and probably mostly, on the capacity of the nucleic acids to exit the endosomes before they are degraded in the lysosomes. This is also proven by the observation that, in cell culture, treatment with compounds that raise the pH of acidic vesicles, such as the anti-malaria drug chloroquine or the macrolide antibiotic bafilomycin A1, a selective inhibitor of the vacuolar-type ATPase (V-ATPase), significantly increase the efficiency of transfection. A few natural proteins and peptides possess natural endosomolytic activity, and can thus be used to form lipoplexes or polyplexes with improved transfection efficiency. These include peptides derived from the HA2 subunit of the hemagglutinin (HA) protein of the influenza virus (which mediates a low-pH-dependent fusion reaction between the viral envelope and the endosomal membrane following cellular uptake of the virus particles by receptor-mediated endocytosis), the envelope of the Sendai virus (a paramyxovirus also called hemoagglutinating virus of Japan, HVJ, which is a powerful inducer of membrane fusion), and a synthetic amphipathic peptide, sensitive to pH, called GALA (named as such because it contains repeat units of glutamic acid-alanine-leucine-alanine). Alternatively, some gene therapy applications have exploited the natural endosomolytic properties of the capsid proteins of adenovirus, by including whole adenoviral virions, inactivated by UV radiation, in the lipoplex.

Finally, other peptides can facilitate the subsequent passage, namely transport of DNA from the cytosol to the nucleus. In particular, peptides carrying a NLS are recognized by the cellular proteins of the importin/karyopherin family, which, in the cells, mediate transport of NLS-containing sequences through the nuclear pores into the nucleus.

3.4.4
Chemical Methods for Gene Transfer: Pros and Cons

Although the chemical methods of gene transfer offer important advantages over viral gene delivery in terms of relative safety and simplicity, and notwithstanding the large amount of research carried out on these methods over the last 20 years, their overall efficiency is still unsatisfactory. Most of the DNA entering the cells remains trapped in the endosomes and is eventually destroyed. In addition, plasmid DNA reaching the nucleus is unprotected from degradation by cellular nucleases and, since it does not integrate into the cellular genome, is progressively lost, thus allowing transgene expression for periods usually shorter than a couple of weeks.

When lipoplexes or polyplexes are administered systemically, additional problems ensue due to their rapid update and elimination by the cells of the reticuloendothelial system (RES) in spleen, liver, and lymph nodes. With analogy to several other conventional drugs, the most widely used system to avoid unspecific interactions is to mask the positive charges of lipoplexes and polyplexes with neutral hydrophilic polymers, such as polyethylene glycol (PEG). On one hand, PEGylation prevents aggregation, thus favoring the formation of smaller complexes (which is usually an advantage for gene transfer), while, on the other hand, it blocks unspecific interaction of the complexes with serum proteins or other extracellular components, thus increasing their persistence in the blood stream. PEGylation, however, in several conditions also leads to a decreased interaction of the complexes with the cells, thus diminishing their biological activity.

Finally it should be noted that, although less immunogenic than viral vectors, lipoplexes and polyplexes are internalized by macrophages and other APCs and can thus elicit an immune response against both the gene transfer molecules and the delivered transgenes. Additionally, cationic liposomes can be toxic, since they rapidly induce the production of pro-inflammatory cytokines, such as TNF-α, IL-6, IL-12, and IFN-γ. In the case of plasmid DNA delivery, part of this response is also due to the presence, within the plasmid, of bacterial, non-methylated CpG sequences, which usually represent a powerful stimulus for immune response through the activation of Toll-like receptor-9 (TLR-9).

In light of these considerations, it is not surprising that most (~70%) of the gene therapy clinical trials, and in particular those aimed at delivering coding genes, exploit viral vectors rather than non-viral methods for gene transfer.

3.5
Viral Vectors

The most efficient system to deliver a gene into a cell is to exploit the properties of vectors derived from the viruses that infect animal cells. In their replicative cycle, viruses make use of very efficient mechanisms to internalize their own genome into the target cells, which have evolved over million years. In the most simplistic view, a viral particle is a tiny object composed of a nucleic acid and a few proteins that impede its degradation in the extracellular environment and mediate its internalization into the target cells. In general terms, the process of viral replication is sustained by the interaction of several proteins of viral or cellular origin with their respective, specific targets on the viral genome. The proteins are said to act *in trans* and the targets *in cis*. Examples of such *cis/trans* interactions are those regulating activation of promoters positioned within the viral genome, transport of the viral nucleic acids from the nucleus to the cytoplasm, or packaging of viral genomes inside the virions.

A viral genome modified in order to accommodate an exogenous sequence of interest (the therapeutic gene in the case of gene therapy) is called a *vector*. The principles according to which the different viral vectors are obtained starting from the parental genomes are common to all systems. They consist in the: (i) removal, from the viral genome, of most genes coding for viral proteins and, in particular, of those that are potentially pathogenic; (ii) maintenance of the *cis*-acting sequences of the viral genomes required for viral replication; in particular, those determining inclusion of the genomes within the viral particles (packaging signal, ψ); (iii) expression of the viral genes required for viral replication within the virus-producing cells (called *packaging cells*) from genes encoded by transiently transfected plasmids, or expressed in the context of a helper virus simultaneously infecting the packaging cells, or directly contained inside the packaging cell genome thanks to previous engineering of these cells.

Five classes of vectors are currently in an advanced stage of clinical experimentation for human gene therapy. These include viruses derived from the *Retroviridae* family (gammaretroviruses and lentiviruses), adenoviruses, AAVs, and herpesviruses. Other viruses, such as vacciniaviruses, the viruses belonging to the spumavirus and alpharetrovirus genera of the *Retroviridae* family, and RNA viruses such as the Semliki Forest Virus are also considered potentially attractive for therapeutic gene transfer, however their use is limited to vaccination (for vacciniaviruses) or they still need vast preclinical development and validation.

The modalities of production and the characteristics of the five main classes of viral vectors are detailed in the following sections, along with the main characteristics of each of the parental viruses.

3.5.1
Vectors Based on Gammaretroviruses

The vast majority of the clinical trials conducted in the 1990s took advantage of the properties of viral vectors based on gammaretroviruses. Among the properties of these viruses are their relative genetic simplicity, their efficiency in infecting a vast series of different cell types, and their peculiar ability to integrate their genetic information into the genome of the infected cells, a characteristic leading to permanent genetic modification.

3.5.1.1
Molecular Biology and Replicative Cycle of Retroviruses

The *Retroviridae* family includes a vast series of enveloped viruses with a positive-strand RNA, having a common genetic structure and replicative cycle. A peculiar characteristic of all members of this family is the presence of an enzyme, reverse transcriptase (RT), which copies the viral RNA into a double-stranded cDNA form, which eventually integrates into the infected cell genome. The integrated DNA form of the viral genome is named *provirus*.

Classification of Retroviruses

The different members of the *Retroviridae* family have been variously classified according to their morphology, natural animal host, type of disease caused, and tropism for different cell types. These characteristics are reported in Table 3.2, along with the indication of a few representative family members. In recent years, the taxonomic classification has changed to consider both the previous parameters and the more recently acquired information on the genetic organization of the viruses. The *Retroviridae* family now consists of 2 subfamilies (*Orthoretrovirinae* and *Spumaretrovirinae*) and 7 genera: alpharetrovirus

3

Table 3.2 Criteria for retrovirus classification

Parameter	Characteristics		
Electron microscopy morphology	*Classification*	*Characteristics*	*Prototype*
	Type A virions	Central nucleocapsid with translucid appearance with one or two concentric layers, without envelope	Immature forms of type B and type D viruses and endogenous retroviruses (intracisternal A particles, IAP)
	Type B virions	Nucleocapsid in eccentric position with prominent surface protrusions	Mouse mammary tumor virus (MMTV)
	Type C virions	Central nucleocapsid, with almost invisible protrusions	Most murine and avian sarcoma and leukemia viruses; for example, Moloney murine leukemia virus (Mo-MLV) and avian sarcoma/leukosis virus (ASLV)
	Type D virions	Oval nucleocapsid with small surface protrusions	Mason-Pfizer monkey virus (MPMV)
	Other types	Similar to type C viruses with different protrusions	Bovine leukemia virus (BLV), human T-cell leukemia virus (HTLV)
		Similar to type C viruses carrying a nucleocapsid with a truncated cone shape	Lentivirus
		Virions with prominent surface protrusions	Spumavirus
Genome organization	Simple retroviruses (*gal*, *pol*, and *env* genes) and complex retroviruses (*gal*, *pol*, *env* plus several accessory genes)		
Host	Murine, avian, feline, bovine, human, etc. retroviruses		
Species tropism	Ecotropic, xenotropic, amphotropic retroviruses		
Pathogenicity	Oncovirus, lentivirus, spumavirus		
Disease caused	Leukemia, sarcoma, myeloblastosis, erythroblastosis, immunodeficiency, anemia, encephalitis, etc.		

(whose prototype species is the avian leukosis virus, ALV), betaretrovirus (mouse mammary tumor virus, MMTV), gammaretrovirus (murine leukemia virus, MLV), deltaretrovirus (bovine leukemia virus, BLV), epsilonretrovirus (Walleye dermal sarcoma virus, WDSV), lentivirus (human immunodeficiency virus type 1, HIV-1), and spumavirus (human foamy virus, HFV). Table 3.3 reports this classification along with the older subgroup denominations and some of the most representative viruses in each genus.

A classification that is sometimes useful for operational purposes divides the *Retroviridae* family into three major subgroups: the oncoretroviruses (which include the

Table 3.3 Taxonomy of the *Retroviridae* family

Subfamily	Genus	Former classifications	Main species	Prototype viruses
Orthoretrovirinae	Alpharetrovirus	Avian type C retroviruses; Avian sarcoma/ leukosis viruses (ASLV)	Avian leukosis virus	ALV
			Rous sarcoma virus	RSV
	Betaretrovirus	Mammalian type B retroviruses; type D retroviruses	Mouse mammary tumor virus	MMTV
			Mason-Pfizer monkey virus	MPMV
	Gammaretrovirus	Mammalian type C retroviruses	Murine leukemia virus	Abelson-MLV, Friend-MLV, Moloney-MLV
			Feline leukemia virus	FeLV
			Gibbon ape leukemia virus	GaLV
			Harvey murine sarcoma virus	Ha-MSV
			Moloney murine sarcoma virus	Mo-MSV
			Simian sarcoma virus	SSV
			Reticuloendotheliosis virus	REV-A, REV-T
	Deltaretrovirus	BLV-HLTV group retroviruses	Bovine leukemia virus	BLV
			Primate T-lymphotropic viruses (human and simian)	HTLV-1, STLV-1, HTLV-2, STLV-2, STLV-3
	Epsilonretrovirus	Fish retroviruses	Walleye dermal sarcoma virus	WDSV
	Lentivirus		Bovine immunodeficiency virus	BIV
			Equine infectious anemia virus	EIAV
			Feline immunodeficiency virus	FIV-O, FIV-P
			Caprine arthritis encephalitis virus	CAEV

(cont→)

3

Table 3.3 (*Continued*)

Subfamily	Genus	Former classifications	Main species	Prototype viruses
			Visna/Maedi virus	VISNA
			Human immunodeficiency virus 1 and 2	HIV-1, HIV-2
			Simian immunodeficiency virus	SIVagm (155), SIVcpz, SIVmac
Spumaretrovirinae	Spumavirus		Simian foamy virus	SFVmac (SFV-1 and SFV-2), SFVagm (SFV-3), SFVcpz, and SFVcpz(hu)
			Bovine foamy virus	BFV
			Equine foamy virus	EFV
			Feline foamy virus	FFV
			Human foamy virus	HFV or HSRV

first five genera), the lentiviruses (*lenti-*: Latin for "slow", since these viruses cause diseases characterized by a long incubation period and slow evolution), and the spumaviruses (*spuma-*: Latin for "foam", from the cytopathic effect induced in monkey kidney cells, which is characterized by the formation of large vacuoles). Given their complex genetic organization, viruses belonging to the latter groups are also collectively named as "complex retroviruses" (see also Figure 3.7).

Genome Organization

The genomes of prototypic members of the *Retroviridae* family, such as the gammaretrovirus Moloney-murine leukemia virus (Mo-MLV), are 9–11 kb long and consist of 3 essential genes (*gag*, *pol*, and *env*) flanked, in their integrated, proviral DNA forms, by two identical sequences of 400–700 bp at the 3' and 5' extremities, named long terminal repeats, LTRs (Figure 3.6). Each LTR consists of three regions: U3, R, and U5. The 5' LTR U3 region contains a promoter driving expression of all viral transcripts. Transcription starts in correspondence with the first nucleotide of the 5' R region and proceeds for the entire length of the genome; the R region contains a polyadenylation signal, which, at the 3' end, drives cleavage and polyadenylation of the mRNA in correspondence with the 3' R-U5 boundary; the ensuing mRNA corresponds to the genome that is eventually packaged into the virions. Thus, in contrast to the provirus, which is flanked by two complete LTRs, the viral RNA genome initiates with the R-U5 sequence at its 5' extrem-

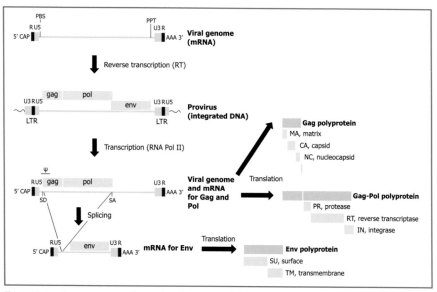

Fig. 3.6 Retrovirus genome and proteins. Structure of the viral genome mRNA, proviral DNA, and major viral transcripts (full-length genomic mRNA and single-spliced transcript) are shown on the *left side*, as indicated from top to bottom. The proteins obtained by transcription of the viral mRNAs are indicated on the *right side*

ity and ends with the U3-R sequence at its 3' extremity. A complete LTR sequence is only generated during reverse transcription.

Three essential genes – *gag, pol,* and *env* – are present between the two LTRs (Figure 3.6). The *gag* gene codes for proteins associated with the viral genome and essential for packaging; these include the matrix (MA), capsid (CA), and nucleocapsid (NC) proteins. The *pol* gene codes for the three essential viral enzymes that characterize all members of the retrovirus family: reverse transcriptase (RT), protease (PR), and integrase (IN). RT is responsible for the process of reverse transcription, converting the RNA genome into its cDNA form. PR catalyzes cleavage of the polyproteins that are generated by translation of the viral genes into the individual viral proteins, as detailed in the discussion of the viral replication cycle. IN catalyzes integration of the viral cDNA into the host cell genome to generate the provirus. The *env* gene codes for two proteins that are displayed first on infected cell surfaces and later, after budding, on the viral envelope. These are the TM (trans-membrane) protein, which positions itself across the membrane, and the SU (surface) protein, which is anchored onto the TM outside the membrane and mediates recognition of cellular receptors.

The viral RNA transcript, besides constituting the retroviral genome that is packaged inside the virions, also acts as the mRNA from which all the viral proteins are translated (Figure. 3.6). In particular, the full-length primary transcript codes for two long polypeptides corresponding to the Gag and Gag-Pol polyprotein precursors; the same transcript undergoes splicing to generate a processed mRNA coding for the Env polyprotein. In all cases, functional retroviral proteins are eventually generated by cleavage of these polyproteins into the final individual peptides.

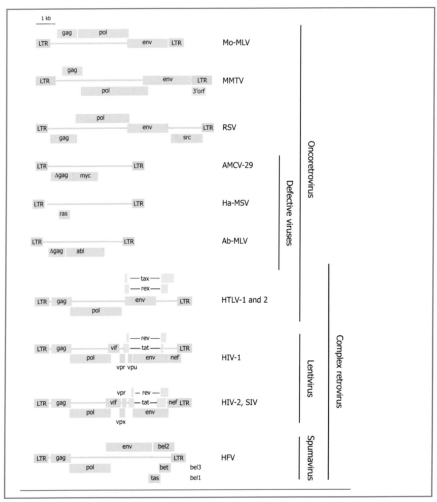

Fig. 3.7 Retroviral genomes. The genetic organization of some common retroviral genomes is shown, along with the indication of the common retroviral genes (*lilac*) and of the accessory genes proper of the individual groups (*other colors*)

Presence of the LTRs and of the *gag*, *pol*, and *env* genes is a hallmark of all retroviruses, since integrity of these genetic elements is essential for viral replication. Some family members, however, contain additional genetic information or display variations of this general genetic structure. The genetic organization of a few prototypic retroviruses is shown in Figure 3.7. In particular:

(i) members of the group of complex retroviruses (including HTLV-1, all lentiviruses, and spumaviruses), in addition to *gag*, *pol*, and *env*, also contain an extra series of genes, encoded from the 3' half of the viral genome. These genes, named "accessory genes" are indeed fundamental for these retroviruses to efficiently infect the respective target cells. For example, HIV-1 contains 6 accessory genes (*tat*, *rev*, *nef*, *vpr*, *vpu*, and *vpr*), which are essential for different

Table 3.4 Examples of retroviruses carrying viral oncogenes (*v-onc*)

Parental/helper virus	Retrovirus	Acronym	*v-onc*
	Rous sarcoma virus	RSV	*src*
Avian leukosis virus (ALV)	Avian myeloblastosis virus	AMV	*myb*
	Avian erythroblastosis virus	AEV	*erbA, B*
	Avian myelocytomatosis virus 29	AMCV-29	*myc*
	Y73 sarcoma virus	Y73SV	*yes*
	Avian sarcoma virus 17	ASV-17	*jun*
Moloney-Murine leukemia virus (Mo-MLV)	Abelson murine leukemia virus	Ab-MLV	*abl*
	Harvey murine sarcoma virus	Ha-MSV	*ras*
	Moloney murine sarcoma virus	Mo-MSV	*mos*
	Finkel-Biskis-Jinkins murine sarcoma virus	FBJ-MSV	*fos*
Feline leukemia virus (FeLV)	Snyder-Theilen feline sarcoma virus	ST-FeSV	*fes*
	Gardner-Arnstein feline sarcoma virus	GA-FeSV	
	Susan McDonough feline sarcoma virus	SM-FeSV	*fms*
	Hardy-Zuckerman 4 feline sarcoma virus	HZ4-FeSV	*kit*
Simian sarcoma virus (SSV)	Woolly monkey sarcoma virus	WMSV	*sis*

steps of the viral life cycle, including transcription (*tat*), transport of viral mRNAs outside of the nucleus (*rev*), cell cycle regulation (*vpr*), and modulation of virion infectivity (*vif*).

(ii) Some alpharetroviruses and gammaretroviruses contain a peculiar gene, derived from the host cell genome. This is an oncogene and represents the activated form of a cellular gene that is normally devoted to the control of cell cycle progression or cell differentiation. The viral versions (v-*onc*) of these normal cellular genes (also named proto-oncogenes, c-*onc*) are devoid of introns and thus similar to the cellular gene cDNAs, and are constitutively active, since they contain mutations that activate the encoded proteins or are continuously transcribed at high levels. Some of the v-*onc*-carrying retroviruses are listed in Table 3.4. Of note, several of the cellular proto-oncogenes have been discovered thanks to the presence, in one of the retroviruses, of their activated counterparts. The viruses containing an oncogene have the ability to transform the cells they infect (that is, to induce an oncogenic behavior) very efficiently and with a very rapid kinetics, since the constitutively active oncogene drives the cells into proliferation and negatively regulates their terminal differentiation. The Rous sarcoma virus (RSV), an alpharetrovirus, is the only retrovirus in which the v-onc is additional to intact *gag*, *pol*, and *env* genes. In all other cases, the oncogene-carrying retroviruses show more or less broad deletions in their genome – see, for example, Figure 3.7 for the avian myelocytomatosis virus 29 (AMCV-29), an alpharetrovirus, or the Abelson murine leukemia virus (Ab-MLV) and the Harvey murine sarcoma virus (Ha-MSV), two gammaretroviruses. Since the presence and integrity of the *gag*, *pol*, and *env* genes are essential, these viruses are defective for replication and can only be propagated if the cell they infect is also superinfected with a replication-competent virus of the same family. The viruses allowing replication of the defective retroviruses are named *helper* viruses. A cell simultaneously infected with a helper virus and a defective virus produces virions with the characteristics of the helper virus but containing genomes corresponding to either the defective or the helper virus. For example, in the gammaretrovirus genus, at least eleven defective retroviruses causing fibrosarcomas in cats have been insolated, possessing seven different oncogenes. All these viruses (named feline sarcoma viruses, FeSVs) have arisen as recombinants from the feline leukemia viruses (FeLV), in which a vast part of the genome was replaced by cellular oncogenes. The FeSVs are defective for replication and their propagation can only occur if the animal is superinfected with a replication-competent FeLV.

The retroviruses containing an oncogene induce tumors efficiently and rapidly after infection. However, the replication-competent retroviruses (RCRs) not carrying an onco-gene – including FeLV itself, Mo-MLV, the mouse mammary tumor virus (MMTV), the ALV, and the human T-lymphotropic virus type-1 (HTLV-1) – can also induce tumors in their natural hosts, albeit with very different kinetics and mechanisms. In these cases, transformation requires several weeks or months (for example, for Mo-MLV) or even decades (for HTLV-1). In the case of Mo-MLV, transformation is due to insertional muta-genesis, that is to activation of a cellular proto-oncogene or inactivation of a tumor sup-pressor gene due to retroviral integration within, or in close proximity to, these genes. This will be further discussed in the section on 'Gene Therapy of Hematopoietic Stem Cells', since insertional mutagenesis was the cause of the occurrence of leukemia in a few patients treated by gene therapy using gammaretroviral vectors. In the case of HTLV-1, cellular transformation is subsequent to the activity of the virus accessory genes, in par-ticular, of the *tax* gene, which interfere with multiple cellular functions and facilitate mutation of cellular proto-oncogenes.

Structure of Virions

Retroviral particles have a diameter of 80–100 nm and consist of an envelope, composed of the host cell plasma membrane with the addition of glycosylated viral TM and SU pro-teins, linked by disulfide bonds. Inside the virions, the viral proteins MA, CA, and NC associate with two identical copies of the viral mRNA genome to form the nucleocapsid, along with the viral enzymes RT, PR, and IN (Figure 3.8). By electron microscopy, the virions appear with an outer ring, corresponding to the envelope, surrounding an electron-dense core, corresponding to the nucleocapsid; the TM and SU proteins protrude as spikes from the envelope. Variations in the morphological structure of the virions represent cri-teria for classification of the different members of the family.

Replicative Cycle

The replicative cycle of retroviruses can be subdivided into a series of subsequent steps (Figure 3.9).

Adsorption. Binding of virions to the cell surface is mediated by the interaction of SU with a cellular plasma membrane protein, acting as a receptor. The different members of

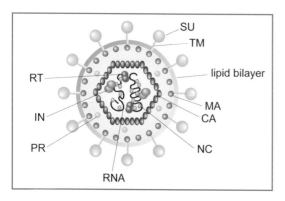

Fig. 3.8 Retroviral virion. The structure of a prototype retroviral virion is shown, with the indication of the proteins contained inside and on the surface. *SU*: surface; *TM*: transmembrane; *MA*: matrix; *CA*: capsid; *NC*: nucleocapsid; *PR*: protease; *IN*: integrase; *RT*: reverse transcriptase

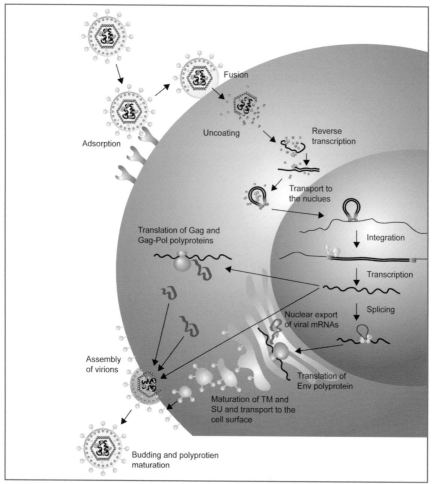

Fig. 3.9 Retroviral replication cycle. The main steps of the retroviral replication cycle are shown. The viral genome RNA is shown in black and the cDNA formed upon reverse transcription in red. The LTR elements are boxed. See text for description

Table 3.5 Examples of cellular proteins acting as membrane receptors for different retroviruses

Retrovirus	Acronym	Receptor	Function
Moloney-murine leukemia virus – ecotropic	Mo-MLV eco	Rec-1 (mCAT-1)	Basic amino acid transporter
Moloney-murine leukemia virus – amphotropic	Mo-MLV anfo	Ram-1	Phosphate transporter
Gibbon ape leukemia virus	GaLV	GLVR-1	Phosphate transporter
Feline leukemia virus	FeLV		
Simian sarcoma virus	SSV		
Avian leukosis virus – subgroup A	ASLV-A	Tv-a	Low-density lipoprotein (LDL) receptor-related protein
Avian leukosis virus – subgroups B, D, and E	ASLV-B, -D, and -E	Tv-b	Member of the tumor necrosis factor receptor (TNFR) family, most likely the avian homologs of mammalian TRAIL receptors
		Tv-c	Member of the immunoglobulin superfamily; most closely resembles the mammalian butyrophilins
Human immunodeficiency virus-1	HIV-1	CD4	T-cell receptor
		CXCR4 or CCR5	Chemokine receptor

the *Retroviridae* family have evolved SU proteins with very different receptor specificity (Table 3.5). For example, the alpharetrovirus ALV encompasses at least 10 different subgroups based on the capacity to bind different cellular receptors. In general, the physiological function of each receptor is different and not necessarily related to the biology of viral infection or the pathogenesis of disease induced.

Fusion and *uncoating*. The entry step is activated by binding of SU to the cellular receptor and mediated by a conformational change of TM, resulting in the fusion of the viral envelope with the plasma membrane. Following this event, the contents of the virions are found inside the cell cytosol and uncoating of the genome RNA from the capsid proteins takes place.

Reverse transcription. This process occurs in the infected cell's cytosol and is catalyzed by the viral RT enzyme (Figure 3.10). It can be subdivided into 9 subsequent steps. (i) A specific sequence immediately downstream of the 5' LTR (*primer binding site*, PBS) hybridizes, thanks to its complementarity, to a cellular tRNA – different retroviruses use different tRNAs; for example, Mo-MLV and HTLV-1 use tRNA[Pro], Visna-, Spuma-, and Mason-Pfizer monkey viruses use tRNA[Lys1,2], and HIV-1 uses tRNA[Lys3]. The 3'-OH end of the tRNA functions as a primer for the synthesis of a complementary DNA, which thus

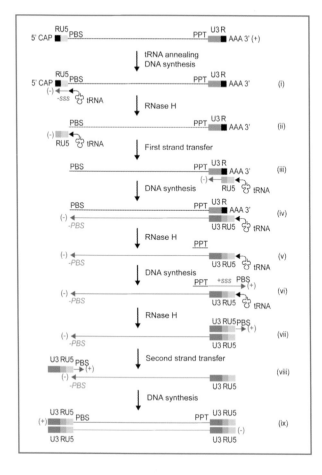

Fig. 3.10 Reverse transcription. See text for description. *PPT*: poly-purine tract; *PBS*: primer binding site

corresponds to the negative (–) strand of the genome. Polymerization progresses to include the R region at the 5' end of the genome. The DNA intermediate formed in this first step is named *minus-strand strong-stop DNA* (*–sss*). (ii) The RT protein is endowed with an enzymatic activity additional to DNA-dependent RNA synthesis, mapping in the C-terminal portion of the protein, consisting in the capacity to digest the RNA moiety in a DNA:RNA hybrid (RNase H activity). Thanks to this activity, the enzyme removes the RNA hybridized to the newly synthesized cDNA, thus exposing the single-stranded R region within the newly synthesized –*sss* cDNA. (iii) This –*sss* cDNA then translocates and hybridizes to the 5' of the viral genome (first strand transfer), thanks to the complementarity between the R regions; the tRNA primer is thus brought into this new position. (iv) Synthesis of the (–) strand cDNA continues starting from the 3'-OH of the –*sss* DNA to reach the PBS region. (v) In the meantime, the RNase H activity of RT continues to digest the viral RNA at the 3' end of the genome, until it stops in correspondence with a central region carrying a short (~10 nt), purine-rich sequence, the *poly-purine tract* (PPT). (vi) The PPT RNA acts as a primer for the synthesis of the positive (+) strand cDNA towards the 3' end of the viral genome. The (+) cDNA that is synthesized includes the entire LTR and the PBS region downstream of the LTR (*plus-strand strong-stop DNA,* *+sss*). (vii) The RNase H activity of RT removes the tRNA primer at the 5' end of the (–) strand DNA. (viii) Removal of the tRNA exposes the *+sss* DNA PBS region, which can then hybridize to the (–) strand 3' end (second strand transfer). This hybridization, which probably occurs by forming a circular intermediate, generates a template having the 3'-OH extremities on both strands available as primers for polymerization. (ix) RT elongates both strands up to the two LTRs, eventually displacing the previously hybridized regions in the circular intermediate. In this manner, a linear proviral cDNA is generated.

Transport to the nucleus. The newly synthesized, viral cDNA is part of a nucleoprotein complex also containing, besides RT, IN, various and still poorly characterized cellular proteins and, in the case of complex retroviruses, other viral proteins (e.g., Vpr and MA for HIV-1). This complex is called *Pre-Integration Complex* (PIC). In oncoretroviruses, the PIC, which is not smaller than 50 nm, cannot directly enter the nuclear pores, and can thus only have access to the cellular DNA when the infected cells undergo mitosis, after disintegration of the nuclear membrane. In contrast, the lentiviral PIC contains some proteins (including IN, probably Vpr, and perhaps MA in the case of HIV-1) that are able to interact with the nuclear pore proteins and mediate nuclear transport. As a consequence, lentiviruses, but not gammaretroviruses, can infect both replicating and quiescent cells, a property which is of paramount interest for gene therapy applications. Transport of the HIV-1 PIC to the nucleus occurs by sliding on actin microfilaments.

Integration. The process of integration is mediated by the viral protein IN with the assistance of various cellular proteins, only a few of which have been characterized. IN recognizes the extremities of the newly synthesized viral cDNA and removes two terminal nucleotides at the 3' ends on both strands. The processed nucleophilic 3'-OH ends of the viral cDNA are then inserted into the backbone of the target DNA through a transesterification reaction. Integration is random in terms of sequence specificity, however it commonly occurs in regions containing actively transcribed cellular genes. The reason for the selection of these "hot spots" is still unclear. It might be related to the presence, within these regions, of a relaxed chromatin structure that is more accessible to PICs compared to heterochromatic regions, or be due to the specific interaction of some still uncharacterized components of PICs with cellular factors involved in transcription.

Transcription. After integration, the proviral DNA genome can be considered as an additional protein-coding gene of the cell. In fact, transcription of the provirus is carried out by cellular RNA polymerase II and involves the same set of transcription factors that control expression of cellular genes (general transcription factors, mediator, chromatin modification and remodeling factors). The promoter controlling transcription corresponds to the U3 region of the 5' LTR. This sequence is very different in the various members of the *Retroviridae* family, since it consists of a *bricolage* of transcription factor binding sites of the host cell. The presence of these binding sites confers the provirus the property of being expressed in specific cell types or upon specific cell stimulation. As an example, Figure 3.11 displays a schematic representation of the HIV-1 LTR with the indication of the cellular transcription factors binding to this sequence. In some complex retroviruses, besides cellular transcription factors binding the LTR, transcription is controlled by an accessory protein encoded by the viral genome (e.g., Tat in the case of HIV-1 and Tax in the case of HTLV-1). In particular, HIV-1 Tat binds a highly structured RNA sequence

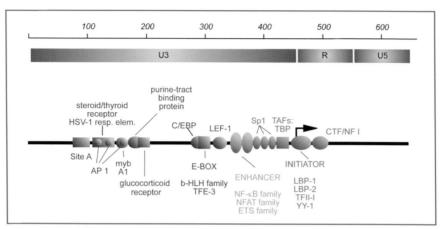

Fig. 3.11 Structure of the HIV-1 long terminal repeat (LTR). The major proteins binding the basal promoter (*green*), enhancer (*orange*), and upstream promoter elements (*blue*) are indicated. The arrow indicates the transcription start site

positioned at the 5' end of the viral mRNA named TAR (*trans-activating response*). From that region, Tat mediates transcriptional activation by recruiting, to the viral LTR promoter, on one hand cellular proteins possessing histone-acetyltransferase (HAT) activity and thus inducing chromatin relaxation and, on the other hand, the cellular P-TEFb kinase, which phosphorylates the carboxy-terminal tail of RNA polymerase II, a modification required for transcriptional elongation. Since transcriptional control is an essential step in the replicative cycle of the retroviruses, it also represents, together with receptor binding, a major determinant of the tropism of the different retroviruses for specific cell types.

Splicing of viral mRNAs. RNA polymerase II generates a single transcript, starting in correspondence with the first nucleotide of the R sequence at the 5' LTR and ending at the polyadenylation site in the U5 sequence of the 3' LTR. This transcript is both the viral genome that eventually becomes packaged inside the virions and the mRNA for the synthesis of all viral proteins. Since, in all eukaryotic cells, mRNAs are monocystronic (that is, each one codes for a single polypeptide), the primary proviral mRNA must undergo splicing to generate different shorter mRNAs, each one devoted to translation of a specific protein. In particular, in cells infected with simple retroviruses, such as Mo-MLV, two mRNAs are found, one corresponding to the original, full-length transcript and a shorter one, in which a large intron in the 5' half of the primary mRNA has been removed (Figure 3.7). The primary mRNA is used for translation of *gag* and *pol*, which occurs in the cytosol, and the shorter for *env*, which is translated by ribosomes associated with the ER. In the case of deltaretroviruses (e.g., HTLV-1), lentiviruses (e.g., HIV-1), and spumaviruses the situation is more complex, since these viruses also code for a series of accessory proteins, for each of which at least one specific mRNA is required. These shorter mRNAs are generated through a multiple splicing process making use of different 5' (splicing donor, SD) and 3' (splicing acceptor, SA) splicing sites. For example, in cells infected with HIV-1, over 35 different mRNAs are generated, which can be classified into 3 different classes according to their length (Figure 3.12). The longer class has ~9 kb and includes a single transcript, corre-

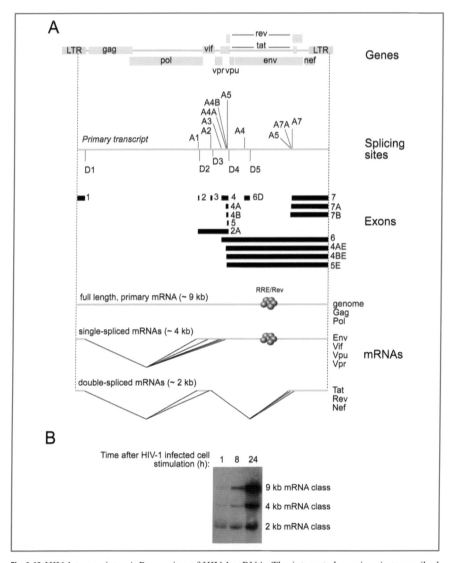

Fig. 3.12 HIV-1 transcripts. **A** Processing of HIV-1 mRNA. The integrated provirus is transcribed to generate a single RNA, containing multiple 5' (splice donor, D) and 3' (splice acceptor, A) sites. These splice sites encompass several exons, containing the open reading frames for all viral proteins. Splicing generates over 35 different mRNAs, which can be grouped into one of three classes: the longest (~9 kb) corresponds to the full-length mRNA; the intermediate (~4 kb) groups all constructs spliced in the 5' portion of the genome; the shortest (~2 kb) includes transcripts spliced twice or more. The first two classes of mRNAs contain the RRE element binding Rev. **B** Northern blotting showing expression of HIV-1 mRNAs in the U1 monocytic cell line, containing two copies of HIV-1 proviral DNA, where transcription is latent under basal conditions and can be activated by a variety of stimuli, including antibodies or cytokines. Upon activation, the shortest transcripts accumulate first, followed by the intermediate mRNAs and, last, by the full-length, genomic mRNA

sponding to the primary mRNA generated by proviral transcription; this mRNA codes for two polyproteins corresponding to Gag and Gag-Pol (see below). The second class includes a series of ~4-kb-long mRNAs, generated by splicing of an intron corresponding to the *gag-pol* sequences. This splicing event utilizes a 5' SD upstream of *gag* and different 3' SA sites downstream of *pol*. The mRNAs generated by this process code for Env (the majority), or the accessory proteins Vif, Vpr, and Vpu. The third class of transcripts includes shorter mRNAs (~2 kb) that, besides removal of the above-described intron, undergo additional splicing events that use different SD and SA sites to remove an intron in the *env* region. These mRNAs code for Tat, Rev, and Nef.

Transport of viral mRNAs. The production of multiple mRNAs by alternative splicing generates the problem of how to transport, into the cytosol, mRNAs that are not fully processed and still contain introns. Different solutions to this problem have been evolutionarily found by the different retroviruses (Figure 3.13A). In the oncoretroviruses, the primary mRNA contains specific sequences (*constitutive export elements*, CTE) that promote nuclear export by binding cellular proteins. In complex retroviruses, a virus-encoded protein (Rev and Rex in the case of HIV-1 and HTLV-1 respectively) binds a structured RNA sequence located in the primary mRNA in correspondence with the *env* gene (*Rev-responsive element*, RRE and *Rex-responsive element*, RXRE, respectively), and thus contained in a potential intron present in the partially spliced mRNAs (Figure 3.13B). Both Rev and Rex bind the cellular protein Crm-1 at the nuclear pores and thus promote translocation, into the cytosol, of partially spliced transcripts retaining the RRE or RXRE sequences. In the case of HIV-1, the RRE sequence plays an important role in the design of lentiviral vectors – see below.

Translation of viral mRNAs. The differentially spliced retroviral mRNAs coding for the essential genes *gag, pol,* and *env* are translated into polyproteins (Gag, Gag-Pol, and Env), which in turn are cleaved to generate the final polypeptides. The Gag polyprotein produces the MA, CA, and NC proteins (plus additional small polypeptides in some retroviruses); the Gag-Pol polyprotein generates the RT, PR, and IN enzymes; the Env polyprotein generates the SU and TM proteins (Figure 3.6). In the case of Gag and Pol, proteolytic cleavage of the polyproteins is carried out by the viral PR enzyme; in the case of Env, a furin protease of cellular origin cleaves the polypeptides inside the Golgi apparatus in parallel with protein glycosylation and before the protein is exposed onto the plasma membrane. While Env is produced from one or more specific mRNAs, Gag and Gag-Pol are usually translated from the same mRNA. In gammaretroviruses (the most frequently used to generate retroviral vectors; e.g., Mo-MLV), the *gag* and *pol* genes are on the same open reading frame (orf), separated by a single Stop codon. The ribosome starts translation from the first AUG to synthesize the Gag polyprotein; once it reaches the Gag Stop codon, sometimes it ignores it (in-frame suppression of termination), inserts an additional amino acid instead of this codon, and then continues translation, thus generating a polyprotein also containing Pol. In alpharetroviruses (ASLV) and lentiviruses (HIV), Gag and Pol are on different orfs. Thus, formation of the Gag-Pol fusion polypeptide requires a ribosomal frameshift during translation, by which the ribosome, once in correspondence with the *gag* 3' extremity, moves one nucleotide backward and changes reading frame. Since both suppression of termination and ribosomal frameshifting have an efficiency of 5–10%, the amounts of Gag proteins synthesized are about 10–20 times higher than those

3

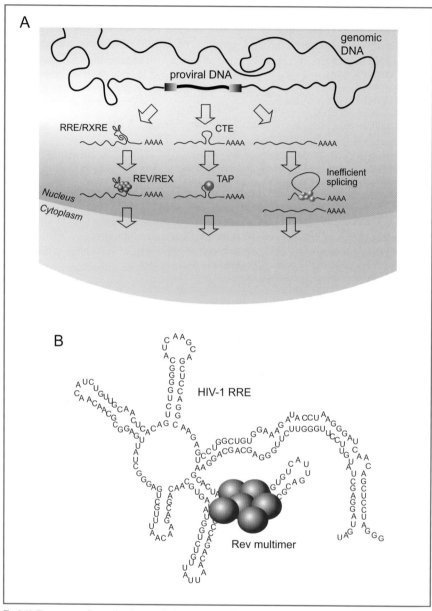

Fig. 3.13 Transport of unspliced retroviral RNAs into the cytoplasm. **A** Different strategies for retro-viral RNA export. Complex retroviruses have evolved binding of a viral protein to a cognate site in the unspliced RNAs (HIV-1: RRE/Rev; HTLV-1: RXRE/Rex) to promote export of unspliced mRNAs (*left part*). In oncoretroviruses, the primary mRNA contains specific sequences (consti-tutive export elements, CTE) that promote nuclear export by binding cellular proteins (*middle*). Other simpler retroviruses exploit inefficient splicing to allow export of incompletely spliced tran-scripts (*right*). **B** Sequence and secondary structure of the HIV-1 RRE RNA region, with the indi-cation of the Rev-binding site. The protein binds as a multimer

of Pol. As discussed above, in complex retroviruses, each of the accessory genes is usually translated from one or more specific mRNAs, generated from the primary transcript by
alternative splicing.

Assembly. Assembly of virions occurs in correspondence with the plasma membrane (for
type C viruses) or in the cytosol (for type B and D viruses); in both cases, assembly is mainly
driven by the Gag polypeptide. Inclusion of the genome-length viral mRNA into the virions is
due to the interaction of a specific RNA sequence, the *packaging signal* (ψ), located in correspondence with the 5' of the *gag* gene, with the Gag polypeptide in the portion corresponding
to the NC protein. Each virion includes 1200–1800 Gag and 100–200 Gag-Pol polyproteins.

Budding and virion maturation. The Env polyprotein is independently translated
inside the ER, becomes glycosylated and matures into TM and SU in the Golgi apparatus,
and is then exposed on the cell membrane of the infected cells. In the regions of viral budding, the protein becomes associated with the virions thanks to its interaction with the N-
terminus of Gag. Once outside the cells, the virions undergo maturation, by which the Gag
and Gag-Pol polyproteins are cleaved to generate the respective individual peptides.
Proteolytic cleavage is mediated by PR, which first excides itself from Gag.

3.5.1.2
Structure of Gammaretroviral Vectors

A prototype member of the gammaretrovirus genus used as a vector for gene therapy, the
Mo-MLV contains, besides the three essential *gag*, *pol*, and *env* genes, at least 5 genetic elements that are necessary for the completion of its replicative cycle and thus essential for the
construction of vectors. These are (listed from the 5' to the 3' of the genome; Figure 3.14):

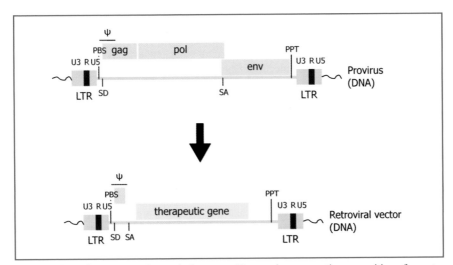

Fig. 3.14 Schematic structure of retroviral vectors. *Upper scheme*: genetic composition of a prototype gammaretroviral vector (e.g., Mo-MLV), with the indication of the most relevant genetic elements (*LTR*: long terminal repeat; *PBS*: primer binding site; *PPT*: poly-purine tract; *SD*: 5' splice
site; *SA*: 3' splice site). *Lower scheme*: retroviral vector

(i) the **LTRs**, of which the 5' U3 region is the promoter for mRNA transcription, the R region is required for reverse transcription, and the 3' U5 region contains the polyadenylation site;

(ii) the **primer binding site** (PBS), positioned immediately downstream of the 5' LTR, which is required for cellular tRNA binding to prime reverse transcription;

(iii) the **5' and 3' splice sites** (SD and SA respectively), which are essential to generate the spliced mRNA used for the translation of *env*; the sequence starting from the U3/R region and ending at the SD site is named the *leader* sequence and is common to all transcripts;

(iv) the **packaging signal** (ψ), which includes a structured RNA region at the 5' of the *gag* gene, partially extending toward the SD site; this is the sequence binding to Gag that is required for the inclusion of the viral genome mRNA inside the virions during assembly; and

(v) the **polypurine tract** (PPT), positioned at the 3' end of the genome upstream of the 3' LTR, which is required for reverse transcription.

Retroviral vectors must contain these five genetic elements, while the rest of the genome is dispensable and can be removed and be substituted by the therapeutic gene, including the sequences coding for the viral proteins (Figure 3.14). Thus, in the simplest version of oncoretroviral vectors, transcription of the therapeutic gene is directly controlled by the viral 5' LTR.

3.5.1.3
Production of Gammaretroviral Vectors

Viral vectors based on gammaretroviruses are produced in cultured mammalian cells. A plasmid containing the proviral DNA, having a structure similar to that depicted in Figure 3.14, is first obtained by standard cloning procedures, amplified in bacteria, and purified. This plasmid is then transfected into a *packaging cell line*, that is a cell line, usually of murine origin, that expresses the retroviral *gag*, *pol*, and *env* genes, which are no longer present in the retroviral vector plasmid but are nevertheless required for virion production. A packaging cell line is usually generated by stable transfection of the DNA sequences coding for Gag-Pol and Env, and thus constitutively expresses the respective proteins. Transfection of the Gag-Pol and Env DNAs to generate a packaging cell line is usually carried out in two subsequent steps, to avoid the possibility that the two constructs might integrate into contiguous regions of the genome, which would favor recombination of the two sequences with that of the retroviral vector plasmid (or with sequences corresponding to endogenous retroviruses (ERVs)), with the consequent generation of infectious RCRs. As a matter of fact, regions of homology no longer than 10 bp between the packaging sequences and the retroviral vectors are sufficient to drive recombination between the two constructs, leading to the generation of infectious viruses able to replicate autonomously. If the Gag-Pol and Env sequences are integrated far apart in the genome, the likelihood of recombination is significantly diminished. Over the first ten years of development of gene therapy, a vast series of packaging cell lines have been generated starting from the *gag*, *pol*, and *env* genes of different murine and avian retroviruses. In particular, since the tropism and efficiency of infection are mainly due to the properties of the Env proteins, these packaging cell lines vary in their capacity to generate retroviral vectors with the capacity to transduce different cell targets – see also below.

Once transfected into a packaging cell line, the plasmid containing the retroviral vector is transcribed starting from the 5' LTR and thus generates an mRNA that encompasses the whole proviral construct and contains the packaging signal (ψ). Presence of this signal permits recognition of the vector mRNA by Gag, followed by its inclusion into a virion (Figure 3.15). A virion generated in this manner is indistinguishable from a wild-type virion, and is thus fully infectious. After infection of a target cell, thanks to RT (which is present inside the virion) and the cis-acting PBS and PPT sequences, the vector genome

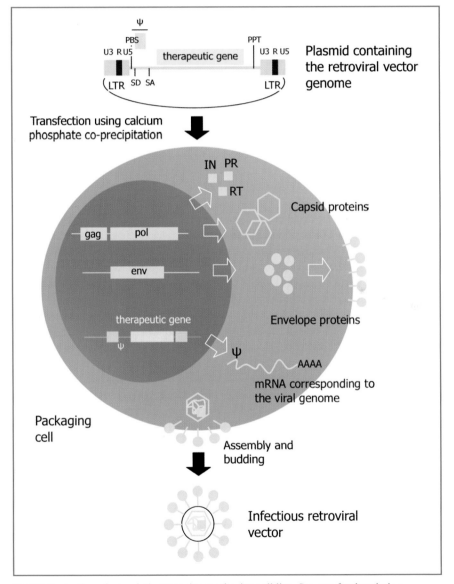

Fig. 3.15 Production of retroviral vectors in a packaging cell line. See text for description

is reverse transcribed. The proviral cDNA is then integrated into the host cell genome by IN, which is also present in the virion. Once integrated, the mRNA expressed by the vector provirus is no longer infectious, since none of the retroviral proteins are present. Thus, retroviral vectors are only capable of a single cycle of infection.

As anticipated above, the properties of the available packaging cell lines are mainly related to the characteristics of the *env* gene they express, since the SU protein dictates tropism of the virions towards different cell types. In particular, the Mo-MLV retrovirus, which is the prototype of several gene therapy retroviral vectors, naturally presents two SU variants. One exclusively binds the murine receptor Rec-1 (also called mCAT-1), a basic amino acid transporter, while the other one binds Ram-1, a phosphate transporter (Table 3.5). While the former protein is only expressed by murine cells, the second is common to cells of several species, including humans. Therefore, Mo-MLV displaying the Rec-1-binding SU variant only infects murine cells, while that displaying the Ram-1-binding SU infects cells of all species. According to their tropism, the former viruses are called *ecotropic* (indicating that they only replicate in cells of the same species in which they were isolated) and the latter *amphotropic* (*ampho-*: Greek for "both", indicating that they replicate well in cells of both of the species from which they were isolated and other species). Finally, *xenotropic* (*xeno-*: Greek for "foreign") retroviruses are endogenous to one species, but can only be propagated well in cells from a species foreign to the normal host (e.g., ERVs of mice that replicate well in rat or hamster cells). Amphotropic, but not ecotropic, vectors can be used for gene therapy of human cells.

Transfection of the retroviral vector-containing plasmid into the packaging cells is commonly performed using conventional methods, such as calcium phosphate precipitation or lipofection, and is thus relatively inefficient, since most of the internalized DNA is degraded or integrates into the cellular DNA in a random manner, often interrupting the continuity of the retroviral vector sequence. Therefore, the amount of retroviral particles found in the packaging cells' supernatant is limited, and thus the titer of the preparation (measured as the concentration of infectious particles) is relatively low. In contrast, if this supernatant is used to infect a second packaging cell line and the retrovirus contains a gene allowing the selection of the transduced cells, it is possible to obtain a population of homogenously transduced packaging cells, which release significant amounts of vectors in their supernatants. The first retroviral vectors designed for gene therapy, therefore, also contained, in addition to the therapeutic gene, a selectable gene, such as the *neo* gene, which confers resistance to the antibiotic geneticin or G418. These vectors are produced by a two-step procedure, the first entailing calcium-phosphate transfection of retroviral plasmid DNA into an ecotropic packaging cell line and the second using the supernatant produced by these cells to transduce an amphotropic packaging cell line, followed by selection of individual producer clones and analysis of the viral titers obtained in each case (Figure 3.16A). Following this procedure, the retroviral vector titers that can be achieved are in the order of $\sim 1 \times 10^6 - 1 \times 10^7$ infectious particles/ml of supernatant. Application of this procedure is very useful to obtain a retroviral vector-producing cell clone, which can be used for the continuous production of retroviral particles for a given gene therapy application without having to rely on transient transfection each time, which is cumbersome and has variable efficiency.

During assembly of retroviral particles, the viral genomes are packaged thanks to the interaction of the Gag polyproteins with the ψ sequence, occurring in the cytosol, while the

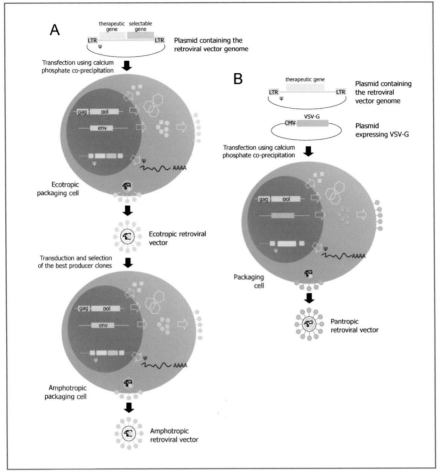

Fig. 3.16 Production of retroviral vectors. **A** Dual-step selection of retroviral vector producer cell clones. The procedure entails a first transfection of an ecotropic packaging cell line, followed by transduction of an amphotropic cell line using the supernatant obtained from the first transfectants. **B** Pseudotyping of retroviral vectors. A packaging cell line only expressing Gag and Pol is transfected with the retroviral vector plasmid and a second plasmid encoding VSV-G

TM and SU Env proteins are independently brought to the cell membrane through the ER-Golgi route. In the course of the study of retrovirus biology it was noticed that, if a cell is infected with a retroviral vector with a given specificity but, at the same time, it expresses an Env gene with a different specificity, some of the viral particles that are produced have the infectious property dictated by this Env gene, however carry a genome with different characteristics. This is known in virology with the term *pseudotyping* (*pseudo-*: Greek for "false"). Thanks to this property, it is possible to package the same viral genomes inside particles having efficiency and specificity of infection that are different from those encoded by the genome itself and are instead dictated by the Env proteins displayed by the virions.

3

The most efficient possibility for retroviral vector pseudotyping is through the use of the G protein encoded by the vesicular stomatitis virus (VSV). VSV is an enveloped virus with a negative sense RNA genome belonging to the family of *Rhabdoviridae*. The virus is of veterinary interest and is a cause of concern, since it can infect different animal species in an epidemic manner, including cattle. In the infected animals, the disease is characterized by the appearance of vesicles in different organs, including mouth and tongue, hence the name (*stoma-*: Greek for "mouth"). In humans, the virus can cause a flu-like syndrome, with the occurrence of vesicles on the lips resembling herpetic infection. The VSV envelope displays the virally encoded G protein (VSV-G), which mediates infection by binding, very efficiently, the phospholipids present on virtually all mammalian cell membranes and triggering endocytosis of the viral particles. Once in the endocytic compartment, lowering of the pH activates the fusogenic properties of VSV-G, which determines fusion of the viral envelope with the endosomal membrane and release of the virion content into the cytosol. Thanks to these properties, the VSV-G protein, once incorporated in a retroviral envelope, mediates viral infection at high efficiency and broad specificity.

Due to its fusogenic properties, it is however not possible to permanently express VSV-G in a packaging cell line. VSV-G-pseudotyped retroviral vectors are thus obtained by the transient transfection of packaging cell lines that only express Gag-Pol with a plasmid expressing VSV-G under the control of a strong promoter (such as the promoter of the cytomegalovirus immediate-early genes), in addition to the plasmid containing the retroviral vector DNA (Figure 3.16B). In contrast to virions containing retroviral Env proteins, VSV-G-pseudotyped virions can be purified by high-speed centrifugation without significant loss of infectivity. By using a single packaging passage followed by centrifugation, titers in the order of $\sim 1 \times 10^8$–1×10^9 infectious particles/ml of supernatant can routinely be obtained. The VSV-G-pseudotyped retroviral vectors have broad species specificity and cell-type range.

3.5.1.4
Variants in the Design of Gammaretroviral Vector Genomes

The simplest retroviral vector maintains all the essential genetic elements in cis (LTR, SD/SA, PBS, PPT, and ψ) and contains the therapeutic gene cloned within the two LTRs (Figure 3.17A). In principle, since the vector contains a single gene, the SA/SD sites are also dispensable, however their presence confers stability to the mRNA and thus allows higher titers and expression levels.

Starting from this relatively simple genetic design, over the last several years a number of modifications have been proposed, with the main purpose of allowing delivery of additional genes or driving expression of the therapeutic gene from a promoter different from the LTR. The main classes of variant vectors are described as follows.

(1) **Vector with an internal promoter**. One of the most frequent conditions in gene therapy is that expression of the therapeutic gene is driven by a promoter different from the vector LTR (e.g., a strong constitutive, or an inducible, or a tissue-specific promoter). In these cases, it is possible to insert the promoter of choice upstream of the therapeutic gene and thus downstream of the 5' LTR (Figure 3.17B). The cells containing such a construct, however, express two mRNAs, one starting from the 5' LTR and the other one from

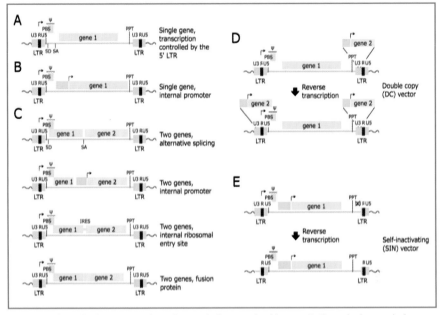

Fig. 3.17 Variants in the construction of retroviral vector backbones. **A** Canonical retroviral vector in which therapeutic gene expression is driven by the viral LTR. **B** Retroviral vector containing an internal promoter driving expression of the therapeutic gene. **C** Retroviral vectors containing two genes. **D** Double copy (DC) vector, before and after reverse transcription. **E** Self-inactivating (SIN) vector, before and after reverse transcription

the internal promoter, both ending in correspondence of the polyadenylation site of the 3' LTR. This condition is far from being ideal, since transcriptional interference is likely to ensue between the two promoters, due to a conflict in the assembly of the transcriptional machineries and to the disturbance that read-through transcription exerts on the downstream promoter. In addition, viral production exclusively relies on LTR-driven transcription: should the internal promoter be very strong, the titers of the viral preparations obtained from the packaging cells are usually low. The combination of two promoters is only effective when the internal promoter is weak or inactive in the packaging cell lines and becomes strong in the final target cells, as might be the case for a tissue-specific promoter, while the opposite is true for the viral LTR.

(2) **Vectors expressing two genes**. To simultaneously express two genes (as is the case, for example, of retroviral vectors containing a selectable gene in addition to the therapeutic gene), at least four different strategies can be followed (Figure 3.17C), detailed as follows.

(i) The first approach is to maintain the SA and SD sequences and clone one of the genes between these sites and the other one downstream of the SD sequence. This arrangement recapitulates that of wild-type simple retroviruses, with expression of the downstream gene relying on splicing of the primary mRNA.

(ii) A second possibility is to clone one gene under the control of the LTR and the other one under the control of an internal promoter. In this case, however, transcriptional competition might ensue between the two promoters, as discussed above.

(iii) A third option is to clone both genes under the control of the LTR and separate the two sequences by the insertion of an internal ribosomal entry site (IRES); this is by far the most efficient solution to both obtain high virus titers and permit high-level gene expression. In this arrangement, the vector produces a single transcript that is used for translation of two proteins from two different AUG codons. The IRESs that are most used for this purpose are those derived from viruses belonging to the Picornaviridae family or from the hepatitis C virus (HCV), since the RNA genomes of these viruses are not capped at their 5' ends and thus require the IRES to direct translation of their own proteins.

(iv) Finally, a fourth possibility is to clone the coding sequences of two genes of interest in frame, in order to obtain a single fusion protein; this strategy obviously requires that both proteins retain their function when fused to their respective partners.

(3) **Double copy vectors**. A very interesting strategy that can be used for the expression of short therapeutic nucleic acids (typically: ribozymes, shRNAs; cf. section on 'Modes of Delivery or Intracellular Synthesis of Small Regulatory RNAs') is to clone the transcriptional cassette expressing these genes within the U3 sequence of the 3' LTR, without interfering with the transcriptional elements contained in this region (Figure 3.17D). In the packaging cells, such a construct is transcribed starting from the 5' LTR and generates an mRNA originating in the 5' R region and ending in correspondence of the 3' U5 region; this mRNA thus contains the modified U3 sequence. During reverse transcription in the target cells, RT also transfers this modified U3 to the 5' LTR, thus duplicating the therapeutic gene. Inside the DC vector LTRs, another gene can be present (e.g., a selectable gene), the transcription of which is normally controlled by the 5' LTR.

(4) **Self-inactivating (SIN) vectors**. As reported above, the 5' LTR U3 region, which controls proviral transcription, is generated during the process of reverse transcription, when the 3' U3 sequence jumps to the 5' end of the genomic RNA. It is thus possible to construct retroviral vectors that, in their plasmid form, contain an intact 5' LTR and a 3' LTR that is mutated or almost entirely deleted. In the packaging cells, these vectors generate a transcript that contains this modified U3 sequence, which will become duplicated at the 5' LTR during reverse transcription in the target cells (Figure 3.17E). The provirus generated in this manner will be incapable of driving transcription of a therapeutic gene cloned within its LTRs, unless a promoter is inserted upstream of this gene. This strategy is very useful to avoid transcriptional interference between the LTR and an internal promoter. In addition, since the viral LTRs often activate expression of cellular genes neighboring the proviral integration sites (see section on 'Gene Therapy of Hematopoietic Stem Cells'), this strategy is currently considered as a possible means to minimize this problem.

3.5.1.5
Properties of Gammaretroviral Vectors

Viral vectors based on gammaretroviruses were the most utilized vectors in the gene therapy clinical trials until the early 2000s. Their popularity was due to a number of reasons, including the relative simplicity of use, high efficiency of transduction of replicating cells (e.g., *ex vivo* cultured cells), the immunogenicity, and ability to integrate their proviral cDNA form into the host cell genome, with the potential to render transduction, and thus therapeutic gene expression, permanent.

The clinical trials so far conducted, however, have highlighted a series of important problems, which have strongly limited the use of these vectors in more recent years.

(i) Some of the problems are related to the construction of the vectors themselves and the modalities of their production. Most of these problems can be circumvented by better vector design, as already outlined above. For example, transcriptional interference between two promoters, which lowers viral titers on one hand and therapeutic gene expression on the other, can be avoided by the use of simple vectors in which a single gene is directly controlled by the viral LTR. Potential formation of infectious recombinant viruses can be controlled by using packaging constructs coding for Gag-Pol that lack any homology stretch with the retroviral vector sequence. Finally, low titers and limited tropism of amphotropic preparations can be circumvented by VSV-G pseudotyping.

(ii) An additional technical issue related to the use of retroviral vectors is caused by their size, which usually permits cloning of therapeutic genes no longer than 6–7 kb (a retroviral particle can only package mRNAs no longer than a total of 9–10 kb). This prevents the delivery of native genes and is an impediment for the use of these vectors for very long cDNAs.

(iii) By far more important in applicative terms is the absolute need of gammaretroviral vectors that their target cells are in active replication. In fact, the pre-integration complex (PIC) of these viruses, which includes the viral cDNA and a series of proteins of cellular and viral origin including IN, remains in the cytosol and does not have access to the nucleus unless during mitosis, when the nuclear membrane breaks down. Since most of the cells in our body, including neurons, skeletal muscle cells, cardiomyocytes, endothelial cells, and the vast majority of peripheral blood lymphocytes, rarely divide or do not divide at all, the use of gammaretroviral vectors is essentially restricted to *ex vivo* applications on cells actively maintained in the cell cycle.

(iv) A fourth essential limitation of gammaretroviral vectors relates to the progressive silencing of therapeutic gene expression in the transduced cells. This occurrence is a consequence of methylation of cytosines in the context of the CpG di-nucleotide at the level of vector LTR promoter region. Methylated cytosines are recognized by various methyl-cytosine-binding proteins, which eventually promote chromatin deacetylation and compaction, eventually leading to silencing of gene expression. Methylation of retroviral DNA is believed to provide an evolutionary response aimed at preserving the integrity of the cellular genetic information against the insertion of transposable elements; in this respect, it is worth considering that over 8% of the human and mouse genomes indeed consists of endogenous retroviral sequences (ERVs), corresponding to over 30,000 proviruses per genome, divided into at least 50 different families, some of which are capable of autonomous replication.

(v) Finally, and probably most important of all, a very serious problem that limits the use of gammaretroviral vectors is the possibility that integration into the transduced cell genome might be mutagenic, leading to the inactivation of a tumor suppressor gene or the activation of an oncogene, thus contributing to oncogenic transformation. While, in theory, this event appears unlikely since retroviral vectors do not replicate and are capable of a single integration event, it has already occurred in at least two clinical trials for gene therapy of SCID-X1, a severe inherited immunodeficiency. Since then, inappropriate gene activation has also been observed in a series of other experimental studies in cultured cells and animal models; this issue will be discussed in more detail in the section on 'Gene Therapy of Hematopoietic Stem Cells'). In light of these problems, the regulatory agencies in Europe no longer approve the use of Mo-MLV-based retroviral vector for stem cell gene therapy of non-lethal disorders.

3.5.2
Vectors Based on Lentiviruses

One of the most striking characteristics that distinguishes lentiviruses from gammaretroviruses is the ability of the former to infect non-replicating cells. For example, one of the relevant cell types infected by HIV-1 *in vivo* are macrophages, which are terminally differentiated cells that have exited the cell cycle. As discussed above, this property is due to the capacity of the lentiviral PIC, which forms in the cytosol, to actively cross the nuclear membrane thanks to the interaction of some of the PIC proteins (IN, MA, Vpr) with proteins of the nuclear pore. This property appears of paramount interest for gene therapy, since it allows a significant extension of the range of cell types in which gene transfer might be of therapeutic benefit, especially because most of the cells in our body are non-replicating. In addition, in the case of hematopoietic stem cells, *ex vivo* transduction with lentiviral vectors appears efficient also in the absence of growth factor stimulation, a condition permitting the preservation of their pluripotency (cf. section on 'Gene Therapy of Hematopoietic Stem Cells').

For these reasons, starting from the late 1990s, the possibility to obtain vectors based on HIV-1 and the other lentiviruses has appeared very appealing for *in vivo* and *ex vivo* gene transfer applications.

3.5.2.1
Structure and Production of Lentiviral Vectors

The lentivirus from which most of the currently available vectors have been generated is HIV-1, mainly because of the vast amount of available information concerning the molecular biology and the properties of this virus. Over the last 10 years, at least three different generation of HIV-1-based lentiviral vectors have been produced, each bearing significant improvements over the preceding ones (Figure 3.18).

In the **first-generation** lentiviral vectors, recombinant viral particles are generated through cell transfection with 3 plasmids.

The *first plasmid* contains, in its proviral DNA form, the gene transfer vector, which carries the therapeutic gene. This proviral DNA contains, in 5' to 3' orientation: (i) the wild-type 5' viral LTR; (ii) the leader region, containing the PBS sequence and the 5' splice site-SD; (iii) ~350 bp of the *gag* gene in the region corresponding to the packaging signal ψ – the gene open reading frame is closed by the insertion of a Stop codon to block translation; (iv) ~700 bp of the *env* gene containing the RRE region, to allow export of the viral transcript from the nucleus, and the 3' splice site-SA; (v) a promoter driving the expression of

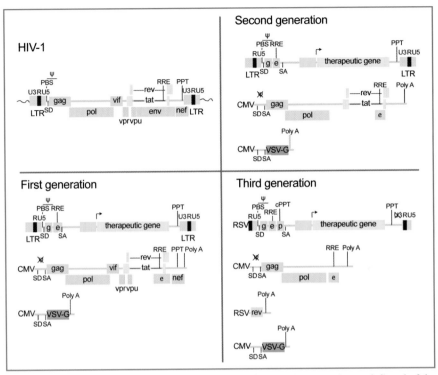

Fig. 3.18 Lentiviral vectors. Schematic representation of the HIV-1 genome (*upper left*) and of the plasmids required to obtain first-, second-, and third-generation lentiviral vectors. See text for description

the therapeutic gene – this is essential, since the natural HIV-1 promoter, consisting in the 5' LTR U3 region, is almost silent in the absence of the viral Tat protein, which is not present in the vector due to safety reasons; (vi) the 3' viral LTR, with the immediately upstream located PPT sequence.

The *second plasmid* is a packaging plasmid, also derived from the HIV-1 genome, which contains all viral genes with the exception of *env*. Besides *env*, this plasmid also carries a mutation in the ψ region to prevent packaging of the encoded mRNA into the viral particles and lacks the 3' LTR, which is substituted by a heterologous polyadenylation sequence. Expression of this plasmid is driven by the strong constitutive promoter of the cytomegalovirus immediate early (CMV IE) genes.

The *third plasmid* codes for the VSV-G protein; to avoid limitations in tropism and circumvent the relative low infectivity of the natural HIV-1 envelope, lentiviral vectors are usually pseudotyped with VSV-G.

Production of the vectors is carried out by the transient transfection of human embryonic kidney 293 T (HEK 293T, expressing the SV40 T antigen protein) cells with these three plasmids; the virions containing the retroviral vector RNA are then found in the cell culture supernatant, similar to gammaretroviral vectors.

First-generation lentiviral vectors elicit important safety concerns, related to both their production strategy and their clinical utilization. At the levels of production, a recombinant event occurring between the packaging plasmid and the plasmid containing the lentiviral vector can generate a replication competent lentivirus (RCL), the infectivity of which can be even extended by the presence of the VSV-G protein. A similar recombination event could also occur at the moment of reverse transcription in the transduced cells, should the virion carry two RNA genomes, one corresponding to the transfer vector and other to the packaging plasmid. At the level of clinical application, once used for gene therapy of HIV-1 infected patients, first-generation lentiviral vectors might recombine with the wild-type virus infecting the patients, thus potentially leading to the creation of novel viruses, with unpredictable potential for diffusion and pathogenicity. Finally, should first-generation lentiviral vectors be used in an HIV-1-infected patient, it is also possible that vector replication is stimulated by infection of the transduced cells with wild-type HIV-1: in this case, the wild-type virus would act as a helper for vector replication, since the vector construct contains all the sequences necessary for replication, including transcription (LTR), packaging (ψ), reverse transcription (R, PBS, PPT), and integration (U3). Thus, superinfection of cells carrying an integrated lentiviral vector with wild-type HIV-1 would determine the mobilization of the vector inside the organism.

To try and overcome these safety issues, further deletions have been introduced into the HIV-1 backbone to progressively remove all genes that are not strictly necessary for the production of viral particles and the transduction of the target cells. The **second-generation** lentiviral vectors entail the use of a similar three-plasmid design as the first-generation vectors, however, in the packaging plasmid, besides *gag* and *pol*, all accessory genes are removed, with the exception of *tat* and *rev*. In this manner, the probability of recombination between the vector and the packaging plasmid is significantly reduced in the packaging cells. However, since such a second-generation vector carries intact LTRs and ψ region, on one hand it can still recombine with wild-type HIV-1 while, on the other hand, it can be mobilized by wild-type HIV-1 if used in an HIV-1-infected patient, similar to first-generation vectors.

A **third-generation** of lentiviral vectors was designed to definitely prevent both the possibility of recombination with wild-type HIV-1 and of vector mobilization inside the organism. Production of these vectors, which have now entered clinical experimentation, now requires four plasmids. The *first plasmid* corresponds to the transfer vector, which is now obtained using the SIN approach (cf. gammaretroviral vectors above) to modify the LTR region. In particular, the 3' U3 LTR region is deleted, to inactivate transcription of the proviral DNA after reverse transcription. In the packaging cells, the vector is transcribed from a constitutively active heterologous promoter, positioned upstream of the R region. In addition, recent evidence indicates that the inclusion, inside the vector proviral DNA, of an HIV-1 sequence located within the *pol* gene significantly increases viral titers. This sequence, named *central polypurine tract/central termination sequence* (cPPT/CTS), would function by enhancing both reverse transcription – acting as an additional PPT to drive synthesis of the *plus-strand strong-stop DNA* (cf. replicative cycle of retroviruses above) – and PIC nuclear transport.

Since transcription of the transfer plasmid depends on a heterologous promoter, the presence of the *tat* gene, coding for the transactivator acting on the viral LTR becomes superfluous. Thus, the *packaging plasmid* now only contains the *gag* and *pol* genes, while the *rev* gene is expressed from a *third plasmid*. Presence of the Rev protein is still necessary, since it allows proper transport into the cytosol of the mRNA expressed from the packaging plasmid. Finally, as in the previous generations, a *fourth plasmid* encodes VSV-G.

This third-generation lentiviral vector production system only requires 3 of the 9 HIV-1 genes, thus offering a safety profile that is definitely more reassuring than that of first- and second-generation vectors.

Based on an analogous design, first-, second-, and third-generation lentiviral vectors have also been obtained from the genomes of other non-human lentiviruses, including FIV, SIV, and BIV. Utilization of these vectors rather than those based on HIV-1 would have the advantage of increased safety, since the viruses from which these vectors are derived do not infect humans.

3.5.2.2
Properties of Lentiviral Vectors

Compared to gammaretroviral vectors, the main advantage of lentiviral vectors is their property to transduce non-replicating cells. This paves the way to the possibility of using these vectors *in vivo*, to transduce organs such as brain or retina, which are mainly composed of quiescent cells. In a similar manner, lentiviral vectors can be used for *ex vivo* transduction of hematopoietic stem cells without the need to induce their replication. In this respect, however, it should be observed that, while it holds true that lentiviral vectors are able to successfully transduce cells that are out of the cell cycle, these cells still need to be metabolically active. This requirement is well exemplified by the observation that, during the natural history of HIV-1 infection, the virus very efficiently transduces resting, however metabolically active, macrophages and much less efficiently transduces resting, and metabolically quiescent, peripheral blood T lymphocytes. This appears to be of particular relevance when lentiviral vectors are considered for gene transfer into stem cells of different derivation, since the activation state of these cells is usually low.

The main concerns elicited by lentiviral vectors relate to safety and, in particular, to the generation of RCLs, to the mobilization of vectors by the wild-type virus in HIV-1-infected patients, and to the potential for insertional mutagenesis.

RCL generation can occur either during vector preparation, by recombination of the transfer vector with the packaging plasmid, or *in vivo*, after superinfection of a transduced cell with wild-type HIV-1. The third-generation lentiviral vectors seem to have a significantly better safety profile compared to the previous generations, due to the limited sequence homology with wild-type HIV-1 they present.

As far as vector mobilization by wild-type HIV-1 is concerned, this is more than a theoretical possibility since it has already been observed in HIV-1-infected patients in the first gene therapy clinical trial exploiting a lentiviral vector (see section on 'Gene Therapy of HIV-1 Infection'). This vector, however, contained an intact LTR, which was transcriptionally activated upon HIV-1 infection. In the third-generation vectors, removal of the LTR U3 region using the SIN technology should prevent the possibility of mobilization, since viral replication requires vector transcription starting from the 5' R region, in order to ensure the inclusion, in the viral mRNA genome, of the packaging signal and the R region itself.

Finally, it still remains to be understood whether integration of lentiviral vectors into the host cell genome might lead to the inappropriate activation of cellular genes through insertional mutagenesis, similar to gammaretroviruses. Several studies are currently addressing this issue experimentally, however only scanty primary data are available in humans due to the very limited number of patients treated with lentiviral gene therapy so far. *Ex vivo* cell transduction indicates that these vectors, similar to wild-type HIV-1, also integrate in correspondence with cellular transcribed genes. However, the region where integration occurs corresponds to the whole gene transcription unit, in contrast to gammaretroviruses, which preferentially integrate in correspondence with the transcription start site, including the gene promoter and first intron (Figure 3.19). Since aberrant transcriptional activation of the gene where proviral integration has occurred is likely to depend on either the interaction of the viral LTR elements with the cellular gene promoter or read-through transcription of the gene from an upstream integrated provirus, this is less likely to occur with lentiviral compared to gammaretroviral vectors. In addition, the HIV-1 LTR promoter, in the absence of the Tat protein, is extremely weak, thus rendering the possibility of aberrant transcriptional activation of a neighboring gene less probable. The SIN technology should, in principle, render the probability of cellular gene activation even less likely to occur, due to the removal of the U3 region. Indeed, various experiments are currently ongoing to comparatively assess the mutagenic potential of third-generation, SIN lentiviral vectors with the former lentiviral and gammaretroviral vectors.

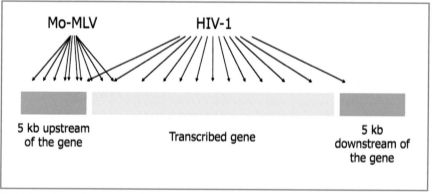

Fig. 3.19 Regions of integration of Mo-MLV and HIV-1 vectors into the cellular genome. Both Mo-MLV and HIV-1 vectors integrate in correspondence with actively transcribed cellular genes. However, Mo-MLV integration occurs in close correspondence with the gene transcription start site, within a few kilobases both upstream or downstream. In contrast, HIV-1 integrates into the whole gene region. Alpharetroviruses, exemplified by the avian sarcoma–leukemia virus (ASLV) group, show fairly random integration, with only weak favoring of transcription units

3.5.3
Vectors Based on Adenoviruses

The first adenovirus was isolated in 1953 from the adenoid tissue (hence the name) recovered during tonsillectomy from a child. Currently, over 100 members of the *Adenoviridae* family are known, able to infect man and various animal species, including non-human primates, mouse, dog, pig, frog, different species of birds, and even some types of snakes. The human adenoviruses are responsible for 5–10% of acute respiratory diseases of children and a variable number of conjunctivitis and gastroenteritis epidemics.

The marked capacity of this virus to infect epithelial cells initially inspired the idea to use adenoviral vectors for gene therapy of diseases of the lung and the airways, typically of cystic fibrosis. However, the natural tropism of adenoviruses for the respiratory epithelium and the conjunctiva is mainly due to its modality of transmission rather than to the molecular characteristics of the virus. Indeed, the receptor mediating cell infection by adenoviruses is ubiquitously expressed and most of the cell types can sustain adenoviral replication independent from the replicative state of the cells, thus opening the way to the possible utilization of these viruses for gene transfer into virtually any organ. Additionally, an intrinsic property of adenoviruses is the great efficiency at which they exploit the cellular machinery to drive synthesis of viral mRNAs and translation of viral proteins: a cell infected with adenovirus produces extremely high levels of viral proteins and thus, in the case of the vectors, of the therapeutic gene they contain. All these properties are of obvious interest for gene therapy; it is thus not surprising that, since the second half of the 1990s, these vectors have been the focus of a vast series of both animal and clinical experimentations.

3.5.3.1
Molecular Biology and Replicative Cycle of Adenoviruses

Based on the capacity of different human sera to neutralize adenoviral infection in cell culture, more than 50 serotypes of adenoviruses capable of infecting humans can be distinguished. The neutralizing antibodies mainly recognize epitopes in the exon protein of the virion and the fiber knob (see below). The different serotypes are classified into 6 subgroups (A–F) on the basis of their capacity to determine human red blood cell agglutination; subgroup C includes serotypes 2 and 5 (Ad2 and Ad5), from which most of the gene therapy vectors are derived.

Structure of Virions

The virion consists of a capsid showing icosahedral symmetry, without an envelope, having a diameter of 70–100 nm, and surrounding the viral nucleic acid.

The capsid has 20 facets, each formed by an identical equilateral triangle, 12 vertexes, and 30 edges (T=4). Each facet of the icosahedron is composed of 240 proteins, named hexons since each of them has contacts with 6 other proteins. Each of the 12 vertexes is instead formed by a different protein, named pentons since each of them has contact with another 5

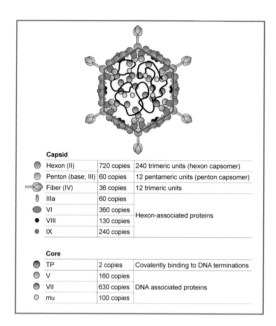

Capsid		
Hexon (II)	720 copies	240 trimeric units (hexon capsomer)
Penton (base; III)	60 copies	12 pentameric units (penton capsomer)
Fiber (IV)	36 copies	12 trimeric units
IIIa	60 copies	
VI	360 copies	Hexon-associated proteins
VIII	130 copies	
IX	240 copies	

Core		
TP	2 copies	Covalently binding to DNA terminations
V	160 copies	
VII	630 copies	DNA associated proteins
mu	100 copies	

Fig. 3.20 Schematic representation of an adenoviral virion. The lower part of the figure lists the 11 proteins taking part in the formation of the virion, also indicating their abundance

neighboring proteins. Each penton is formed by a base, which is part of the capsid surface, and a fiber projecting outward, which has different lengths in the various serotypes.

At least 11 different proteins take part in the formation of the virion (Figure 3.20). Each **exon** is formed by 3 subunits of protein II, which is therefore the most abundant protein of the virions; a protein II trimer forming an exon is also called a hexon capsomer. Proteins VI, VIII, and IX are associated with the exon and probably stabilize the interactions among the different protein II monomers and between the exon and the inner proteins of the virion. The **base** of each penton is formed by 5 subunits of protein III, which associate with protein IIIa, while the **fiber** is composed of 3 subunits of protein IV; the combination of the penton base and the fiber is also called a *penton capsomer*.

The *inner part* of the virion contains four different proteins and the viral genome. The terminal protein (TP) is covalently attached to the extremities of the linear genome DNA, while the basic proteins V, VII, and μ (mu) bind the genome and promote its condensation. Additionally, protein V forms a bridge between the virion core and the pentons, thanks to its binding to protein VI. The virion also contains a protease (Pr), encoded by the viral genome, which is necessary for the maturation of some of the structural proteins of the virion and is thus required for proper infectivity.

Genome Organization

The adenoviral genome consists of a double-stranded, linear DNA molecular of 36 kb in the case of Ad2 and Ad5, bearing at the two extremities two identical sequences in reverse orientation (inverted terminal repeats, ITRs; 103 bp in the case of Ad2 and Ad5); these regions act as origins of DNA replication of the entire genome.

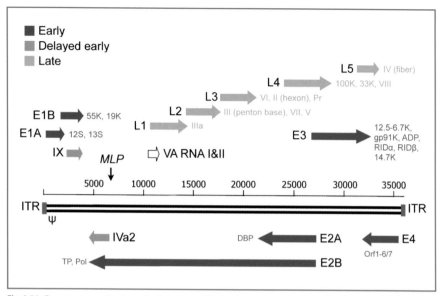

Fig. 3.21 Genome organization of adenovirus. The early genes are shown in *blue*, the delayed early genes in *gray*, and the late genes in *red*. *MLP*: major late promoter. For each gene, the major encoded proteins are indicated. The *arrows* indicate direction of transcription. *ITR*: inverted terminal repeat

The genome contains: (i) five early transcriptional units, which become activated upon cell infection: E1A, E1B, E2 (E2A and E2B), E3, and E4; (ii) two delayed early transcriptional units (IX and IVa2); and (iii) one major late (ML) transcription unit, which is processed to generate 5 families of late mRNAs through post-translational processing (from L1 to L5) (Figure 3.21). All these units are transcribed by RNA polymerase II. The genome also contains 1 or 2 (according to the different serotypes) genes transcribed by RNA polymerase III (virus-associated (VA) RNA genes). Conventionally, the adenovirus genome map is shown with the E1A gene on the "left" side, and thus the E1A, E1B, IX, ML, VA RNA, and E3 genes are transcribed on the "upper" strand and the E4, E2, and IVa2 genes on the "lower" strand. The genomes of all adenoviruses known to date show the same genetic organization.

Functions of the Adenoviral Proteins

E1A is the master gene upon which activation of the whole replicative cycle of adenovirus depends. This gene encodes two proteins (E1A-13S and E1A-12S, generated through alternative splicing of the same mRNA), which exert a variety of functions inside the infected cells, having the ultimate goal to promote viral replication. In particular, E1A binds different cellular proteins controlling the cell cycle, including the tumor suppressor pRb, thus stimulating cell cycle entry; in addition, the protein binds different components of the transcriptional machinery, including the transcriptional coactivators and histone-acetyltransferases p300/CBP and P/CAF, different cellular transcription factors, proteins

3

of the mediator complex, and the TATA-binding protein TBP, thus stimulating transcription of a series of cellular and the majority of viral genes. The presence of the E1A protein inside the cells also activates p53, since E1A stimulates transcription of the tumor suppressor p19^{ARF}, which binds p53 and modulates its activity.

The E1B gene codes for two proteins, of 55 and 19 kDa. E1B-55K binds p53 and inhibits its transcriptional activity, thus blocking the induction of apoptosis that would be elicited by cell infection with the virus. This property is further discussed in the section on 'Oncolytic Viruses' in the context of 'Gene Therapy of Cancer'. E1B-19K has homology with the cellular gene *bcl-2* and also displays anti-apoptotic activity by binding members of the cellular family of Bax proteins.

The E2 region includes two genes coding for factors necessary for viral DNA replication: E2A, coding for the DNA binding protein DBP, and E2B, coding for the terminal protein TP – which binds the extremities of the linear viral genome – and the viral DNA polymerase.

E3 contains a series of genes that are dispensable for adenoviral replication in cell culture however become essential to overcome the host response to viral infection *in vivo*. The encoded proteins include E3-gp19K, which is localized on the membranes of the ER and prevents transport of MHC class I molecules to the cell surface, where they would allow adenoviral antigen presentation and thus infected cell recognition by cytotoxic T lymphocytes; RIDα, RIDβ, and E3-14.7K, which inhibit TNF-α-, Fas- and TRAIL-induced apoptosis; and adenovirus death protein (ADP), which facilitates cytolysis and thus release of virions from the infected cells.

E4 contains a series of genes the main function of which is to facilitate mRNA processing, stimulate viral DNA replication, and switch cellular transcription off. E4orf6 participates in the formation of the complex between E1B and p53, which inhibits p53 transcriptional activity and targets the protein for degradation. E4orf3 determines localization of the adenoviral DNA replication foci inside the infected cell nucleus.

VA RNA is a genetic region transcribed by RNA polymerase III that generates, in Ad2 and Ad5, two small regulatory RNAs (VA-I and VA-II; 160 nt) that are not translated and have the function to inhibit the cellular protein kinase R (PKR). This enzyme is activated by the double-stranded RNA that accumulates in adenovirus-infected cells and blocks phosphorylation of the cellular translation initiation factor eIF2-α. If not inhibited, PKR would thus inhibit translation of viral and cellular mRNAs and abort infection.

The product of gene IX (pIX) is a multifunctional protein that stabilizes the viral capsid and possesses transcriptional activity. In addition, the protein contributes to the reorganization of the infected cell's nuclear structure; in particular, pIX induces the formation of peculiar nuclear inclusions, where the cellular protein PML becomes localized.

IVa2 codes for a protein involved in the transcriptional activation of the viral major late promoter (MLP), which controls transcription of all the late genes of adenovirus.

The L1–L5 genes code for the viral proteins essential for the late phases of the infection, in particular those taking part in the formation of the virions (cf. below).

The functions of the main regulatory genes of adenovirus are summarized in Table 3.6.

Table 3.6 Functions of the main regulatory genes of adenovirus

Gene	Proteins	Function	Major effects
E1A	E1A 13S E1A 12S	Interacts with proteins regulating the cell cycle (e.g., Rb) and proteins controlling gene expression (p300/CBP, P/CAF, transcription factors, TBP, mediator)	Promotes cell entry into the S-phase; activates transcription of cellular and viral genes
E1B	E1B 19K	Bcl-2 homolog	Inhibits apoptosis
	E1B 55K	Binds and inactivates p53; facilitates the selective transport of viral mRNAs from the nucleus to the cytoplasm	
E2A	DBP	Binds viral DNA	Necessary for viral DNA replication
E2B	TP	Binds the extremities of viral DNA	
	DNA polymerase	Synthetizes of viral DNA	
E3	12.5K-6.7K		
	gp19K	Binds MHC class I molecules and blocks their transport to the cell surface; inhibits MHC class I gene transcription	Inhibits infected cell recognition by the immune system
	11.6K (ADP)	Has cytolytic activity	Promotes release of virions from the infected cells
	10.4K and 14.5K (RIDα and RIDβ)	Inhibits TNF-α, Fas-L and TRAIL-induced apoptosis	Inhibits apoptosis
	14.7K	Inhibits TNF-α and Fas-L-induced apoptosis; stabilizes NF-κB	
E4	E4orf6/7	Modulates activity of E2F transcription factors	Modulates E1A and E1B activities; inhibits apoptosis
	E4orf6 (34K)	Cooperates with E1B 55K	
	E4orf4	Inhibits E1A-induced activation of E2F; binds protein phosphatase PP2A	
	E4orf3 (11K)	Binds E1B 55K; determines localization of nuclear foci of adenoviral DNA replication	
	E4orf2		
	E4orf1	Facilitates cell transformation by E1A and E1B	
VA RNA	-	Inhibits cellular kinase PKR	Permits translation of viral mRNAs
IX	pIX	Structural protein of the capsid and transcriptional activator	Activates late gene expression
IVa2	IVa2	Transcriptional coactivator	Contributes to activation of the major late promoter (MLP)

Replicative Cycle

The replicative cycle of adenovirus is conventionally divided into two phases, separated by viral DNA replication. The early events include binding of the virus to the cell surface (adsorption), penetration of the virus inside the cells, transport of the viral DNA into the nucleus and expression of the early genes, starting with E1A. The early gene products allow further expression of the viral genes, stimulate viral DNA replication, induce cell cycle progression, block apoptosis, and antagonize a series of cellular responses with potential antiviral activity. The early phase lasts about 5–6 h, after which replication of the viral genome starts, concomitant with the late phase of gene expression, leading to transcription of the late genes and to virion assembly. The IVa2 and IX genes are expressed with a timing intermediate between early and late. The replicative cycle takes about 20–24 h in HeLa cells to complete; at the end, each cell has generated about 1×10^4–1×10^5 new infectious viral particles.

The adenoviral replicative cycle can be schematically divided into a series of subsequent steps.

Absorption. Adenoviruses absorb to the cell surface thanks to the interaction of the C-terminal portion of the fiber protein, extending outward like a knob, with a cell surface receptor known as CAR (coxsackie/adenovirus receptor). The CAR protein belongs to the immunoglobulin superfamily and acts as a receptor for the adenovirus subgroups A, C, D, E, and F (but not B) and for B-type Coxsackieviruses, hence the name of the protein.

Internalization. After interaction of the fiber with CAR, virion internalization occurs through receptor-mediated endocytosis mediated by clathrin-coated vesicles. During this process, a fundamental role is played by the interaction of the penton base with the $\alpha v\beta 5$ and $\alpha v\beta 3$ integrins on the cell surface.

Exit from the endosomes and transport to the nucleus. More than 90% of the internalized virions exit from the endocytic vesicles at the level of early endosomes, thanks to the endosomolytic property of the penton base, which is stimulated by the progressive acidification of the endosomes. Of interest for gene therapy, the endosomes are physically destroyed after this process, since exogenous protein–DNA complexes entering into the same endocytic vesicles with adenovirus, although not physically linked to the virions, are also released into the cytosol after endosomolysis. Once in the cytoplasm, the viral particles are transported into the nucleus in an active manner, thanks to the interaction of the exon with the cellular microtubuli. Concomitant with internalization, the virion undergoes progressive disassembly, mediated by the dissociation and proteolytic degradation of its protein components, in particular of protein VI, which functions as a glue between the capsid and the inner components of the virion. A complex consisting of viral DNA, with covalently bound protein TP, and the basic proteins VII, V, and mu, then translocates from the cytosol to the nucleus. Once in the nucleus, protein TP transports the complex in the nuclear matrix compartment, an event essential for efficient viral DNA replication.

Transcription of early genes. Immediately after the adenoviral genome enters the nucleus, the early phase of transcription starts. This phase has three primary objectives: (i) to promote entry of the infected cell into the S-phase of the cell cycle, thus generating a cellular environment optimal for viral replication – this activity is exerted by the products of the E1A, E1B, and E4 genes; (ii) to protect the infected cell from the various antiviral defense

mechanisms at the cell and organism levels – E1A, E3, and VA RNA genes; (iii) to synthesize the viral proteins that are necessary for viral DNA replication – the E2 gene. Achieving
all three of these objectives depends on transcriptional activation of the viral genome, which
is mediated by the E1A gene product. Thanks to its interaction with numerous cellular factors (cf. above), E1A alone is indeed capable of stimulating entry of the cell into the S-phase.
At least three proteins directly inhibit cell apoptosis: E1B-55K and E4orf6 bind and inactivate p53, while E1B-19K is a Bcl-2 homologue. In addition, proteins encoded by E3 block
apoptosis by interfering with signaling emanating from the TNF-α receptors and promoting
degradation of Fas on the cell surface (this receptor triggers a death signal after its interaction with the Fas ligand (FasL) expressed by cytotoxic T lymphocytes). Other E3 proteins
block transport of MHC class I molecules to the cell surface and inhibit their transcription,
thus blocking recognition of the infected cells by the immune system. Finally, both E1A and
VA RNAs block inhibition exerted by interferon α and β. E1A exerts this activity by binding and inactivating the STAT proteins, which transduce the signal from the interferon receptors on the plasma membrane. The VA RNAs instead bind and block the PKR kinase; interferon induces transcription of this kinase in an inactive form, while its activation is induced
by the accumulation, in the infected cells, of double-stranded RNA molecules.

Genome replication. The production of the E2 gene-encoded proteins DNA polymerase, DBP, and TP marks the beginning of the viral DNA replication phase. This begins
from the ITRs at the two extremities of the genome and continues in both directions; the
process, which is catalyzed by the viral DNA polymerase, requires the covalent binding
of TP to the genome ends and involves binding of a series of cellular factors (NF-I, NF-
III, and others) to the ITRs. Elongation of the newly synthesized DNA requires the viral
protein DBP, which binds DNA, and the cellular factor NF-II. As a rule, the adenoviral
genome never integrates into the host cell, a property that also characterizes the vectors
derived from the wild-type virus.

Transcription of late genes. At the beginning of the DNA replication phase, transcription of the late genes also starts. These are organized as a single long transcript of about
29,000 nt, which is subsequently processed through the utilization of different polyadenylation and alternative splicing sites to give rise to a series of shorter mRNAs. These can
be grouped into 5 families (L1–L5) on the basis of the utilization of 5 different polyadenylation sites. Expression of all these transcripts is controlled by a single, specific promoter,
the major late promoter (MLP), which is activated by the cellular transcription factor
USF/MLTF and transactivated by E1A. The product of the delayed early gene IVa2 coop-
erates with USF/MLTF in MLP activation. Once replication of viral DNA starts and the
late gene mRNAs are synthesized, the cellular mRNAs are selectively retained in the
nucleus, due to the capacity of E1B 55K and E4orf6 to block their export. In addition to
cytoplasmic transport, translation of the viral mRNAs is also favored at this stage compared to that of cellular mRNAs. These properties have obvious relevance in considering
adenovirus as a vector for high-level expression of therapeutic proteins.

Virion assembly and cell lysis. Translation of the L1–L5 mRNAs leads to the synthesis
of the virion structural proteins; packaging of the viral DNA inside the virions is mediated
by the recognition of the packaging signal ψ, present at about 260 bp from the left side of
the genome and consisting of a series of AT-rich DNA stretches. Release of the virions from
the infected cells is accompanied by disintegration of the plasma membrane during cell lysis.

3

3.5.3.2
Structure of Adenoviral Vectors

Three different classes of vectors based on Ad2 or Ad5 are considered for gene therapy. The first generation consists of viruses in which the therapeutic gene substitutes the E1 and/or E3 regions. The second generation carries additional deletions in the E2 and E4 regions. Finally, the third generation includes the so-called *gutless* or helper-dependent vectors, in which the whole viral genome is substituted with the exception of the ITR and ψ regions (Figure 3.22).

The *first-generation* adenoviral vectors are obtained by substituting the E1, or the E1 and E3, regions with an expression cassette, consisting of the therapeutic gene, a promoter, and a polyadenylation site. As reported above, the E1 region (containing the early genes E1A and E1B) codes for proteins essential for the expression of the other early genes as well as for the late genes of the virus. Since these proteins are required for viral replication, to produce the vector particles, they are supplied in trans by specific cell lines, such as HEK 293, 911, N52.E6, or PER.C6. The E3 region codes for proteins that are important to counteract the host antiviral mechanisms. These products, however, are not required for *in vitro* adenovirus replication, and thus it is not necessary to complement their loss in trans during vector production. However, for some applications, it is desirable to maintain or even increase expression of some of the E3 proteins. For example, E3-11.6K (ADP) facilitates the release of infectious particles from the producing cells, and gp19k reduces the T-cell response against the transduced cells and thus increases persistence of gene expression *in vivo*. Vectors carrying deletions in only E1 can accommodate foreign DNA stretches up to 5.1 kb, while those deleted in E1 and E3 up to 8.3 kb (considering that the maximum length of DNA that can be packaged in Ad2 or Ad5 virions is about 38 kb).

Although the E1-deleted vectors cannot replicate *in vivo*, expression of the several adenoviral genes that are still present stimulates a powerful inflammatory and immune response of the host, which raises important safety concerns, as will be further discussed below. In addition, the immune response limits the duration of therapeutic gene expression driven by these vectors, since the transduced cells are eliminated by cytotoxic T lymphocytes.

Since inflammatory and immune response towards first-generation adenoviral vectors is stimulated by the various vector-encoded proteins, a *second generation* of vectors was obtained, bearing additional deletions in the E2 region – in particular, in the E2A (coding for DBP), E2B (TP), or DNA polymerase genes – or in the whole or vast majority of the E4 region. These vectors can accommodate up to 14 kb of foreign DNA. Despite elimination of these genetic regions, these vectors do not completely solve the issue of adenovirus-induced toxicity, given the immunogenic and inflammatory potential of the residual genes. In addition, expression of the therapeutic gene from these vectors is reduced compared to first-generation vectors, probably because some of the E2 and E4 genes code for proteins that directly or indirectly increase the levels of expression of the virus-encoded genes.

Finally, the *third-generation* adenoviral vectors are characterized by the complete deletion of the adenoviral genome and its substitution with exogenous DNA, with the exception of the regions required in cis for viral DNA replication and packaging (ITRs and ψ respectively). These vectors are named *gutless* or *gutted* or, more appropriately, *helper-dependent* (since their replication entirely depends on the co-infection of the cells in

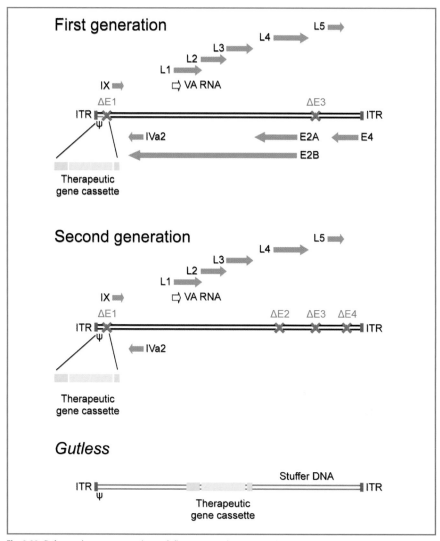

Fig. 3.22 Schematic representation of first-generation, second-generation, and gutless adenoviral vectors. The genes deleted or otherwise inactivated in each type of vector are indicated, along with the site of insertion of the therapeutic gene cassette

which packaging occurs with a helper vector producing in trans all the required proteins) or *high-capacity* (HC, since they can accommodate up to 37 kb of exogenous DNA, thus also allowing delivery of large DNA sequences or multiple genes).

An additional class of adenovirus-derived vectors that are used for cancer gene therapy are the *oncolytic viruses* (or oncolytic vectors), in which only the E1B-55K gene is deleted, thus exclusively allowing viral replication in p53– cells. The properties of these mutated viruses are described in the section on 'Oncolytic Viruses' in the context of cancer gene therapy.

3.5.3.3
Production of Adenoviral Vectors

Production of adenoviral vectors requires a two-step procedure, entailing first the generation of a vector genomic DNA with the sequence of interest and later its replication and packaging to obtain infectious viral preparations.

As far as vector DNA production is concerned, the relative length of the wild-type adenoviral genome (~36 kb) poses an important obstacle to the use of conventional genetic engineering techniques, essentially based on recombinant DNA manipulation *in vitro* followed by amplification of plasmids in simple microorganisms. Therefore, a series of relatively complex protocols have been set up over recent years for the production of first-generation, second-generation, and gutless adenoviral vectors exploiting various alternative approaches.

The second step, namely the production of infectious vector particles, is usually obtained in human cells, named *helper cells*, supplying al the necessary functions in trans (in particular, the products of the E1 gene) in the case of first- and second-generation vectors. The production of gutless vectors is more complex.

Production of First- and Second-Generation Adenoviral Vectors

The production of first- and second-generation vectors is essentially based on the generation of long molecules of linear DNA corresponding to the desired adenoviral vector genome by recombination. The recombinant genomes can be obtained: (i) by direct recombination in helper cells; (ii) by *in vitro* ligation followed by helper cell transfection; (iii) by recombination in bacteria; and (iv) by *in vitro* ligation followed by plasmid transformation of bacteria.

(i) Methods based on recombination in helper cells

The classic method used to generate adenoviral vectors in which the therapeutic gene substitutes the E1 region takes advantage of the recombination events occurring spontaneously in mammalian cells between two homologous DNA sequences (Figure 3.23A). Cells of an E1-expressing cell line (typically, HEK 293 cells, generated in the 1970s by transformation of human embryonic kidney cells with Ad5 DNA) are transfected with: (a) a DNA molecule corresponding to the majority of the "right" arm of the adenoviral genome, obtained starting from the entire purified adenoviral genome (or from a plasmid containing the adenoviral genome), after digestion with a restriction enzyme cutting in the E1 region, followed by the eventual purification of the "right" fragment; (b) a plasmid containing the "left" arm of the adenoviral genome, starting from the ITR down to the region immediately downstream of the restriction site; in this *shuttle* plasmid, an expression cassette containing the gene of interest substitutes the E1 region. Thanks to the region of homology downstream of the restriction site, which is present in both DNA arms, recombination between the two molecules occurs inside the helper cells; this event, although rare, generates a complete viral genome, which is packaged, followed by the release of viral particles into the cell supernatant. This supernatant is then utilized to infect other cells, which are covered by a layer of agar, in order to block diffusion of the virions and only allow infection of the neighboring cells. In this manner, a lysis

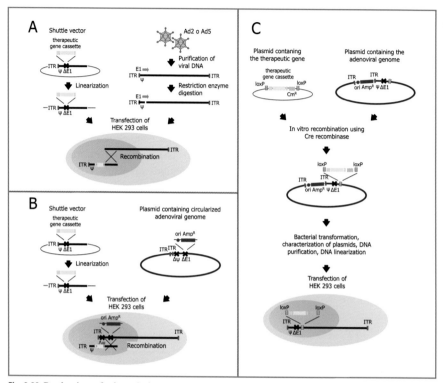

Fig. 3.23 Production of adenoviral vectors. **A** Production of adenoviral vectors by recombination in helper cells transfected with a linearized plasmid carrying the gene of interest and the digested, linear viral DNA. **B** Production of adenoviral vectors by recombination in helper cells transfected with a linearized plasmid carrying the gene of interest and with another plasmid corresponding to the adenoviral genome in which the E1 region is substituted with an antibiotic resistance gene and a prokaryotic DNA replication origin, and in which the packaging signal is deleted. **C** Production of adenoviral vectors by *in vitro* recombination mediated by the Cre recombinase. See text for description

plaque is obtained. Once a plaque containing the desired vector and devoid of parental virus is identified, this is amplified to produce a small quantity of vector, which is then used to infect a larger number of helper cells and produce batch quantities of vector.

The above-described procedure is tedious and labor-intensive (it usually takes 2–4 weeks to complete), since DNA recombination rarely occurs inside mammalian cells. Furthermore, unwanted recombination events from unrestricted, parental DNA molecules also lead to the generation of wild-type virus. Thus, vector production requires multiple subsequent passages of infection and analysis of several lysis plaques.

Different procedures have been developed to improve selection and purification of the desired recombinant vector. For example, as a source of adenoviral DNA, it is possible to use a genome in which a cassette expressing the herpes simplex virus thymidine kinase (HSV-TK) enzyme substitutes the E1 gene. In this manner, cell treatment with gancyclovir (to which cells expressing HSV-TK are sensitive) eliminates all cells infected with a parental rather than a recombinant virus. The same strategy can be followed by using, instead of the HSV-TK gene, the gene coding for *E. coli* β-galactosidase (LacZ) or the *Aequorea victoria* green fluorescent protein (GFP). In the former case, the plaques formed by the parental, unwanted virus are recognized since they become blue after staining with the chromogenic substrate X-gal, while, in the latter case, they are green at the fluorescence microscope. Finally, instead of naked adenoviral DNA, it is possible to use DNA in a complex with the TP protein, a strategy significantly increasing recombination efficiency.

An additional manner to improve recombinant adenoviral production is based on the capacity of circular molecules containing the adenoviral genome to generate infectious virus after HEK 293 cell transfection (Figure 3.23B). These cells are transfected with: (a) a circular viral DNA containing, instead of the E1 region, a prokaryotic origin of DNA replication and the ampicillin resistance gene, and carrying a deletion of the ψ region: this construct can only be propagated and selected in *E. coli* as a plasmid, however cannot be incorporated into virions; (b) a small shuttle plasmid carrying the gene of interest and the "right" arm of the adenovirus genome, similar to the previous method; this plasmid is transfected after linearization. Inside the cells, recombination between the two molecules generates a viral genome that can be packaged. The same technique can be used to also insert expression cassettes into the E3 region, or in both E1 and E3. The main disadvantage of this system is that it requires the use of very large plasmid molecules in *E. coli*, where these are often unstable and potentially deleterious to bacterial growth.

(ii) Methods based on in vitro ligation, followed by helper cell transfection

Instead of transfecting the "right" and "left" arms of the adenoviral vector DNA in the cells relying on their *in vivo* recombination, as in the above-described procedure, it is possible to obtain the desired DNA molecule through canonical ligation *in vitro*. The simplest approach is based on the ligation of the "left" arm of the genome, obtained by restriction enzyme digestion followed by purification, with the "right" arm, recovered from the insert of the shuttle plasmid, as above. The ligation mixture is then transfected into the helper cells to propagate and package the virus. A drawback of this method is that it is particularly subject to contamination from the parental virus, since tiny amounts of uncut parental DNA are inevitably present in the ligation mixture.

(iii) Methods based on recombination in E. coli

Although poorly utilized due to the difficulty of manipulating large molecules in bacteria, some procedures have been developed to generate adenoviral vector genomes by recombination in bacteria. These methods essentially rely on the utilization of two elements: (a) a large plasmid containing the whole adenoviral genome (or at least its "right" arm); (b) a smaller shuttle plasmid containing an expression cassette for the gene of interest flanked by regions of homology to the adenoviral genome region where recombination is sought (i.e., the E1 or E3 regions). In its simplest formulation, the procedure entails transfection of the two plasmids into a recombination-proficient (RecA+) *E. coli* strain, in which the desired molecule is usually produced with an efficiency varying from 20 to 100%. The recombinant plasmid is then transferred to a RecA− strain for propagation, to avoid further recombination events. Once purified, the plasmid DNA is digested to linearize the adenoviral genome and transfected into the helper cells for packaging.

(iv) Methods based on in vitro ligation or recombination, followed by amplification in bacteria

Instead of relying on *in vivo* recombination in helper cells or bacteria, it is possible to obtain the desired recombinant plasmids *in vitro* and then propagate these constructs in *E. coli*. The plasmid DNA is then recovered and transferred into the helper cells for replication and packaging. For the *in vitro* production of the adenoviral vector genomes, it is possible to use large plasmids containing the whole adenoviral genome, previously engineered to contain, in place of the E1 region, recognition sites for rare endonucleases, which can be used for standard cloning of an insert containing the therapeutic gene cassette. Much more efficiently, instead of cloning, this cassette can be inserted into the vector by site-specific *in vitro* recombination (Figure 3.23C). For this purpose, a system based on two plasmids is commonly used: (a) the first plasmid contains the adenoviral DNA circularized in order to contain, in the outside region flanked by the two ITRs, a prokaryotic cassette including an origin of DNA replication and an antibiotic resistance gene; the adenoviral DNA contains an intact ψ region and carries a deletion in E1; downstream of this deletion, a recognition site for a prokaryotic recombinase is inserted; (b) the second plasmid contains the expression cassette for the therapeutic gene and an antibiotic resistance gene different from that contained in the first construct; these sequences are flanked by two recognition sites for the same recombinase targeting the adenoviral genome. The two DNAs are purified from *E. coli*, mixed and incubated together with the purified recombinase: this enzyme, using its target sites on the two molecules, mediates insertion of the fragment containing the therapeutic gene into the adenoviral genome. The reaction product is then used to transfect bacterial cells, which are selected with the antibiotic to which the gene flanking the therapeutic gene confers resistance. After extraction and characterization, this plasmid DNA is then used to transfect helper cells.

The most commonly used recombination procedure is that based on the site-specific recombination system Cre-*loxP*. The P1 bacteriophage produces an enzyme, the Cre recombinase, that recognizes a specific 34 bp sequence, named *loxP* (locus of crossover in P1). When two *loxP* sequences are located far apart, Cre binds both of them and activates their recombination: as a consequence of this process, the DNA segment between the two sequences is removed, only leaving a single *loxP* sequence on site (the P1 bacteriophage uses such a recombination strategy to depolymerize the concatemers that its genome forms during DNA replication in bacteria, since the two *loxP* sites are located at the two extremities of the phage genome). Using the same recognition sequences, Cre is also able to mediate the insertion of a DNA segment flanked by two *loxP* sites (in the case of adenoviral vectors, the DNA segment containing the therapeutic gene) using a third *loxP* site inside the target DNA molecule (in this case, the adenoviral DNA).

The methods based on *in vitro* recombination followed by bacterial transformation are relatively simple and rapid, and nowadays represent the methods of choice to produce adenoviral vectors, also thanks to the commercial availability of kits that facilitate the whole procedure. However, transformation of bacteria with very large molecules can lead to their rearrangement. Therefore, much attention needs to be paid to the careful characterization of the final recombinant adenoviral DNA molecule before helper cell transfection.

Production of Gutless Adenoviral Vectors

Gutless adenoviral vectors only possess the ITRs at the two extremities of the genome and the ψ region inside, while the DNA of interest substitutes the rest of the genome (Figure 3.22). Since adenoviral vectors are able to package linear DNA molecules having a length corresponding to 75–105% of the wild-type genome (about 27–37.5 kb), when a gutless vector is considered for the delivery of a cDNA or of a small gene it is usually necessary to complement the insert with irrelevant DNA sequences acting as *stuffer* DNA. As a source of stuffer, DNA of prokaryotic or yeast origin, or, better, DNA sequences derived from large human introns can be used. The gutless vector DNA is cloned as a plasmid, amplified in *E. coli*, purified and linearized to release the segment flanked by the two ITRs, and, finally, transfected into the cells.

Since gutless vectors are completely devoid of viral genes, all the proteins necessary for vector DNA replication must be provided in trans. This can be obtained by co-infection of the cells with a replication-competent adenovirus acting as a helper. In this case however, both the gutless and the wild-type genomes are packaged, causing significant contamination of the final vector preparations. To selectively avoid packaging of the helper virus, different strategies can be followed, including mutation of the ψ region of the helper virus, elimination of ψ during vector production, or use of helper viruses with genomes significantly longer or shorter than those that can be packaged.

The most effective strategy so far developed to avoid helper virus packaging is based on the use of HEK 293 cells previously selected to express the Cre recombinase (293Cre). These cells are transfected with the linearized gutless vector genome and infected with a first-generation adenoviral vector (E1-deleted), in which the ψ region is flanked by two *loxP* sequences. Inside the cells, the Cre recombinase removes ψ from the helper virus genome, thus selectively preventing its incorporation into the virions (Figure 3.24). Using this strategy, the extent of contamination of helper virus in the final vector preparations is in the order of 0.1–10% of the gutless vector.

Notwithstanding the relative efficiency with which the gutless vectors can be produced using the above-described procedure, the residual levels of contamination by the helper virus pose important safety issues in light of clinical application. For this reason, several laboratories are currently trying to improve the system, using different recombinases (for example, the yeast Flp recombinase, which catalyses recombination between the *frt* sites) in addition or as an alternative to Cre-*loxP*, or using helper viruses carrying a mutation in exon protein IX, which is necessary for packaging. In the presence of this mutation, the genome of a gutless vector of optimal length is packaged much more efficiently than that of a helper virus that is significantly shorter or longer than the optimal packaging range.

Purification and Characterization of Adenoviral Vectors

Starting from the helper cell supernatant, the virions corresponding to the adenoviral vectors are purified by three subsequent centrifugation steps, the first of which is a conventional centrifugation to pellet the virus while the last two are run in a cesium chloride gradient, in

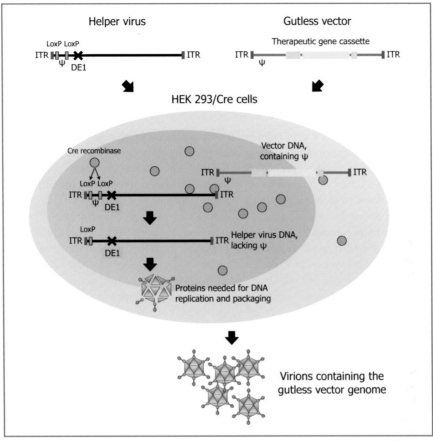

Fig. 3.24 Production of adenovirus gutless vectors. The vectors are produced in 293/Cre cells transfected with the linearized gutless vector DNA and infected with a helper adenoviral vector in which the ψ region is flanked by two *loxP* sequences. Inside the cells, the Cre recombinase removes ψ from the helper virus genome, thus selectively preventing its incorporation into the virions and allowing selective packaging of the gutless vector DNA

which the viral particles are separated according to their specific density, and thus purified.

Before its utilization, an adenoviral vector preparation must be controlled for the possible presence of replication competent adenovirus (RCA). In particular, the HEK 293 cells used by most production protocols contain about 4.5 kb of the Ad5 "left" arm, including the E1 region, integrated into human chromosome 19. This region can thus recombine with the genome of first- and second-generation adenoviral vectors, or with the helper virus in the case of the gutless vectors, thus leading to the formation of RCAs. This becomes progressively more likely should subsequent production steps be carried out, or if very large batches are obtained, since the RCAs replicate more efficiently than the vectors. In case an RCA contamination is detected, for example by PCR, the original vector must be again isolated from lysis plaques generated from the replication of single clones.

3

To lower the probability of RCA formation, helper cells alternative to HEK 293 have been generated, such as 911 or PER.C6, which do not contain stretches of homology between the sequence contained in the vectors and that integrated in the cell genome.

3.5.3.4
Properties of Adenoviral Vectors

Adenoviral vectors are a very efficient tool for gene transfer in mammalian cells, since they infect a vast range of both resting and replicating cells, can be purified and concentrated to reach titers in the order of 1×10^{13} particles/ml, and their genomes do not integrate into the target cells, which might be advantageous for several applications. In addition, the gutless vectors can accommodate large segments of exogenous DNA, up to 35 kb.

As far as first- and second-generation vectors are concerned, these continue to be very interesting for experimental gene transfer in animals. However, their clinical utilization is hampered by the inflammatory and immune response they elicit, which on one hand limits the duration of *in vivo* gene expression and, on the other hand, raises important safety issues. The administration of a first- or second-generation adenoviral vectors, similar to wild-type adenovirus, stimulates both the adaptive and the innate immune response. Immediately after inoculation, expression of a series of inflammatory cytokines is activated, determining recruitment, to the sites of inoculation, of macrophages, neutrophils, and NK cells. For example, in the liver, 80–90% of vector is rapidly eliminated by this inflammatory response within the first 24 h after inoculation. This response is triggered by the adenoviral particle itself and does not require viral gene expression. Subsequently, starting from 4 to 7 days after injection, the humoral and cellular immune response starts to be activated. The inoculation site becomes infiltrated by cytotoxic T lymphocytes, which recognize and eliminate the transduced cells. Furthermore, the immune system mounts a very vigorous antibody response, which, thanks to the production of neutralizing antibodies, prevents any possibility of re-injecting the same vector or vectors based on the same serotype. In this context, it is also important to observe that 30–40% of individuals living in western countries and 80–90% of those living in sub-Saharan Africa naturally possess anti-Ad5 antibodies, which completely prevents utilization of this serotype for gene therapy or vaccination.

The powerful induction of an inflammatory and immune response was the cause of the death of an 18-year-old patient enrolled in a gene therapy clinical trial for the hereditary deficit of ornithine transcarbamylase (OTC), an enzyme of the urea cycle, at the University of Pennsylvania, Philadelphia, PA in 1999. This patient received an injection in the liver, through the hepatic artery, of a second-generation adenoviral vector carrying the OTC cDNA. A few hours after infusion of a relatively high dose of vector, the patient started to show severe symptoms of systemic toxicity, and died after 4 days (cf. section on 'Gene Therapy of Liver Diseases'). Death of this patient was subsequently attributed to a massive, acute inflammatory response to the adenoviral vector injection, probably due to a cytokine storm triggered by the viral capsid.

In light of the above observations, it is thus possible to conclude that the utilization of first- and second-generation adenoviral vectors should now be limited to applications in which prolonged transgene expression is not desirable or required, and in which immune stimulation is instead a requisite. In practical terms, this is the case in two very important applications: gene therapy of cancer and genetic vaccination.

In the case of gutless vectors, the systemic administration of these viruses continues to stimulate the immune response, similar to first- and second-generation vectors, since this depends on the viral capsid proteins. The same proteins also trigger the production of neutralizing antibodies, which prevent re-administration of vectors of the same serotype. After the initial inflammation, however, the gutless vectors do not express any viral genes, and the transduced cells are therefore not recognized and eliminated by the immune system, unless the transgene protein itself is immunogenic. Experiments performed by transduction of various tissues in rodents, dogs, and non-human primates have indeed indicated that the administration of these vectors, in particular to the liver and skeletal muscle, determines a stable transduction, lasting over time and leading to the expression of the transgenes at therapeutic levels. This appears to be of particular relevance for applications in which other vectors that are efficient in these tissues (in particular, those based on AAV) are instead incapable of delivering very large inserts. This is the case, for example, of Duchenne muscular dystrophy, a disease caused by mutations in the dystrophin gene (cf. section on 'Gene Therapy of Muscular Dystrophies'). The dystrophin cDNA is about 14 kb long and thus unfit for cloning into AAV, but very well suited for gutless adenoviral vectors.

Despite the great potential of gutless vectors, their clinical application is however still limited by two major technical issues. The first one consists in their contamination with a clinically still unacceptable proportion of helper virus; the second one relates to the difficulty of obtaining the large batches of vectors that are needed for clinical use, since the procedures so far developed, which are described above, are unsuitable for scaling up.

Cell Targeting

In concluding the discussion on adenoviral vectors it is important to remember that, over the last 10 years, much effort has been put into direct transduction with these vectors towards specific cell types, a property known as *cell targeting*. As reported above, the Ad5 virions bind the CAR receptor thanks to the C-terminal portion of the fiber protein. Following this interaction, a secondary interaction occurs between the penton base protein (displaying the RGD amino acid sequence) and the target cell $\alpha v\beta 3$ and $\alpha v\beta 5$ integrins, leading to virion internalization by clathrin-mediated endocytosis. In addition, the elongated portion of the fiber establishes contacts with the cell surface HSPGs. It is thus possible to modify the terminal amino acid sequence of the fiber (binding to CAR), the penton base (binding to integrins), and the fiber body (binding to HSPGs) in order to modify the tropism of Ad5 *in vitro* and, possibly, *in vivo*.

The modification mainly considered so far consists in the insertion of peptides at the fiber extremity, after deleting the portion of the protein binding to CAR. The majority of the peptides considered for targeting have been isolated using phage display technology, which allows the selection of short amino acid stretches, expressed on the surface of a filamentous phage, for their property of binding to a specific ligand of interest.

An alternative strategy to modify the adenoviral vector tropism is to use antibodies with double specificity, namely capable of binding and inactivating the CAR-binding domain of the fiber protein on one side, and binding a different cellular receptor on the other side (for example the c-Erb-2 receptor to target the adenoviral vectors towards the breast cancer cells that express this receptor, or the FGF-2 or VEGF receptors, to target the vectors towards activated endothelial cells such as those of the tumor vasculature).

Despite the encouraging *in vitro* results obtained with both these approaches, there is no compelling evidence, at the moment, that these strategies aimed at re-targeting adenoviral vectors might be easily applied to the clinic *in vivo*, especially in light of the systemic toxicity of first- and second-generation vectors.

3.5.4
Vectors Based on the Adeno-Associated Virus (AAV)

In sharp contrast to retroviruses, adenoviruses, and herpesviruses, before its entry into the gene therapy arena, not many laboratories were interested in the biology of the AAV, most likely because this virus, despite its wide diffusion in nature, has never been associated to any human disease. As a consequence, different aspects of its life cycle, including the molecular determinants regulating its tropism, are still largely unknown. In contrast, however, now AAV represents one of the most appealing vectors for *in vivo* gene therapy and several clinical experimentations have already been conducted or are ongoing with very encouraging results, especially for incurable disorders of tissues incapable of regeneration, such as brain and retina.

3.5.4.1
Molecular Biology and Replicative Cycle of AAV

The *Parvoviridae* family (*parvo-*: Latin for "small") includes a vast series of small viruses with icosahedral symmetry, without envelope, containing a single-stranded DNA genome, which infect numerous species of mammals, including man. The family is divided into two genera, the erythroviruses and the dependoviruses. The human prototype of the former genus is human parvovirus B19, the etiologic agent of the fifth disease or erythema infectiosum, while the murine prototype is the minute virus of mice (MVM). AAV instead belongs to the Dependovirus genus, the members of which, in contrast to the erythroviruses, are incapable of autonomous replication and depend on the superinfection of the cells with another virus to complete their replicative cycle, hence the name. In particular, AAV owes its name to its original isolation as a contaminant of cell cultures infected with adenovirus.

The members of the Dependovirus genus are very diffuse in nature: in primates alone over 100 AAV variants have been discovered to date and new serotypes are continuously being identified (i.e., variants with different antigenic properties, not recognized by the currently available antisera). More than 80% of adults of 20 years or older show an antibody response against AAV, proving that they have encountered the virus, probably in their infancy. Despite their diffusion, none of the dependoviruses has ever been associated with any human disease to date.

Structure of Virions

AAV virions have a capsid with icosahedral symmetry (T=1) with a diameter of 18–25 nm, composed of 60 proteins. These include 3 proteins derived from the same gene (the Cap gene) and differing in their N-terminus: VP1, VP2, and VP3, with a 1:1:18 ratio (i.e., each virion has 3, 3, and 54 VP1, VP2, and VP3 proteins respectively). The capsid includes the viral genome, consisting of a linear single-stranded DNA, having either positive or negative polarity; in any AAV preparation, about half of the virions have a DNA with positive polarity and the rest a DNA with negative polarity.

Over recent years, at least 12 different AAV serotypes have been isolated (AAV1–AAV12) and well characterized antigenically, while over 100 additional genetic variants have been identified by PCR amplification of DNA from cultured cells infected with adenovirus or derived from human and non-human primate tissues. All these viruses share similar structure, size, and genetic organization and only significantly differ in the amino acid composition of the capsid proteins. The sequence homology between these proteins ranges from 55 to 99%, and is the major determinant dictating the use of the receptors for cell internalization. In general terms, all AAVs use receptors that are ubiquitously and abundantly expressed (Table 3.7). The most utilized serotype both experimentally and clinically is AAV2, which binds to cell surface HSPGs; $\alpha v \beta 5$ integrin and the receptors for fibroblast growth factor (FGFR-1) and hepatocyte growth factor (HGFR)

Table 3.7 Receptors for some parvoviruses

Parvovirus	Receptor
AAV1	Sialic acid ($\alpha 2,3$ N-linked and $\alpha 2,6$ N-linked)
AAV2	Heparan sulfate proteoglycans (HSPGs)
	Co-receptors: $\alpha v \beta 5$ integrin, FGFR-1, HFGR
AAV3	Heparan sulfate proteoglycans (HSPGs)
AAV4	Sialic acid ($\alpha 2,3$ O-linked)
AAV5	Sialic acid ($\alpha 2,3$ O-linked and $\alpha 2,3$ N-linked)
	PDGF receptor (PDGFR)
AAV6	Sialic acid ($\alpha 2,3$ N-linked and $\alpha 2,6$ N-linked)
AAV7	Not known
AAV8	Laminin receptor (LamR)
AAV9	Not known (LamR?)
Parvovirus B19	Red blood cells P antigen
CPV (canine parvovirus)	Transferrin receptor
	Sialic acid (N-glycolyl neuraminic acid, NeuGC)
FPV (feline panleukopenia parvovirus)	Transferrin receptor

function as co-receptors in some cells. Similar to AAV2, AAV3 also binds HSPGs. In contrast, AAV1, AAV4, AAV5, and AAV6 interact with sialic acid (N-acetylneuraminic acid, Neu5Ac) residues, linked with various bonds to the cell surface glycans. AAV8 binds a specific cell surface protein, LamR, which exerts several functions in the cells, including that of receptor for extracellular laminin. AAV2 and AAV5 particles enter cultured cells by clathrin-mediated endocytosis and are found in early endosomes immediately after entry. These cellular compartments are trafficked through the cytoplasm and rapidly approach a perinuclear location, where they mature into late endosomes. In contrast to AAV2, AAV5, in addition to the endosomes, can also be found in the trans-Golgi apparatus, indicating differences in endosomal trafficking between serotypes.

Organization of the Genome

The single-stranded AAV genome has about 4.7 kb and contains two orfs, corresponding to two genes, *rep* and *cap* (Figure 3.25A). *Rep* codes for the proteins necessary for viral replication, and *cap* for the proteins of the viral capsid.

By the use of two different promoters (p5 and p19) and the inclusion or not of an exon, the *rep* gene codes for 4 protein isoforms (Rep78, 68, 52, and 40). The Rep proteins are necessary for replication of the viral DNA, its integration into the host cell genome, and the transcriptional regulation of the viral promoters. They are endowed with single-stranded endonuclease (nickase) and helicase activities. Furthermore, the Rep proteins exert a series of effects on the infected cells, including the inhibition of cellular DNA replication and of transcription of several cellular genes.

The VP1, VP2, and VP3 proteins are generated from the *cap* gene by using three different start sites (AUG codons) for translation. All the AAV transcripts have the same polyadenylation site, located at the 5' end of the genome.

The AAV coding region is flanked by two ~145-nt-long ITRs, having an internal complementarity stretch in their first 125 nt and thus forming a T-shaped hairpin structure, identical at the two viral ends (Figure 3.25B). This palindromic sequence is the only cis-acting genetic element necessary for all AAV functions, including viral DNA replication, site-specific integration into the host cell DNA, and packaging of virions. The first two activities (replication and integration) require the presence of Rep68 or Rep78 proteins, which specifically bind a sequence within the ITR, the Rep binding site (RBS), and cleave in a site- and strand-specific manner at the terminal resolution site (TRS) located 13 nucleotides (nt) upstream of the RBS. An almost identical sequence in human chromosome 19q13.4 represents the minimal sequence necessary and sufficient for AAV site-specific integration – see below. The two ITRs are the only AAV sequences preserved in the vectors, while a transcriptional cassette (promoter+therapeutic gene+polyadenylation site) substitutes the rest of the genome.

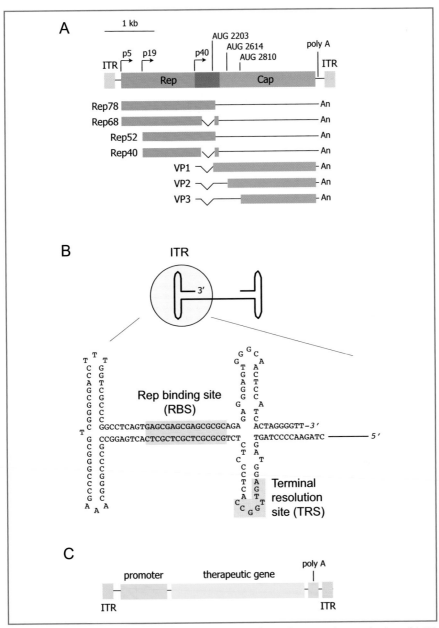

Fig. 3.25 Genetic organization of AAV. **A** The upper part shows the genomic organization of AAV, with the indication of the promoters (p5, p19, p40), the AUG codons for translation of the Cap-encoded proteins, and the polyadenylation site. The bottom part shows the structure of the viral mRNAs, indicating the intron–exons organization, and of the encoded proteins. *ITR*: inverted terminal repeat; *An*: polyA tail. **B** Enlargement of the ITR region, with the indication of the Rep binding site (*RBS*) and the terminal resolution site (*TRS*). **C** Schematic representation of the structure of an AAV vector

3

Replicative Cycle

After binding to cell surface receptors, AAV is internalized by receptor-mediated endocytosis and is thus found inside the endosomal compartment. Although the virus can penetrate a vast series of different cell type, thanks to the ability of its capsid to interact with ubiquitously expressed receptors, the fate of infection strictly depends on the physiological state of the infected cells (Figure 3.26). If the cells are exposed to genotoxic stress (e.g., are treated with X-rays, or γ-rays, or other DNA-damaging agents) or are infected with another virus (typically, adenovirus or herpesvirus), the AAV DNA, once exited from the endosomes, is efficiently transferred to the nucleus and replicated by the cellular machinery with the assistance of the viral protein Rep. In particular, viral DNA replication is carried out by a cellular DNA polymerase (probably DNA polymerase δ) using as a primer the exposed 3'-OH from one of the ITRs. Completion of double-stranded DNA synthesis requires the nickase activity of Rep, which cleaves one strand of the ITR and thus permits elongation of DNA synthesis to reach the end of the template molecules (Figure 3.27). At the end of the replication process, two complete viral genomes are generated, with complementary polarity; both of these are packaged inside the virions at equal efficiency. In a few hours,

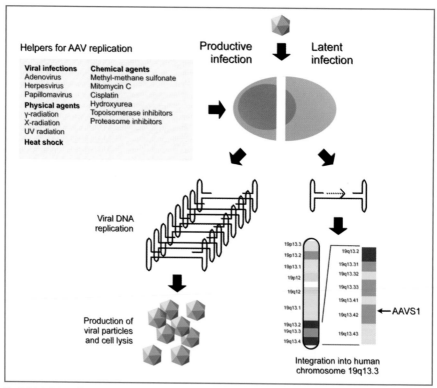

Fig. 3.26 AAV replication cycle. Schematic representation of the AAV replication cycle, under permissive (*left side*) and non-permissive conditions (*right side*). In the latter case, the viral DNA integrates site specifically in the AAVS1 region of chromosome 19q13.3. See text for details

every cell produces $5 \times 10^5 - 1 \times 10^6$ viral particles; the infected cells eventually lyse and the virions are released in the outside environment.

Productive viral infection requires exit of the viral particles from the endosomes, transport to the nucleus, removal of the capsid and release of the nucleic acid, and, most important, conversion of the single-stranded genome into a double-stranded replication intermediate. In physiological conditions, that is in the absence of any treatment with chemical or physical agents or superinfection with another virus, most human cells do not allow productive viral replication, and infection is probably blocked at multiple steps. Under non-permissive conditions, however, in a fraction of the infected cells the AAV genome becomes integrated in a site-specific manner into a specific region of human chromosome 19q13.4, named AAVS1, which contains a 33-bp sequence almost identical to the RBS and TRS sites in the viral ITRs. AAVS1 is positioned immediately upstream of the translation initiation site of the gene coding for the protein phosphatase 1 regulatory inhibitor subunit 12C (PPP1R12C), also known as MBS85 (myosin-bind-

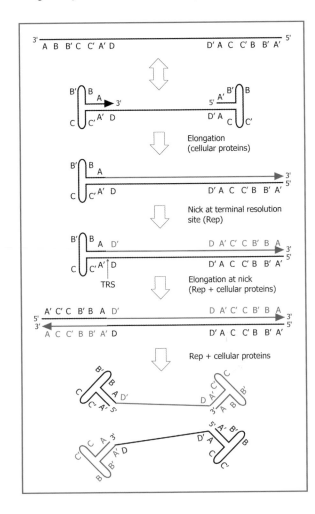

Fig. 3.27 Mechanism for AAV DNA replication. Letters indicate specific sequences in the inverted terminal repeats (A-A', B-B', C-C', D-D' denote complementary sequences). Input viral DNA is in black and neosynthesized DNA in red. The arrow indicates the 3' OH available as a primer for DNA synthesis. *TRS*: terminal resolution site

ing subunit 85), a protein involved in the regulation of actin–myosin fiber assembly. The translation initiation start codon of this gene is located only 17 nt downstream of the chromosomal RBS sequence. The AAVS1 region is located in a centromeric position with respect to the genes coding for the slow skeletal muscle troponin T (TNNT1) and cardiac troponin I (TNNI3), located 15 and 26 kb apart respectively. The capacity of AAV DNA to integrate into the AAVS1 region strictly depends on the availability of the Rep proteins to simultaneously interact with both the ITR and the cellular RBS–TRS motifs, thus mediating a semi-homologous recombination between the two sequences, with the almost certain involvement of still unidentified cellular proteins. Integration of AAV into the human genome is the only event of site-specific integration known to occur in mammalian cells.

It is still largely unclear which molecular determinants govern cell permissivity to productive AAV replication and their relationship with the induction of genotoxic damage or infection of the cells with another virus. These treatments do not directly act on the replication machinery of the virus or on its proteins, but do induce some cellular functions that render the cellular environment permissive for replication. Experimental evidence indicates that this process is controlled by the proteins belonging to the cellular DNA damage repair (DDR) system, namely the cellular machinery that physiologically surveys the integrity of the cellular genome. In particular, proteins of the MRN complex (Mre11, Rad50, and Nbs1) bind the AAV genome which, similar to damaged cellular DNA, is single-stranded and bears imperfectly paired DNA sequences at the level of its terminal hairpins. These proteins block replication of the genome by impeding its conversion to a double-stranded form. Once the cell is treated with chemical or physical agents, the DDR proteins are recruited to other sites of cellular DNA damage, thus permitting the AAV genome to complete its replication. In the case of adenovirus, the helper effect exerted by this virus is mediated by a few known viral genes, which are encoded by the early regions E1A/E1B, E2A, and E4 (in particular, E4or6) and by the gene coding for VA-I. The E1B and E4orf6 proteins are indeed able to induce degradation of the cellular MRN complex.

3.5.4.2
Structure and Production of AAV Vectors

The AAV genome can be converted into a double-stranded DNA form and cloned into a bacterial plasmid. Once transfected into mammalian cells, thanks to its site-specific nickase activity, the Rep protein is able to excide the AAV sequence from the plasmid and initiate its replication. This process only requires integrity of the ITRs in cis and the presence of Rep in trans. If the cells also express the capsid proteins, the single-stranded DNA genome that is formed by the process of DNA replication becomes packaged into the capsids thanks to the interaction of the ITRs with the VP1-3 proteins.

AAV vectors are usually obtained starting from the AAV2 genome, cloned in a plasmid form, by removing all the viral sequences with the exception of the two ITRs (about 145 nt each). Between the ITRs, an expression cassette is cloned containing the therapeutic gene and its regulatory elements (Figure 3.25C). In contrast to retroviruses, the replicative cycle of which involves the generation of both RNA and DNA genomic forms and in which the choice of the promoter is thus critical to determine the vector efficiency, AAV replication only involves DNA intermediates. As a consequence, any promoter can be chosen to direct expression of the therapeutic gene, without interfering with the production of vectors. This includes strong constitutive, inducible, or tissue-specific promoters. The only strict requirement is that the transcriptional cassette cloned between the two ITRs does not exceed 4–4.5 kb.

The expression of Rep proteins exerts toxic effects on the cells, since these proteins interfere with several essential processes, including DNA replication and cellular gene transcription. Therefore, it has not been possible to obtain packaging cell lines stably expressing Rep, and AAV packaging thus occurs upon transient transfection. This is usually achieved by transfecting, using calcium phosphate co-precipitation, HEK 293 cells with one plasmid containing the AAV vector, as described above, and one plasmid containing the AAV *rep* and *cap* genes without the ITRs. To stimulate the induction of cell permissivity to productive AAV replication, the cells are also infected with adenovirus or, more conveniently, treated with a third plasmid bearing the adenovirus helper genes E2A, E4, and VA-I RNA; the E1A and E1B genes are already expressed in the HEK 293 cells. Several laboratories now exploit a single helper plasmid, containing both the AAV2 *rep* and *cap* genes and the adenoviral helper genes; in this case, the production of vectors involves cell transfection with only two plasmids (Figure 3.28).

Forty-eight hours after transfection, the cells start to show a clear cytopathic effect, due to viral replication, and a large quantity of virions is found in both the supernatant and the cell lysates. In contrast to retroviral vectors and similar to adenoviruses, the AAV genomes are very resistant to manipulation and treatment with chemical and physical agents. Thus, they can be easily purified by cesium chloride or iodixanol gradient centrifugation, or by chromatography. The viral preparations obtained using these procedures are sufficiently pure to be used in both experimental animals and in the clinics. The titers can reach or surpass 1×10^{14} viral particles/ml; the concentration of viral particles is thus several orders of magnitude higher than both VSV-G-pseudotyped retroviral vectors and adenoviral vectors.

A standard AAV production protocol entails utilization of the AAV2 ITRs in conjunction with the AAV2 *rep* and *cap* genes. However, the capsid proteins corresponding to any AAV serotype can recognize the AAV2 ITRs and mediate packaging of the AAV2 genome inside the virions. It is thus possible to change the serotype of the vector simply by using, during production, an expression vector for any desired *cap* gene. Vectors with a capsid corresponding to the AAV1–AAV9 serotypes are commonly generated to exploit the different organ tropism of these viruses (cf. below).

3.5.4.3
Properties of AAV Vectors

AAV vectors represent an outstanding tool for *in vivo* gene transfer, for a series of reasons, which are summarized as follows.

(i) AAV vectors do not express any viral protein; therefore, they are not immunogenic and do not cause inflammation (in contrast to first- and second-generation adenoviral vec-

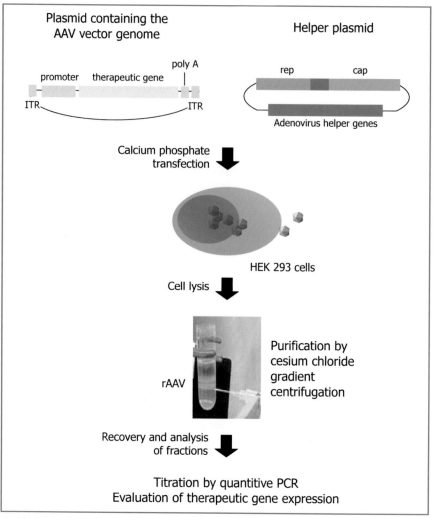

Fig. 3.28 Production of AAV vectors. For AAV vector production, two plasmids are transiently trans-fected into HEK 293 cells. The first plasmid corresponds to the AAV vector itself, in which the ther-apeutic gene cassette is flanked by the inverted terminal repeats (ITRs), and the second codes for Rep and Cap and for the adenoviral proteins providing helper functions. Twenty-four hours after transfection, cells are lysed and the vectors are purified by cesium chloride centrifugation

tors); as a consequence, therapeutic gene expression usually lasts for month- or year-long periods. In this respect, it is however worth mentioning that, while this conclusion certain-ly holds true in mice, pigs, and dogs, which are not the natural hosts of human AAVs, from which most of the vectors are currently derived, it might be different in humans and non-human primates, where the pre-existing immunity against the virus in some cases might determine the elimination of the transduced cells over the first weeks post-inoculation (cf. also the section on 'Gene Therapy of Hemophilia').

(ii) AAV vectors do not integrate into the host cell genome, but persist in an episomal form, probably as head-to-tail or head-to-head extrachromosomal concatemers, in non-replicating cells; therefore, they avoid the problem of insertional mutagenesis (in contrast to retroviral vectors). Of notice, the ability of the wild-type virus to integrate site specifically into the AAVS1 region of human chromosome 19q13.4 strictly requires the AAV Rep protein; since the gene coding for this protein is not present in the vectors, site-specific integration does not occur.

(iii) Possibly as a consequence of the lack of integration into the transduced cell chromosomes, therapeutic gene expression is not subject to significant methylation and silencing (in contrast to retroviral vectors).

(iv) AAV vectors can be generated at high titers, thus allowing the simultaneous expression of different genes from the same cells or tissues. This property could be of great importance in light of the possibility to deliver multiple growth factor coding genes, for example for gene therapy of cardiovascular or neurodegenerative disorders, or for the administration of multiple shRNAs to inhibit different proteins acting along the same metabolic pathway.

(v) AAV vectors do not experience the problem of transcriptional interference from different promoters (in contrast with retroviral vectors). The therapeutic gene can thus be controlled by any promoter of choice, provided that its length is suitable for cloning into AAV.

Taken together, these properties have encouraged, over the last 5 years, the use of these vectors in over 50 clinical trials, which have enrolled several hundred patients. These are Phase I/II trials for various hereditary (in particular hemophilia B, deficit of α1-antitripsin, cystic fibrosis, muscular dystrophies, retinal degeneration) and acquired (rheumatoid arthritis, Parkinson's disease, Alzheimer's disease) disorders. The preliminary results of these trials will be detailed in the respective sections in the chapter on 'Clinical Applications of Gene Therapy'.

Despite the successful utilization of AAV vectors for clinical gene transfer, several of their molecular properties remain poorly understood. Most AAV serotypes use ubiquitous molecules that are expressed at high levels by most cell types as receptors for internalization, such as HSPGs or sialic acid linked to cell surface glycoproteins and gangliosides (Table 3.7); as a consequence, internalization of these vectors occurs in most cells. However, in the majority of cell types, the vector DNA does not reach the nucleus or, most frequently, is not converted from its single-stranded DNA form to double-stranded DNA, a step that is obviously essential for transcription to occur. In vivo, only a few tissues show high-level natural permissivity to AAV transduction. These include the heart (cardiomyocytes), skeletal muscle (skeletal myofibers), brain (neurons), retina (ganglionar cells, pigment epithelium and photoreceptors), and, to a lesser extent, liver (hepatocytes). The reasons why the virus is particularly efficient in these cell types are still largely unknown, but clearly involve molecular events following vector internalization inside the cells. For example, after injection of AAV2 in the brain, the vector transduces neurons very efficiently but does not transduce glial cells at all, even though these cells express much higher levels of HSPGs on their surface. Of note, all the cell types that are naturally permissive to AAV transduction are post-mitotic and will never re-enter the cell cycle; it is thus likely that the DDR proteins that bind single-stranded DNA and block AAV transduction in replicating cells are downregulated in these permissive cell types. Indirect evidence of the importance of single-stranded to double-stranded DNA conversion in determining efficiency of transduction is the significantly higher transduction efficiency of the so-called *self-complementary* AAV vectors (scAAVs), in which the gene cassette is cloned in the form of two complementary copies, positioned in tandem one after the other; the DNA of these vectors is thus capable of spontaneously forming double-stranded DNA by internal self-complementation. Due to their intrinsic design, however, the scAAV vectors have a cloning capacity about half of the already constrained limit of normal AAV vectors.

The use of capsids with serotypes different from AAV2 on one hand increases efficiency of transduction in the already permissive cell types but, on the other hand, extends tropism to a few other organs (Table 3.8). For example, skeletal muscle is transduced with particular efficiency by AAV1 and AAV6 (which differs only 6 amino acids from AAV1); in the retina, photoreceptors are an efficient target of AAV5 and the pigment epithelium of AAV5 and AAV4; finally, AAV8 transduces both the endocrine and exocrine pancreas very well, in addition to the liver. None of the serotypes, however, permits significant transduction of cells physiologically refractory to AAV2 gene transfer, including endothelial cells, fibroblasts, keratinocytes, and several others.

Another very interesting property of some of the most recent AAV serotypes, AAV8 and AAV9, is their capacity to cross the endothelial barrier of blood vessels. Once injected intravenously or intraperitoneally in the experimental animal, these vectors reach the skeletal muscle parenchyma and transduce myofibers highly efficiently. They thus represent potential tools for whole muscle transduction for gene therapy of muscle dystrophies. A molecular reason for the different cell-type selectivity of the various AAV serotypes is still to be found. It is conceivable that the use of different entry molecules routes the vectors towards different pathways, or that the various capsid proteins modulate the interaction of the viral genome with cellular proteins differently.

Table 3.8 Tropism of AAV serotypes for different organs

Organ		Serotype (in order of efficiency)
Liver		AAV8, AAV9, AAV5
Skeletal muscle		AAV1, AAV7, AAV6, AAV8, AAV9, AAV2, AAV3
Central nervous system		AAV5, AAV1, AAV4, AAV2
Eye	Pigment epithelium	AAV5, AAV4, AAV1, AAV6
	Photoreceptors	AAV5
Lung		AAV5, AAV9
Heart		AAV9, AAV8
Pancreas		AAV8
Kidney		AAV2

3.5.5
Vectors Based on the Herpes Simplex Virus (HSV)

Different aspects of the biology of herpes simplex virus type 1 (HSV-1) suggest the potential value of this virus as a gene therapy vector. These include: (i) the ample cellular host range, since most of the receptors the virus exploits for internalization are widely expressed by mammalian cells, including HSPGs and nectin-1; (ii) its high natural infectivity; (iii) its capacity to efficiently infect non-replicating cells; (iv) the dispensability of over half of the 80 natural HSV-1 genes for virus replication in cultured cells – these genes can thus be removed and replaced by the gene(s) of interest; (v) the ability of the virus to establish latent infection lasting for very prolonged periods of time in neurons, a property that can be exploited to selectively express therapeutic genes in these cells; and finally (vi) the possibility to produce vectors at high titers in the absence of contamination by wild-type virus. On the other hand, however, the relative complexity of the viral genome and our still incomplete knowledge of the molecular properties of various viral proteins still hamper wider utilization of this vector system.

3.5.5.1
Molecular Biology and Replicative Cycle of Herpesviruses

The *Herpesviridae* family consists of over 130 different viruses, very diffuse in most animal species, 9 of which infect humans. These are HSV-1 and HSV-2, cytomegalovirus (CMV), varicella-zoster virus (VZV), Epstein-Barr virus (EBV), and human herpesviruses types 6A, 6B, 7, and 8 (HHV-6A, HHV-6B, HHV-7, and HHV-8).

All members of the family share at least three common biological characteristics: (i) the presence, in the viral genome, of a vast series of genes coding for enzymes involved in nucleic acid metabolism, including thymidine kinase (TK), which is used in several gene therapy applications, especially as a suicide gene (cf. section on 'Clinical Applications of Gene Therapy'); (ii) the nuclear localization of the sites of genome replication and capsid assembly; and (iii) the capacity to establish two modalities of infection, one leading to production of new viral particles and eventual lysis of the infected cells, and the other one in which the viral genome is maintained in a circular, double-stranded DNA form in the nucleus of the infected cells, with only a minority of the viral genes being expressed.

The various members of the family can be divided into one of three sub-families: the alpha-herpesviruses, characterized by broad host range, rapid replicative cycle, and capacity to establish latent infections mainly in the sensory ganglia (this subfamily includes, among the human herpesviruses, HSV-1, HSV-2, and VZV); the beta-herpesviruses, which have a more restricted host range, longer replicative cycle, and capacity to establish latent infections in the salivary glands, the lymphoreticular system, and the kidney (this subfamily includes CMV, HHV-6, and HHV-8); and the gamma-herpesviruses, which prevalently show tropism for lymphoid cells (this subfamily includes EBV and HHV-8).

3

Structure of Virions

All *Herpesviridae* are characterized by a common virion structure, consisting of a central core containing the linear genomic DNA, an icosahedral capsid consisting of 162 capsomers, an apparently amorphous structure surrounding the capsid named *tegument* and containing about 20 different types of proteins, and an envelope derived from the plasma membrane during budding; the envelope displays a series of glycoproteins of viral derivation (Figure 3.29A). In the case of HSV-1, at least 11 different glycoproteins (gB-gN) are present on the virion envelope, for a total of over 1000 copies. The virion has an overall diameter varying from 120 to 300 nm.

Genome Organization

The genome of herpesviruses consists of a large molecule of double-stranded linear DNA (120–250 kb). In particular, the HSV-1 genome has 152 kb and encodes over 80 proteins. Approximately half of these genes are essential for viral replication in cell culture; the other half encode accessory functions, which contribute to the virus life cycle in specific tissues (e.g., post-mitotic neurons) and can be removed without significantly affecting the capacity of the virus to replicate in cell culture. The genome is composed of unique long (U_L) and unique short (U_S) segments which are both flanked by inverted repeats (Figure 3.29B). The HSV-1 genes fall into one of three classes depending on the kinetics of their transcription in the viral replication cycle: *immediate early* (IE), *early* (E), or *late* (L) genes. The IE genes code for regulatory proteins, the E genes for factors necessary for viral replication, and the L genes for the structural proteins of the virions. Replication of the genome occurs through a rolling circle mechanism leading to formation of head-to-tail concatemers.

Replicative Cycle of HSV-1

Cellular infection starts with binding of the virus to the cell surface glycosaminoglycans (GAGs), in particular with heparan sulfate and dermatan sulfate GAGs. This initial interaction, which is mediated by the C and B glycoproteins (gC and gB) of the viral envelope, is then followed by a more specific binding between gD and some membrane receptors, including HveA (herpes virus entry A, also known as HveM, a member of the tumor necrosis factor (TNF) receptors) and HveC (also known as nectin-1, a transmembrane protein of the immunoglobulin superfamily expressed at high levels in sensory neurons). These interactions lead to fusion of the viral envelope with the cell membrane, followed by entry of the viral capsid into the cytosol together with the tegument proteins. Entry is mediated, by not completely understood mechanisms, by the gB, gH/gL, and gD viral glycoproteins.

Once in the cytosol, the viral DNA is transported to the nucleus through the nuclear pores and a productive, lytic cycle starts; the whole process of viral replication, from transcription to packaging, takes places in the nucleus. The viral genes are expressed in a temporally regulated manner: immediately after entry of the viral DNA into the nucleus, and in the absence of *de novo* synthesis of viral proteins, five IE genes are transcribed (infect-

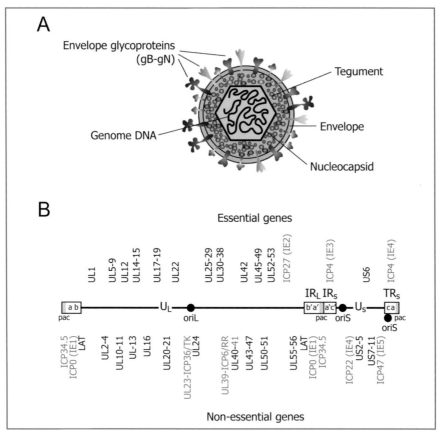

Fig. 3.29 HSV-1 and HSV-1 vectors. **A** Schematic representation of the structure of an HSV-1 virion. **B** HSV-1 genome organization. The HSV-1 genome consists of a linear, double-stranded DNA molecule of 152 kb containing more than 80 genes. The genome is composed of unique long (U_L) and unique short (U_S) segments which are flanked by inverted repeats. These are designated as TR_L and IR_L (terminal and internal repeat of the long segment, respectively) and TR_S and IR_S (terminal and internal repeat of the short segment). The repeats surrounding U_L are designated ab and b'a', while those surrounding U_S are designated a'c' and ca. There are two different origins of replication, oriL in the long segment and oriS in the short segment. OriS is duplicated, along with ICP4, because it is found in the inverted repeats surrounding the long segment. Approximately half of the genes are essential for viral replication in cell culture (*listed on top*); the other half are non-essential for viral replication in cultured cells (*bottom*). Genes in blue are non-essential genes that are mutated in the replication-competent viruses so far developed and described in the text; genes in red are immediate early (IE) genes that are mutated in the replication-defective viruses. The genome contains three pac signals (shown in *yellow*) that assist in packaging the viral genome DNA into virions

ed cell protein-0 (ICP0), ICP4, ICP22, ICP27, and ICP47). The proteins encoded by these genes activate expression of the E genes, coding for factors necessary for viral DNA replication. Once DNA replication is complete, the IE proteins activate transcription of the L genes. Expression of the IE genes is increased by VP16, a structural protein present in the teguments, which acts in concert with several cellular transcription factors that bind the IE gene promoters. The L gene products include the structural proteins of the virus, which allow packaging of new viral particles and completion of the lytic cycle.

During primary infection, HSV-1 initially replicates in the epithelial cells close to the site of exposure. The virus then enters the sensitive nervous terminations and the capsid is convoyed by retrograde axonal transport along the axon cytoskeleton to the nuclear body of neurons in the sensitive ganglia. Once entered the nucleus of these cells, latent infection ensues. This is characterized by the presence of the viral genome in the form of circular, double-stranded DNA molecules, or as concatemerized multimers, which persist in the nucleus episomally (that is, not integrated inside the host cell genome). In this latent state, all genes proper of the lytic phase are transcriptionally silent; expression is active only for a single family of non-polyadenylated transcripts named *latency-associated transcripts* (LATs), which remain localized in the nucleus. The exact function of LATs, which have a structure similar to *lariats* (i.e., the RNA products generated by intron processing during splicing), is not known; however, their presence, which can persist for the whole life of the host, can be used as a marker of latent herpesviral infection.

3.5.5.2
Structure and Production of HSV-1 Vectors

One of the major limitations imposed by HSV-1-derived vectors originates from the high pathogenicity of the wild-type virus; for example, the intracerebral injection of HSV-1 typically causes lethal encephalitis. Removal of all the pathogenic genes from the vectors is therefore imperative. Three strategies are currently considered for this purpose:

(i) removal of all genes dispensable for *in vitro* replication however essential for *in vivo* replication; this generates vectors that are still capable of replicating, however with attenuated virulence (**attenuated, replication-competent vectors**);

(ii) removal of all genes necessary for all types of replication (**replication-defective vectors**);

(iii) deletion of the entire viral genome with the exception of an origin of DNA replication and the packaging signal (**amplicon vectors**).

The characteristics and modalities for construction for these three categories of vectors are described in the following sections.

Attenuated, Replication-Competent Vectors

Limited capacity of replication of an HSV-1 vector *in vivo* can be useful as a means to transfer the transgene to cells neighboring those originally transduced, thus leading to amplification of the therapeutic efficacy. Deletion of some non-essential genes allows the generation of HSV-1 mutants that are still capable of replicating *in vitro* but severely impaired *in vivo*. These genes include those coding for proteins necessary for DNA replication, or mediating virulence, or conferring the infected cell the capacity to evade immune recognition (Figure 3.29B). Examples of these proteins are TK and ribonucleotide reductase (RR), two enzymes dispensable in cell culture but essential for viral DNA replication in neurons, where cellular DNA replication proteins are no longer expressed; the *vhs* (*virion-host shut off*), the product of the UL41 gene, which rapidly destabilizes and blocks translation of the infected cell mRNAs; and the neurovirulence factor ICP34.5, which allows continuation of translation in the infected cells notwithstanding the activation of the cell kinase PKR, which would otherwise phosphorylate and inactivate translation initiation factor eIF2α as an antiviral defense mechanism. Different studies have revealed that the attenuated, replication-competent HSV-1 vectors are not only capable of replicating autonomously but, when inoculated into the brain, also circulate to areas different from those of the original injection, similar to the wild-type virus.

Replication-Defective Vectors

Wild-type HSV-1 replication in neurons begins immediately after entry of viral DNA into the infected cell nucleus. An essential role in this process is played by the IE genes, which activate transcription first of the E genes, involved in DNA replication, and later, once DNA replication is complete, of the L genes, coding for the structural proteins of the virions. Deletion of the IE genes generates mutants that only replicate in cells in which the missing functions are supplied in trans. Once inoculated *in vivo*, these defective viruses are not able to activate the cascade of events leading to lytic infection and thus remain in the cells in a state that is similar to viral latency, persisting for prolonged periods in both neurons and non-neuronal cells *in vivo*.

The first generation of replication-defective HSV-1-based vectors consisted of mutants deleted in the single essential IE gene encoding ICP4. Although these vectors show reduced pathogenicity and could be used to efficiently transfer and transiently express reporter genes in brain, they were nonetheless cytotoxic for neurons in culture. Further improvements involved the introduction of deletion in additional genes, first in ICP27 and later in various combinations of other IE genes (Figure 3.29B). Besides prolonged persistence *in vivo*, an additional advantage of these multiply deleted viruses is their capacity to provide enough space to introduce distinct and independently regulated expression cassettes for different transgenes.

Amplicon Vectors

Amplicon vectors are viral particles identical to wild-type HSV-1 virions, in which the genome is however made of a concatemeric form of a plasmid, the amplicon. This consists of a conventional *E. coli* plasmid carrying an origin of DNA replication (usually ori-S) and a packaging signal (pac), both derived from the HSV-1 genome (Figure 3.30). The remaining portions of the amplicon contain the transgenic sequence(s) of interest; given the capacity of HSV-1 virions to package long DNA molecules, these can extend to over 150 kb. This represents the largest cloning capacity of all currently available gene transfer systems.

Both replication-defective and amplicon vectors are packaged into complete HSV-1

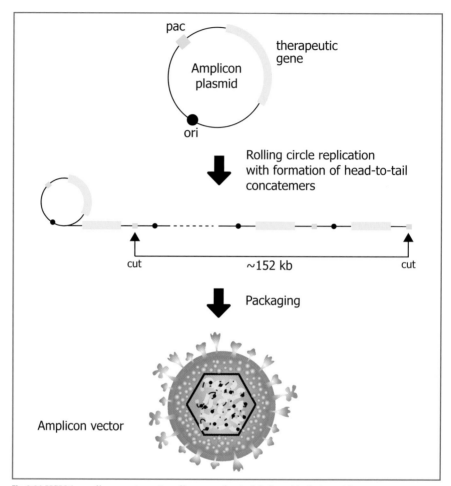

Fig. 3.30 HSV-1 amplicon vectors. Amplicons are bacterial plasmids that contain one or more trans-gene cassettes and two non-coding viral sequences, an origin of DNA replication (*ori*) and a DNA cleavage/packaging signal (*pac*). Upon transfection into a cell line providing HSV-1 helper functions in trans, an amplicon is replicated by a rolling circle mechanism, which generates head-to-tail concatemers which are packaged into HSV-1 particles as approximately 152-kb linear DNA

particles to infected target cells; however, their major difference is that, afterwards, the amplicons persist in the cells without expressing any viral proteins. Thus, while replication-defective vectors are always subject to the potential risk of reactivation and virulence, the amplicon vectors are devoid of this problem. On the other hand, the production of high-titer preparations of amplicons is considerably more difficult than that of replication-defective vectors.

The production of amplicon vectors was initially obtained in cells transfected with an amplicon plasmid (produced in bacteria) and superinfected with a defective HSV-1 virus providing helper function, i.e., supplying in trans all the factors necessary for amplification and packaging. In this manner, however, the probability of contaminating the vector preparations with helper virus was exceedingly high. This problem was recently solved by co-transfecting the amplicon with a set of 5 partially overlapping cosmids, expressing all the required viral proteins; the system can be further improved by using a bacterial artificial chromosome (BAC) as a source of viral proteins.

3.5.5.3
Properties of HSV-1 Vectors

Each of the three types of HSV-1 vectors has different characteristics and, as a consequence, possible modes of application in gene therapy.

One of the major applications of attenuated, *replication-competent vectors* is for the oncolytic therapy of cancer (cf. also the section on 'Oncolytic Viruses'). Different modified viruses have been produced so far for this purpose. The first generation contained mutations in a single gene, aimed at limiting viral replication in actively replicating cells. The genes considered included the gene UL23 (ICP36), coding for TK, or the gene UL39 (ICP6), coding for the major subunit of RR, or the gene coding for the neurovirulence factor ICP34.5. While use of the first two mutants did not extend beyond animal experimentation due to the risk of toxicity, different mutants of ICP34.5, which show considerable anti-tumor effect in animal models, are currently the subject of different clinical trials. Given the success of this first set of attenuated viruses, additional series of vectors were obtained. A second generation includes viruses bearing multiple mutations, in particular, ICP34.5 plus ICP6, currently also in clinical experimentation. Viruses of the third generation, in addition to deletions in the above-mentioned genes, also function as real gene transfer vectors, since they contain genes coding for various cytokines (IL-4, IL-12, IL-10, GM-CSF) or for the co-stimulatory molecule B7.1, with the ultimate purpose of increasing tumor immunogenicity in addition to oncolysis.

An additional interesting property of attenuated herpetic vectors is their use as live, attenuated vaccines for immunization against wild-type HSV-1. Different combinations of mutants in the genes coding for the surface glycoproteins or the IE genes have been obtained. The ultimate purpose of these modifications is to generate an ideal strain that is able to propagate in a limited manner without inducing neurotoxicity, albeit eliciting a strong immune response. Experimentation with such mutants is currently limited to the preclinical phase.

Finally, numerous studies have indicated that the HSV-1 tropism can be changed through the deletion or modification of the surface glycoproteins of the virions. In particular, deletion of gB and/or gC, substitution of gD with VSV-G, and creation of chimeras between gC and receptor-specific ligands could allow, alternatively, broadening of the cell tropism of the virus or its restriction to cell types expressing defined receptors.

The *replication-defective vectors* and the *amplicon vectors* have instead been utilized, so far at the preclinical level, to express a variety of genes in different tissues. In the brain, these genes include those coding for proteins having toxic or pro-apoptotic (for gene therapy of gliomas), neurotrophic (such as NGF or BDNF for gene therapy of neurodegenerative disorders), or enzymatic activity (such as tyrosine hydroxylase for gene therapy of Parkinson's disease). Other studies have shown that both the replication-defective and the amplicon vectors can be used in tissues different from the nervous tissue, including muscle, heart, and liver, or for genetic vaccination. It is likely that a few of these studies will reach clinical experimentation stages in the near future.

3.5.6
Viral Vectors for Gene Therapy: Fields of Application and Comparative Evaluation

A synopsis of the properties, pros, and cons of the five major classes of viral vectors for gene therapy is reported in Table 3.9. The main parameters distinguishing the different vectors can be summarized as follows.

Table 3.9 Pros and cons of the main viral vectors for gene therapy

Vector	Pros	Cons
Gammaretroviral vectors	Efficient transduction Integration into the host cell genome	Low titers (if not pseudotyped) Insertional mutagenesis Silencing of gene expression Exclusively transduce actively replicating cells
Lentiviral vectors	Transduction of quiescent cells *in vitro* and *in vivo* Integration into the host cell genome	Need pseudotyping Possible generation of RCLs Possible mobilization of vector in HIV-infected patients Potential for insertional mutagenesis
Adenoviral vectors (first generation)	Very efficient transduction High-level transgene expression Production at high titers Transduction of both quiescent and replicating cells Broad host range	Transient transduction Stimulation of strong inflammatory and immune responses
AAV vectors	Derived from a non-pathogenic virus Production at high titers Infection of quiescent cells *in vivo* Very long persistence and gene expression	Limited cloning capacity (<5 kb) Lack of a packaging cell line Tropism limited to specific cell types
Herpesviral vectors	Persistence in latent form Large cloning capacity Tropism for neuronal cells	Difficult to manipulate Poor knowledge of several biological features Pathogenic genes difficult to identify and eliminate

Cloning capacity. The currently available vectors significantly differ in their capacity to accommodate DNA fragments of different lengths, with a spectrum ranging from 3–4 kb for AAV vectors, to 8 kb for retroviruses, to 30–40 kb for gutless adenoviral vectors, to 150 kb for HSV amplicons. In evaluating these lengths, one has to consider that the coding portion of human genes has an average size of ~1.5 kb; as a consequence, even the relatively small size of AAV allows the delivery not only of short regulatory RNAs but also of the vast majority of therapeutic cDNAs. The size of vectors becomes substantially limiting in two specific situations, namely when the cDNAs to be transferred are exceedingly long (for example, in the case of dystrophin –9.7 kb, or coagulation Factor VIII >8 kb), or when transcription of the therapeutic gene needs to be strictly regulated, a condition usually requiring very long regulatory elements (for example, for gene therapy of thalassemias or diabetes). In these situations, the possibility of using gutless adenoviral vectors or herpetic amplicons would be most desirable. In particular, the latter class of vectors appears suitable to accommodate entire genetic loci, composed of the whole gene (exons plus introns) and its regulatory elements.

Simplicity of production. The production systems for the different vectors are quite different. In the case of amphotropic and ecotropic gammaretroviral vectors, the possibility to utilize packaging cell lines offers an obvious advantage in terms of simplicity and cost. In contrast, production of AAV, lentiviral, and herpetic amplicon vectors is based on transient transfection of plasmids in the producer cells. Transient transfection is however also used for gammaretroviral vectors when VSV-G pseudotyping is required. Another issue related to vector production concerns the purity of the preparations and, in particular, the contamination of vectors with autonomous replicating or helper viruses.

Efficiency of transduction. The gammaretroviral and lentiviral vectors, both pseudotyped with VSV-G, are capable of transducing a vast number of cell types. However, one of the strict requirements of gammaretroviral vectors is that the infected cells are in active replication. This characteristic substantially prevents the use of these vectors in most cell types *in vivo* and restricts their utilization to *ex vivo* cultured cells. In contrast, lentiviral vectors can transduce non-replicating cells, albeit provided that they are metabolically active. These vectors can thus successfully be used for gene transfer in neurons *in vivo* and in non-stimulated hematopoietic stem cells *ex vivo*. Adenoviral vectors are also very efficient in transducing both replicating and non-replicating cells and, by virtue of the utilization of the ubiquitously expressed CAR receptor, are capable of transducing most cell types, both *in vivo* and *ex vivo*. These vectors are probably the most efficient delivery system currently available both in terms of number of transduced cells and levels of therapeutic gene expression. However, in replicating cells, their efficacy is limited by the lack of integration into the genome: during cell proliferation and expansion, they progressively become diluted or lost. AAV vectors also use ubiquitous receptors for internalization (HSPGs and sialic acid) and are thus internalized by most cells. However, the tropism of these vectors is essentially restricted by events occurring after internalization and involving vector transport to the nucleus and, mostly, single-stranded to double-stranded DNA conversion. The efficiency of these events essentially restricts AAV transduction to postmitotic tissues such as brain, retina, skeletal muscle, and heart. AAV vectors thus represent the vectors of choice for transduction of these tissues. The recent identification of an array of different AAV serotypes now extends their use to other organs, such as liver, pancreas, and lung. Finally, while gammaretroviruses and lentiviruses transduce cells at low multiplicity (one or maximum two vector copies per cell), adenoviruses and AAVs infect at high multiplicity, with several copies of the vector commonly being found to be non-integrated in the nucleus of the transduced cells.

Persistence. The five viral vectors differ substantially in terms of persistence of their genome in the transduced cells and duration of therapeutic gene expression. Gammaretroviruses and lentiviruses integrate into the host cell genome and are thus permanently inherited at every cell division. These are the vectors of choice for the treatment of inherited disorders with recessive monogenic transmission, in which permanent correction of the molecular defect is the therapeutic goal. However, in several circumstances, gammaretroviral vectors especially undergo progressive silencing of gene expression, due to irreversible proviral methylation. In contrast, albeit not being integrated into the genome, AAV transduces long-living and non-replicating cells, and thus persists for month- or year-long periods in these tissues; probably because it remains episomal, methylation-induced transcriptional silencing does not occur and therapeutic gene expression persists. In contrast, the duration of gene expression is very short with first-generation adenoviral vectors, not because of any intrinsic property of the vector itself, but due to the recognition of the transduced cells by the immune system: in immunocompetent animals, expression of the therapeutic effect usually does not last longer than 10–14 days after inoculation. This can still be useful in conditions in which persistence of expression is not required or desirable: this is the case, for example, of gene therapy of cancer or genetic vaccination, or when gene therapy is used to transiently express a growth factor, for example to stimulate therapeutic angiogenesis. Finally, the vectors based on HSV-1 might become very useful for the prolonged expression of genes in the brain, thus exploiting the property of the wild-type virus to persist in episomal form in the neurons for the whole life of the infected organism.

Induction of undesired effects. The use of viral vectors is still strongly impacted by the fear of the possible induction of undesired or frankly pathologic effects. First-generation adenoviral vectors are strongly pro-inflammatory and induce a powerful immune response. Gammaretroviral vectors and, potentially, lentiviral vectors can induce insertional mutagenesis. Both replication-defective and oncolytic HSV vectors raise concerns for their possible reactivation and consequent neurovirulence. Finally, the production of adenoviral, gammaretroviral, lentiviral, and replicating herpesviral vectors is fraught with the possible generation of autonomously replicating wild-type or recombinant viruses. Of note, the only vector system not raising important safety concerns at this moment is that based on AAV.

3.1 Cellular Barriers to Gene Delivery

Further Reading

Conner SD, Schmid SL (2003) Regulated portals of entry into the cell. Nature 422:37–44
Doherty GJ, McMahon HT (2009) Mechanisms of endocytosis. Annu Rev Biochem 78:857–902
Sandvig K, van Deurs B (2005) Delivery into cells: lessons learned from plant and bacterial toxins. Gene Ther 12:865–872

Selected Bibliography

Kerr MC, Teasdale RD (2009) Defining macropinocytosis. Traffic 10:364–371

Kirkham M, Parton RG (2005) Clathrin-independent endocytosis: new insights into caveolae and non-caveolar lipid raft carriers. Biochim Biophys Acta 1745:273–286

Medina-Kauwe LK (2007) "Alternative" endocytic mechanisms exploited by pathogens: new avenues for therapeutic delivery? Adv Drug Deliv Rev 59:798–809

Nichols B (2003) Caveosomes and endocytosis of lipid rafts. J Cell Sci 116:4707–4714

Pelkmans L, Puntener D, Helenius A (2002) Local actin polymerization and dynamin recruitment in SV40-induced internalization of caveolae. Science 296:535–539

Plemper RK, Wolf DH (1999) Retrograde protein translocation: ERADication of secretory proteins in health and disease. Trends Biochem Sci 24:266–270

Roth MG (2006) Clathrin-mediated endocytosis before fluorescent proteins. Nat Rev Mol Cell Biol 7:63–68

Torgersen ML, Skretting G, van Deurs B, Sandvig K (2001) Internalization of cholera toxin by different endocytic mechanisms. J Cell Sci 114:3737–3747

3.2 Direct Inoculation of DNAs and RNAs

Selected Bibliography

Braun S (2008) Muscular gene transfer using nonviral vectors. Curr Gene Ther 8:391–405

Herweijer H, Wolff JA (2003) Progress and prospects: naked DNA gene transfer and therapy. Gene Ther 10:453–458

3.3 Physical Methods

Further Reading

Frenkel V (2008) Ultrasound mediated delivery of drugs and genes to solid tumors. Adv Drug Deliv Rev 60:1193–1208

Hynynen K (2008) Ultrasound for drug and gene delivery to the brain. Adv Drug Deliv Rev 60:1209–1217

Wells DJ (2004) Gene therapy progress and prospects: electroporation and other physical methods. Gene Ther 11:1363–1369

Selected Bibliography

Andre F, Mir LM (2004) DNA electrotransfer: its principles and an updated review of its therapeutic applications. Gene Ther 11[Suppl 1]:S33–42

Bigey P, Bureau MF, Scherman D (2002) In vivo plasmid DNA electrotransfer. Curr Opin Biotechnol 13:443–447

Hagstrom JE (2003) Plasmid-based gene delivery to target tissues in vivo: the intravascular approach. Curr Opin Mol Ther 5:338–344

Heller LC, Heller R (2006) In vivo electroporation for gene therapy. Hum Gene Ther 17:890–897

Lewis DL, Wolff JA (2007) Systemic siRNA delivery via hydrodynamic intravascular injection. Adv Drug Deliv Rev 59:115–123

3

Mennuni C, Calvaruso F, Zampaglione I et al (2002) Hyaluronidase increases electrogene transfer efficiency in skeletal muscle. Hum Gene Ther 13:355–365

Mir LM (2008) Application of electroporation gene therapy: past, current, and future. Methods Mol Biol 423:3–17

Newman CM, Bettinger T (2007) Gene therapy progress and prospects: ultrasound for gene transfer. Gene Ther 14:465–475

Reed SD, Li S (2009) Electroporation advances in large animals. Curr Gene Ther (*in press*)

Stein U, Walther W, Stege A et al (2008) Complete in vivo reversal of the multidrug resistance phenotype by jet-injection of anti-MDR1 short hairpin RNA-encoding plasmid DNA. Mol Ther 16:178–186

Walther W, Siegel R, Kobelt D et al (2008) Novel jet-injection technology for nonviral intratumoral gene transfer in patients with melanoma and breast cancer. Clin Cancer Res 14:7545–7553

Walther W, Stein U, Fichtner I et al (2001) Nonviral in vivo gene delivery into tumors using a novel low volume jet-injection technology. Gene Ther 8:173–180

Walther W, Stein U, Fichtner I et al (2002) Intratumoral low-volume jet-injection for efficient nonviral gene transfer. Mol Biotechnol 21:105–115

3.4 Chemical Methods

Further Reading

Beerens AM, Al Hadithy AF, Rots MG, Haisma HJ (2003) Protein transduction domains and their utility in gene therapy. Curr Gene Ther 3:486–494

Elouahabi A, Ruysschaert JM (2005) Formation and intracellular trafficking of lipoplexes and polyplexes. Mol Ther 11:336–347

Fittipaldi A, Giacca M (2005) Transcellular protein transduction using the Tat protein of HIV-1. Adv Drug Deliv Rev 57:597–608

Giacca M (2004) The HIV-1 Tat protein: a multifaceted target for novel therapeutic opportunities. Curr Drug Targets Immune Endocr Metabol Disord 4:277–285

Park TG, Jeong JH, Kim SW (2006) Current status of polymeric gene delivery systems. Adv Drug Deliv Rev 58:467–486

van Dillen IJ, Mulder NH, Vaalburg W et al (2002) Influence of the bystander effect on HSV-tk/GCV gene therapy. A review. Curr Gene Ther 2:307–322

Vile RG, Russell SJ, Lemoine NR (2000) Cancer gene therapy: hard lessons and new courses. Gene Ther 7:2–8

Wasungu L, Hoekstra D (2006) Cationic lipids, lipoplexes and intracellular delivery of genes. J Control Release 116:255–264

Selected Bibliography

Dass CR (2004) Lipoplex-mediated delivery of nucleic acids: factors affecting in vivo transfection. J Mol Med 82:579–591

Dincer S, Turk M, Piskin E (2005) Intelligent polymers as nonviral vectors. Gene Ther 12[Suppl 1]:S139–145

Dufes C, Uchegbu IF, Schatzlein AG (2005) Dendrimers in gene delivery. Adv Drug Deliv Rev 57:2177–2202

Duncan R, Izzo L (2005) Dendrimer biocompatibility and toxicity. Adv Drug Deliv Rev 57:2215–2237

Fittipaldi A, Giacca M (2005) Transcellular protein transduction using the Tat protein of HIV-1. Adv Drug Deliv Rev 57:597–608

Hoekstra D, Rejman J, Wasungu L et al (2007) Gene delivery by cationic lipids: in and out of an endosome. Biochem Soc Trans 35:68–71

Lange, A, Mills RE, Lange CJ et al (2007) Classical nuclear localization signals: definition, function, and interaction with importin alpha. J Biol Chem 282:5101–5105

Pedroso de Lima MC, Simoes S, Pires P et al (2001) Cationic lipid-DNA complexes in gene delivery: from biophysics to biological applications. Adv Drug Deliv Rev 47:277–294

Rainov NG (2000) A phase III clinical evaluation of herpes simplex virus type 1 thymidine kinase and ganciclovir gene therapy as an adjuvant to surgical resection and radiation in adults with previously untreated glioblastoma multiforme. Hum Gene Ther 11:2389–2401

Ram Z, Culver KW, Oshiro EM et al (1997) Therapy of malignant brain tumors by intratumoral implantation of retroviral vector-producing cells. Nat Med 3:1354–1361

Svenson S (2009) Dendrimers as versatile platform in drug delivery applications. Eur J Pharm Biopharm 71:445–462

Trask TW, Trask RP, Aguilar-Cordova E et al (2000) Phase I study of adenoviral delivery of the HSV-tk gene and ganciclovir administration in patients with current malignant brain tumors. Mol Ther 1:195–203

Wolff JA, Rozema DB (2008) Breaking the bonds: non-viral vectors become chemically dynamic. Mol Ther 16:8–15

Zuhorn IS, Engberts JB, Hoekstra D (2007) Gene delivery by cationic lipid vectors: overcoming cellular barriers. Eur Biophys J 36:349–362

3.5 Viral Vectors

Further Reading

Bessis N, GarciaCozar FJ, Boissier MC (2004) Immune responses to gene therapy vectors: influence on vector function and effector mechanisms. Gene Ther 11[Suppl 1]:S10–17

Bestor TH (2000) Gene silencing as a threat to the success of gene therapy. J Clin Invest 105:409–411

Chang AH, Sadelain M (2007) The genetic engineering of hematopoietic stem cells: the rise of lentiviral vectors, the conundrum of the ltr, and the promise of lineage-restricted vectors. Mol Ther 15:445–456

Coffin JM, Hughes H, Varmus HE (1997) Retroviruses. Cold Spring Harbor Laboratory Press, Cold Spring Harbor, New York, NY, USA

Daniel R, Smith JA (2008) Integration site selection by retroviral vectors: molecular mechanism and clinical consequences. Hum Gene Ther 19:557–568

Kay MA, Glorioso JC, Naldini L (2001) Viral vectors for gene therapy: the art of turning infectious agents into vehicles of therapeutics. Nat Med 7:33–40

Knipe DM, Roizman B, Howley PM et al (2006) Fields' virology, 5th edn. Lippincott Williams & Wilkins, Philadelphia, PA, USA

Schambach A, Baum C (2008) Clinical application of lentiviral vectors: concepts and practice. Curr Gene Ther 8:474–482

St George JA (2003) Gene therapy progress and prospects: adenoviral vectors. Gene Ther 10:1135–1141

Thomas CE, Ehrhardt A, Kay MA (2003) Progress and problems with the use of viral vectors for gene therapy. Nat Rev Genet 4:346–358

Wu Z, Asokan A, Samulski RJ (2006) Adeno-associated virus serotypes: vector toolkit for human gene therapy. Mol Ther 14:316–327

Zentilin L, Giacca M (2008) Adeno-associated virus vectors: versatile tools for in vivo gene transfer. Contrib Nephrol 159:63–77

3

Selected Bibliography

Aghi M, Martuza RL (2005) Oncolytic viral therapies: the clinical experience. Oncogene 24:7802–7816

Aiken C (1997) Pseudotyping human immunodeficiency virus type 1 (HIV-1) by the glycoprotein of vesicular stomatitis virus targets HIV-1 entry to an endocytic pathway and suppresses both the requirement for Nef and the sensitivity to cyclosporin A. J Virol 71:5871–5877

Alba R, Bosch A, Chillon M (2005) Gutless adenovirus: last-generation adenovirus for gene therapy. Gene Ther 12[Suppl 1]:S18–27

Argnani R, Lufino M, Manservigi M, Manservigi R (2005) Replication-competent herpes simplex vectors: design and applications. Gene Ther 12[Suppl 1]:S170–177

Barnard RJ, Elleder D, Young JA (2006) Avian sarcoma and leukosis virus-receptor interactions: from classical genetics to novel insights into virus-cell membrane fusion. Virology 344:25–29

Barquinero J, Eixarch H, Perez-Melgosa M (2004) Retroviral vectors: new applications for an old tool. Gene Ther 11[Suppl 1]:S3–9

Berges BK, Wolfe JH, Fraser NW (2007) Transduction of brain by herpes simplex virus vectors. Mol Ther 15:20–29

Berns KI, Linden RM (1995) The cryptic life style of adeno-associated virus. BioEssays 17:237–245

Berto E, Bozac A, Marconi P (2005) Development and application of replication-incompetent HSV-1-based vectors. Gene Ther 12[Suppl 1]:S98–102

Brunetti-Pierri N, Ng P (2008) Progress and prospects: gene therapy for genetic diseases with helper-dependent adenoviral vectors. Gene Ther 15:553–560

Buning H, Ried MU, Perabo L et al (2003) Receptor targeting of adeno-associated virus vectors. Gene Ther 10:1142–1151

Burton EA, Bai Q, Goins WF, Glorioso JC (2002) Replication-defective genomic herpes simplex vectors: design and production. Curr Opin Biotechnol 13:424–428

Cereseto A, Giacca M (2004) Integration site selection by retroviruses. AIDS Rev 6:13–20

Cervelli T, Palacios JA, Zentilin L et al (2008) Processing of recombinant AAV genomes occurs in specific nuclear structures that overlap with foci of DNA-damage-response proteins. J Cell Sci 121:349–357

Chirmule N, Propert K, Magosin S et al (1999) Immune responses to adenovirus and adeno-associated virus in humans. Gene Ther 6:1574–1583

Danthinne X, Imperiale MJ (2000) Production of first generation adenovirus vectors: a review. Gene Ther 7:1707–1714

Dull T, Zufferey R, Kelly M et al (1998) A third-generation lentivirus vector with a conditional packaging system. J Virol 72:8463–8471

Dutheil N, Shi F, Dupressoir T, Linden RM (2000) Adeno-associated virus site-specifically integrates into a muscle-specific DNA region. Proc Natl Acad Sci U S A 97:4862–4866

Epstein AL (2005) HSV-1-based amplicon vectors: design and applications. Gene Ther 12[Suppl 1]:S154–158

Flotte TR (2004) Gene therapy progress and prospects: recombinant adeno-associated virus (rAAV) vectors. Gene Ther 11:805–810

Flotte TR, Berns KI (2005) Adeno-associated virus: a ubiquitous commensal of mammals. Hum Gene Ther 16:401–407

Gao G, Vandenberghe LH, Wilson JM (2005) New recombinant serotypes of AAV vectors. Curr Gene Ther 5:285–297

Gregorevic P, Blankinship MJ, Allen JM et al (2004) Systemic delivery of genes to striated muscles using adeno-associated viral vectors. Nat Med 10:828–834

Grimm D, Kern A, Rittner K, Kleinschmidt JA (1998) Novel tools for production and purification of recombinant adenoassociated virus vectors Hum. Gene Ther 9:2745–2760

Hendrie PC, Russell DW (2005) Gene targeting with viral vectors. Mol Ther 12:9–17

Jooss K, Chirmule N (2003) Immunity to adenovirus and adeno-associated viral vectors: implications for gene therapy. Gene Ther 10:955–963

Linden RM, Ward P, Giraud C et al (1996) Site-specific integration by adeno-associated virus. Proc Natl Acad Sci U S A 93:11288–11294

Liu Q, Muruve DA (2003) Molecular basis of the inflammatory response to adenovirus vectors. Gene Ther 10:935–940

Manganaro L, Lusic M, Gutierrez MI et al (2009) Concerted action of cellular JNK and Pin1 restricts HIV-1 genome integration to activated CD4+ T lymphocytes. Nat Med

Marconi P, Argnani R, Berto E et al (2008) HSV as a vector in vaccine development and gene therapy. Hum Vaccin 4:91–105

McCarty DM, Young SM Jr, Samulski RJ (2004) Integration of adeno-associated virus (AAV) and recombinant AAV vectors. Annu Rev Genet 38:819–845

Merten OW, Geny-Fiamma C, Douar AM (2005) Current issues in adeno-associated viral vector production. Gene Ther 12[Suppl 1]:S51–61

Miller AD (1990) Retrovirus packaging cells. Hum Gene Ther 1:5–14

Miller AD (1996) Cell-surface receptors for retroviruses and implications for gene transfer. Proc Natl Acad Sci U S A 93:11407–11413

Miller DG, Wang PR, Petek LM et al (2006) Gene targeting in vivo by adeno-associated virus vectors. Nature Biotech 24:1022–1026

Mueller C, Flotte TR (2008) Clinical gene therapy using recombinant adeno-associated virus vectors. Gene Ther 15:858–863

Naldini L, Blomer U, Gallay P et al (1996) In vivo gene delivery and stable transduction of nondividing cells by a lentiviral vector. Science 272:263–267

Nienhuis AW, Dunbar CE, Sorrentino BP (2006) Genotoxicity of retroviral integration in hematopoietic cells. Mol Ther 13:1031–1049

Russell DW, Hirata RK (1998) Human gene targeting by viral vectors. Nat Genet 18:325–330

Russell WC (2000) Update on adenovirus and its vectors. J Gen Virol 81:2573–2604

Russell WC (2009) Adenoviruses: update on structure and function. J Gen Virol 90:1–20

Samulski RJ, Berns KI, Tan M, Muzyczka N (1982) Cloning of adeno-associated virus into pBR322: rescue of intact virus from the recombinant plasmid in human cells. Proc Natl Acad Sci USA 79:2077–2081

Schroder AR, Shinn P, Chen H et al (2002) HIV-1 integration in the human genome favors active genes and local hotspots. Cell 110:521–529

Sinn PL, Sauter SL, McCray PB Jr (2005) Gene therapy progress and prospects: development of improved lentiviral and retroviral vectors: design, biosafety, and production. Gene Ther 12:1089–1098

Vasileva A, Jessberger R (2005) Precise hit: adeno-associated virus in gene targeting. Nat Rev Microbiol 3:837–847

Vihinen-Ranta M, Suikkanen S, Parrish CR (2004) Pathways of cell infection by parvoviruses and adeno-associated viruses. J Virol 78:6709–6714

Wang Z, Zhu T, Qiao C et al (2005) Adeno-associated virus serotype 8 efficiently delivers genes to muscle and heart. Nat Biotechnol 23:321–328

Yamamoto T, Tsunetsugu-Yokota Y (2008) Prospects for the therapeutic application of lentivirus-based gene therapy to HIV-1 infection. Curr Gene Ther 8:1–8

Zentilin L, Qin G, Tafuro S et al (2000) Variegation of retroviral vector gene expression in myeloid cells. Gene Ther 7:153–166

Zuffery R, Dull T, Mandel RJ et al (1998) Self-inactivating lentivirus vector for safe and efficient in vivo gene delivery. J Virol 72:9873–9880

Clinical Applications of Gene Therapy

4

Since 1989, the year of the first gene therapy clinical trial (cf. section on 'Genes as Drugs'), over 1500 clinical studies have been conducted involving several tens of thousands of patients. If objectively evaluated, the overall clinical success of these trials has been modest so far. With some remarkable exceptions, most of the trials have encountered unanticipated technological and biological problems. In this evaluation, however, it should be taken into account that the vast majority of the diseases faced by gene therapy are life-threatening conditions, for which no conventional medical therapy exists, and that gene therapy is a completely new discipline, both conceptually and technically. Indeed, 20 years after the first application, the possibility of success of gene therapy now appears much closer. This is a consequence of the significant improvements made in the development of both *in vivo* and *ex vivo* systems for gene delivery and the identification of novel classes of therapeutic genes. The recent results obtained by gene therapy of inherited blindness and of some neurodegenerative disorders, as well as the progress made in several other clinical trials, now encourage informed and firm optimism on the eventual success of this discipline.

4.1
Clinical Applications of Gene Therapy: General Considerations

An updated source of information on gene therapy clinical trials is available on the website http://www.wiley.co.uk/genetherapy/clinical/. The trials conducted in the United States and United Kingdom are included in the databases maintained by the Genetic Modification Clinical Research Information System (GeMCRIS®) (http://www.gemcris.od.nih.gov/) and the Gene Therapy Advisory Committee (GTAC) of the Department of Health (http://www.dh.gov.uk/ab/GTAC/index.htm), respectively.

4.1.1
Number of Clinical Trials

Up to July 2009, over 1500 clinical trials had been conduced or were currently ongoing. In the years immediately following 1989, the numbers of trials rose very rapidly, to reach an average of over 100 trials per year, especially in the United States, the country in which more than two thirds of the clinical studies are still conducted. At the end of the 1990s, a peak was reached, with an average of more than 10 patients recruited in one or another gene therapy trial every week.

After this initial enthusiasm, the death of a patient enrolled in a clinical study for the deficit of ornithine transcarbamylase (OTC) using an adenoviral vector (in 1999; cf. section on 'Gene Therapy of Liver Diseases'), the development of leukemia in two patients with SCID-X1 treated with a gammaretroviral vector (in 2002; cf. section on 'Gene Therapy of Hematopoietic Stem Cells'), and, more generally, the growing perception of the general inefficacy of the protocols so far developed determined a decline in the number of clinical trials in the subsequent years, in favor of more efforts in improving the available gene delivery technologies and identifying more suitable disease targets.

Over the last 5 years, however, the enthusiasm for gene therapy has been strengthened by the initial success of a growing number of applications exploiting the properties of AAV vectors for gene therapy of a few neurodegenerative disorders. Starting from 2005, over 100 clinical trials were again registered in the regulatory databases worldwide, but with a spectrum of applications significantly different from those at the beginning, and now including a series of degenerative disorders of the central nervous system, cardiovascular system, and retina, which are described in the respective sections.

4.1.2
Phases of Clinical Trials

Similar to conventional pharmaceutical trials, clinical experimentation of gene therapy involves the successful completion of a series of sequential phases. After first proving success of the therapeutic approach in a small animal model of disease (typically mouse or rat) and later, when feasible and appropriate, in large animals (pigs, dogs, non-human primates), the first human experimentation is a *Phase I* trial. The purpose of this kind of investigation is to evaluate the safety and determine the pharmacokinetics of often different doses of the compound under investigation (in the case of traditional drugs: adsorption, duration, metabolism, route of excretion; in the case of gene therapy: fate of vector or engineered cells, levels of therapeutic nucleic acid expression, duration of effect, etc.). Phase I studies are commonly conducted in a small number (10–20) of patients or healthy volunteers, for example to experiment on genetic vaccination. In order to move to the subsequent phases, the drug (or the genetic treatment) must prove safe or at least show a tolerable toxicity in relation to the foreseen therapeutic benefit.

The *Phase II* trials start to evaluate efficacy, using a defined dosage and modality of administration. In the case of gene therapy, these trials are still conducted on a very small number of patients.

When a Phase II trial provides encouraging results, a large *Phase III* trial is organized. These studies involve a larger number of patients, usually one or a few hundred individuals, and their purpose is to confirm the efficacy data, identify the most appropriate dosage, and monitor the appearance of adverse effects in a statistically significant sample. Most Phase III trials are randomized (i.e., patients are randomly assigned to the treatment or placebo groups), double-blinded (i.e., neither the patient nor the treating doctor know whether the patient is receiving the drug or the placebo), and multicentric (i.e., they involve different hospitals and clinical investigators).

Should a drug or a treatment modality pass Phase III, it becomes commercially available. At this point, clinical experimentation continues with *Phase IV*, or post-marketing pharmaco-surveillance, trials. These are aimed at confirming the safety, tolerability, and efficacy of the drug or treatment in a very large number of patients, usually internationally recruited.

Figure 4.1A reports the distribution of the gene therapy clinical trials so far conducted divided by phase of experimentation. It is immediately evident that almost 80% of the investigations so far conducted have been Phase I (evaluation of safety only) or I/II (safety plus preliminary efficacy) studies; these studies were usually conducted on a very small number of patients, sometimes involving only one or two individuals. Only less than 17%

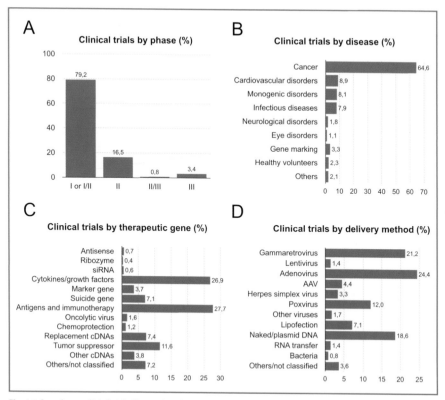

Fig. 4.1 Gene therapy clinical trials. The number of trials are shown by phase (panel **A**), type of disease (**B**), action of therapeutic gene used (**C**), and gene delivery method (**D**)

of trials were Phase II (efficacy), also conducted in a relatively small number of patients. About 50 trials have reached Phase III, including 6 vast trials for anti-HIV genetic vaccination (which have produced very frustrating results) and over 30 trials for cancer, based on the delivery of cytokines to stimulate the immune system, or on anti-cancer genetic vaccination, or, more recently, exploiting the oncolytic potential of replicating adenoviruses and herpesviruses. In the past, a few large trials have explored the efficacy of thymidine kinase as a suicide gene for glioblastoma, with poor results (see section on 'Gene Therapy of Cancer'). Finally, 7 Phase III clinical trials are aimed at inducing therapeutic angiogenesis in patients with myocardial or critical limb ischemia (see section on 'Gene Therapy of Cardiovascular Disorders').

The limited number of clinical studies having reached phases more advanced than Phase I is strongly indicative of the problems that gene therapy has so far encountered at the clinical level. In particular, the results of several trials have not conformed to the expectations, especially concerning the efficacy and duration of gene transfer, and therefore have not proceeded forward. As an example, there have been more than 30 clinical trials on gene therapy of cystic fibrosis by CFTR gene transfer; although half of these trials started before the end of 1995, none has yet reached Phase III, due to the discouraging results so far generated.

4.1.3
Diseases Addressed by Gene Therapy Clinical Trials

Table 4.1 reports a list of diseases addressed by the already conducted or ongoing gene therapy clinical trials. It is immediately evident that the fields of applications include a vast series of conditions, which comprise virtually all the types of human disorders.

When the over 1500 clinical trials are divided by disease addressed (Figure 4.1B), it becomes apparent that the hereditary monogenic diseases, which initially inspired the concept of gene therapy itself, have been the objective of only less than 10% of the trials (Table 4.1). As will be discussed later, the weak interest in these diseases has essentially two explanations. On the one hand, treatment of these diseases by gene therapy is very challenging, since it often requires the permanent modification of all the relevant cells in the body, a task difficult to accomplish in most cases. On the other hand, the monogenic hereditary disorders only represent a limited portion of human disease (less than 2%). The gene therapy clinical trials for the most studied monogenic disorders (cystic fibrosis, to which one third of the studies for hereditary diseases are devoted, hemophilia, and muscle dystrophies) are discussed in specific sections. The trials for hereditary diseases affecting specific tissues (blood, brain, eye, liver, skin) are instead treated in the sections on gene therapy for the respective organs.

Over 60% of the gene therapy clinical trials so far conducted have been in the cancer field, having as their target a vast series of neoplasias and exploiting various strategies, including anticancer vaccination. Of interest, a growing number of trials now have the objective of treating cardiovascular disorders. This field of application has grown very rapidly in recent years. In 1998, less than 10 trials aimed to treat these disorders, while now they include over 130 studies; most of these trials have the objective to induce ther-

Table 4.1 Diseases addressed by gene therapy clinical trials

Pathological condition	Disease
Monogenic hereditary disorders	Cystic fibrosis (CF)
	Hemophilia A and B
	Duchenne muscular dystrophy (DMD)
	Limb-girdle muscular dystrophy (LGMD 2C)
	Severe combined immunodeficiency syndromes (SCID-X1, ADA-SCID, PNP-SCID)
	Wiskott-Aldrich syndrome (WAS)
	Lysosomal storage disorders (Hurler's disease, Hunter's disease, Gaucher's disease, Fabry's disease, Sly syndrome)
	Defects of phagocytes (chronic granulomatous disease, leukocyte adhesion deficiency)
	Fanconi anemia (FA)
	Canavan disease
	Late infantile neuronal ceroid lipofuscinosis (CLN2 or LICLN)
	Huntington's disease (HD)
	Familial amyotrophic lateral sclerosis (ALS)
	Familial hypercholesterolemia (FH)
	Ornithine transcarbamylase (OTC) deficiency
	Junctional epidermolysis bullosa
	Lipoprotein lipase (LPL) deficiency
	Leber's congenital amaurosis (LCA) due to RPE65 deficiency
	Gyrate atrophy
Cancer	Breast cancer
	Ovary cancer
	Cervix cancer
	Glioblastoma
	Leptomeningeal carcinoma
	Colorectal cancer
	Pancreatic cancer
	Hepatocarcinoma
	Prostate carcinoma
	Kidney carcinoma
	Melanoma
	Nasopharyngeal carcinoma
	Small-cell lung carcinoma
	Non-small-cell lung carcinoma (NSCLC)
	Pleuric mesothelioma
	Leukemia
	Lymphoma
	Multiple myeloma
	Retinoblastoma
	Soft tissue sarcoma
Infectious disorders	HIV/AIDS
	Hepatitis B
	Hepatitis C

(cont→)

Table 4.1 *(continued)*

Pathological condition	Disease
Infectious disorders	Influenza
	Tetanus
	Adenovirus infection
	Cytomegalovirus infection
	Epstein-Barr (EBV) infection
Neurological disorders	Alzheimer's disease (AD)
	Parkinson's disease (PD)
	Carpal and cubital tunnel syndromes
	Epilepsy
	Multiple sclerosis (MS)
	Myasthenia gravis
	Diabetic neuropathy
Cardiovascular disorders	Ischemic cardiomyopathy
	Myocardial infarction (MI)
	Heart failure (HF)
	Peripheral artery disease (PAD)
	Pulmonary hypertension
Eye diseases	Age-related macular degeneration (AMD)
	Diabetic macular edema
	Glaucoma
	Retinitis pigmentosa (RP)
	Superficial corneal opacity
Other diseases	Inflammatory disease of the gut
	Rheumatoid arthritis (RA)
	Anemia
	Bone fractures
	Erectile dysfunction
	Intractable pain
	Hypofunction of salivary glands after radiotherapy
	Type I diabetes
	Vesical hyperactivity
	Graft-versus-host disease (GvHD)
	Peanut allergy

apeutic angiogenesis for myocardial and peripheral ischemia. In 2008, the first clinical trial for heart failure was initiated.

Two other fields of application of gene therapy that appear very promising and thus attract vast interest are the degenerative disorders of the eye and the retina. In particular, these include Parkinson's and Alzheimer's diseases in the brain and Leber's congenital amaurosis due to defects of the RPE65 protein in the eye. In all these cases, the initial success of these clinical studies is essentially due to the efficacy of AAV vectors for gene transfer to neurons and retina.

Finally, it is worth commenting on the over 120 trials for infectious disorders, of which the majority were for HIV-1 infection. Most of these trials entailed genetic vaccination, both preventive and therapeutic, using different combinations of viral genes delivered by means of various gene delivery systems. Unfortunately, none of these trials has been successful so far, parallel to the lack of success of the studies aimed at eliciting protective immunization against HIV-1 by vaccination, using strategies other than gene transfer. Other clinical trials are aimed at inducing cellular resistance to HIV-1 or redirecting cytotoxic T lymphocytes (CTLs) towards HIV-1-infected cells.

4.1.4
Therapeutic Genes Used in the Clinical Trials

Until recent years, gene therapy has mainly been based on the transfer of cDNAs or their coding regions. In most cases, these cDNAs encoded cytokines and growth factors, antigens against which an immune response is elicited, or proteins replacing missing cellular functions (Figure 4.1C). Over the last few years, a growing number of trials have started in which the therapeutic genes are small regulatory nucleic acids, including ribozymes, siRNAs, and antisense oligodeoxynucleotides. Finally, over 20 cancer gene therapy trials now take advantage of replication-competent, defective adenoviral or herpesviral vectors with oncolytic properties.

4.1.5
Modality of Therapeutic Gene Delivery in Gene Therapy Clinical Trials

More than two thirds of the clinical studies so far conducted have exploited the properties of viral vectors for gene administration, given the overall higher efficacy of viral compared to non-viral methods (Figure 4.1D). Gammaretroviral vectors, which were the delivery system of choice in the first half of the 1990s, are now much less considered due to their incapacity to transduce resting cells, the shut-off of gene expression over time, and, most importantly, the induction of insertional mutagenesis, leading to aberrant cellular gene activation. Overall, only 21% of the trials have used these vectors so far. In 1996 lentiviral vectors entered the gene therapy arena and since then have gained progressive popularity especially due to their property to transduce non-replicating cells, both *in vivo* and *ex vivo*. Due to safety considerations, the first clinical trial using a lentiviral vector was approved only in 2003 for gene therapy of HIV-1 infection; since then, over 20 other trials are now ongoing or awaiting approval, including studies for different monogenic disorders.

The first- and second-generation adenoviral vectors have been used in about 25% of the trials, especially in the second half of the 1990s. But now important safety concerns have been raised because of their propensity to elicit significant inflammatory and immune responses in the injected individuals. Appropriate use of these vectors now appears to be restricted to gene therapy of cancer and genetic vaccination.

Finally, very promising results are currently being obtained using AAV vectors, which are now being considered more and more for applications entailing gene transfer in muscle, heart, brain, and retina. Over 65 trials have exploited these vectors so far, especially in more recent years. Among the most promising studies, those for RPE65 deficiency, hemophilia, and neurodegenerative disorders deserve mention. In particular, in the last field of application, almost 7 different clinical trials are showing initial promising results, including studies for Alzheimer's disease, Parkinson's disease, and inborn errors of metabolism with prominent neuronal involvement, such as Batten disease and Canavan disease.

Besides viral vectors, a considerable number of trials (almost 20%) are based on the use of naked DNA, in the form of oligodeoxynucleotides or plasmids. In the latter case, these are mostly studies based on the capacity of a few cell types to spontaneously internalize naked nucleic acids, including antigen-presenting cells (APCs) and skeletal muscle fibers. This property has been mostly exploited for genetic vaccination or induction of therapeutic angiogenesis. Formulations based on cationic lipids have been used in less than 8% of the clinical studies so far.

4.2
Gene Therapy of Hematopoietic Stem Cells

One of the gene therapy applications offering broad therapeutic potential is gene therapy of hematopoietic stem cells (HSCs), namely the cells residing in the bone marrow and acting as precursors of all the cells circulating in blood and lymphoid organs (erythrocytes, lymphocytes, granulocytes, monocytes/macrophages, dendritic cells, platelets). Similar to other stem cells, HSCs are capable of proliferating while keeping their stem cell property intact; at the same time, a fraction of them differentiate into the various terminal cell types. Elegant experiments indicate that a single HSC is sufficient to repopulate the whole hematopoietic and immune system in a mouse in which all endogenous HSCs have been killed by high-dose radiation. Given the broad regenerative capacity of HSCs, it is intuitive that their stable genetic modification is very appealing, since it offers the opportunity to correct a genetic defect or modify the phenotype in the different cell types of their progeny.

4.2.1
Bone Marrow and Hematopoiesis

Bone marrow is a soft tissue contained in the trabecular ("spongy") bone that constitutes about 4% of the total weight of the body. Besides HSCs and their progeny, it contains fibroblasts, adipocytes, osteoblasts, a rich network of blood vessels, and mesenchymal stem cells (MSCs), potentially able to differentiate into various cell types of mesodermal derivation.

In humans, within the hematopoietic compartment a small fraction of cells is positive for the CD34 antigen. This is an integral membrane protein, with an apparent mass of 105–120 kDa corresponding to a heavily glycosylated 40-kDa polypeptide, which belongs to the sialomucin family. The function of this protein is still unclear; other cells abundantly expressing CD34 are blood vessel endothelial cells. In the bone marrow, the fraction of

CD34$^+$ cells contains both the true HSC and a series of already committed progenitors derived from these cells, in different stages of differentiation. The phenotype CD34$^+$, CD38$^-$, HLA-DR$^-$, and CD90(Thy1)$^+$ defines a subset of cells containing the most primitive HSCs, able to self-renew. Expression of the CD38 and loss of the CD90 antigens occurs when these primitive cells are activated *in vitro* and their differentiation towards defined cell types starts. Differentiation can be further verified by evaluating the expression of antigens specific for the different cell types that eventually derive from the HSC, including CD33 (myeloid cells), CD13 (granulocytes and monocytes), CD7 (T lymphocytes), CD10 (all lymphoid cells), CD19 (B lymphocytes), CD56 (natural killer cells), CD41a (platelets), and glycophorin A (erythroblasts and erythrocytes).

Maintenance of the stem properties is tightly controlled by a series of growth factors and their receptors. In particular, two tyrosine kinase receptors play an essential role: c-kit (CD117), which binds the stem cell factor (SCF) cytokine and is commonly used as a stem cell marker in both the hematopoietic system and other tissues, and Flt-3/Flk-2 (CD135), which binds the Flt-3L (FL) ligand.

Approximately 1–4% of bone marrow cells express the CD34 antigen; only 1% of these cells, however, has the characteristics of a HSC (CD34$^+$/CD38$^-$). As a consequence, in the bone marrow, only 1 cell out of 1×10^4 mononuclear cells is a true HSC. The minimum number of HSCs considered sufficient for a successful HSC transplantation in patients is 1–2×10^6 CD34$^+$ cells (corresponding to ~10 ml bone marrow) per kg of the recipient's body weight (or 1–2×10^4 CD34$^+$/CD38$^-$ cells/kg).

4.2.2
Hematopoietic Stem Cell Transplantation

In its canonical formulation, conventional BMT consists in the recovery of bone marrow from the posterior iliac crest of a donor (allogeneic transplant) or the patient themself (autologous transplant), followed by its intravenous infusion into the patient, who has previously been treated with high-dose chemotherapy and/or radiation to destroy endogenous HSCs. Once infused, the transplanted HSCs populate the bone marrow and start the complete regeneration of the patient's hematopoietic system.

In addition to the bone marrow, HSCs are also present in the peripheral blood and are enriched in the umbilical cord blood, hence the more appropriate term *stem cell transplantation* (SCT) to substitute that of BMT. HSCs can be recovered from peripheral blood by productive apheresis, a procedure entailing the processing of several liters of blood to recover cells with the density proper of mononuclear cells, including lymphocytes, monocytes, and HSCs. The percentage of circulating HSCs is usually very low (0.1% of all circulating mononuclear cells), however their number can be increased by mobilizing these cells from the bone marrow by treating the donor with chemotherapy (in patients undergoing autologous transplantation for malignant disorders) or growth factors such as granulocyte colony stimulating factor (G-CSF). These procedures can lead to a 10-fold expansion in the number of HSCs in peripheral blood. Another source of HSCs is cord blood (about 40–70 ml per placenta). Thanks to the development of cord blood banks, the use of this source of allogeneic HSCs for transplantation is destined to increase.

4

Independent of the source of the sample (bone marrow, peripheral blood, umbilical cord blood), HSCs can be separated from the other mononuclear cells using antibodies recognizing the CD34 surface antigen. Purification is usually carried out using a straightforward procedure based on anti-CD34$^+$ monoclonal antibodies coupled to magnetic beads: once added to a mononuclear cell sample, the bead-antibody-HSC complexes are separated from the other cells using a magnet.

After various attempts starting in the 1950s, the first successful BMTs were performed between the end of the 1960s and the beginning of the 1970s. Since that moment, the clinical indications for SCT have progressively broadened and now several otherwise lethal or highly invalidating disorders can take advantage of this technology, with a success rate ranging from 30 to 90%, according to the indication and the type of transplantation.

SCT can be allogeneic, autologous, or syngeneic. In **allogeneic SCT**, the hematopoietic system of the recipient is first destroyed (myeloablation) with high-dose chemotherapy or radiotherapy and then reconstituted by HSC infusion from an unrelated, normal donor. As in all transplants, this procedure requires that the donor and the recipient are matched, i.e., that they share similar HLA antigens. In particular, identity of the HLA-A, HLA-B, and HLA-DR loci is considered critical for the success of transplantation, that of HLA-C and HLA-DQ highly desirable. The extent of similarity between the donor's and recipient's HLA determines, on one hand, the frequency and severity of the rejection and, on the other hand, the extent of activation of the donor immune cells against the recipient's tissues, an occurrence known as *graft-versus-host disease* (GvHD). Part of GvHD, however, appears to be beneficial in the settings of SCT for malignancies, since the donor immune cells also recognize the recipient's tumor cells persisting despite myeloablation (*graft-versus-tumor, GvT*; in the case of SCT for leukemia: *graft-versus-leukemia, GvL*). About 30–40% of patients needing SCT have a compatible sister or brother; if no compatible individuals are found among the close relatives, international bone marrow donor registries are searched. In these databases, the probability of finding a compatible donor ranges, according to the frequency of the patient's HLA in the population, from a maximum of 1:5000 to a minimum of 1:1 million individuals.

In **autologous SCT**, the HSCs recovered from the patient's bone marrow are cryopreserved and then infused back into the patient after myeloablation with high-dose chemotherapy or radiotherapy. This is the commonest form of transplant, which is performed at a frequency that is twice that of the other types of SCTs; since the donor and the recipient are the same, most common complications associated with allogeneic transplant, in particular GvHD, are avoided. This situation is similar to **syngeneic SCT**, in which the donor is an identical twin of the recipient; this is, however, a rare condition (<1%), as few individuals have identical twins. While in both these types of transplantations patients do not experience GvHD, there is also no GvT. Therefore, in the field of SCT for malignant disorders, allogeneic or syngeneic SCT only find application when chemotherapy or radiotherapy are expected to eradicate all tumor cells from the recipient prior to transplantation.

Over 30,000 autologous SCTs and 15,000 allogeneic SCTs are performed every year in the world; the commonest indications for these procedures are reported in Table 4.2. Allogeneic SCT finds its major indications in the treatment of different neoplastic disorders of the hematopoietic systems (leukemia, lymphoma, and multiple myeloma) and a series of non-neoplastic conditions, mostly consisting of inherited monogenic disorders of

Table 4.2 Major indications for stem cell transplantation (SCT)

Autologous SCT		Allogeneic SCT	
Cancer	**Other disorders**	**Cancer**	**Other disorders**
Acute myeloid leukemia (AML)	Autoimmune disorders	Acute myeloid leukemia (AML)	Severe combined immunodeficiencies (SCID)
Acute lymphocytic leukemia (ALL)	Amyloidosis	Acute lymphocytic leukemia (ALL)	Aplastic anemia
Multiple myeloma (MM)		Chronic myeloid leukemia (CML)	Thalassemia
Non-Hodgkin's lymphoma (NHL)		Chronic lymphocytic leukemia (CLL)	Fanconi anemia
Hodgkin's lymphoma (HL)		Myelodysplasia	Diamond-Blackfan anemia
Neuroblastoma		Multiple myeloma (MM)	Sickle cell anemia
Medulloblastoma		Hodgkin's lymphoma (HL)	Osteopetrosis
Germ cell tumors		Non-Hodgkin's lymphoma (NHL)	Wiskott-Aldrich syndrome
			Inborn errors of metabolism
			Autoimmune disorders

blood cells and autoimmune disorders with a severe prognosis such as lupus erythematosus systemicus or rheumatoid arthritis. In addition to neoplastic disorders of the HSC, autologous SCT also finds an indication for autoimmune disorders or as an adjuvant to high-dose chemotherapy for the treatment of different solid tumors (in particular, neuroblastoma, medulloblastoma, and germ cell tumors).

4.2.3
Gene Therapy of Hematopoietic Stem Cells: Major Applications

All cells of the hematopoietic and immune systems are a progeny of the HSC: therefore, the individual properties of each cell type in the two systems can be modified by gene transfer into the HSC. Gene therapy of HSCs exploits the technologies available in the context of autologous SCT, with the essential difference that, while in conventional autologous SCT the HSCs are usually cryopreserved after recovery awaiting reinfusion, for gene therapy they are cultured *ex vivo* and transduced with a viral vector to achieve their permanent modification (Figure 4.2).

The main applications of HSC gene therapy are reported in Table 4.3 and discussed below.

(1) **Gene therapy of hereditary disorders of the hematopoietic and immune systems**. One of the major fields of application of gene therapy of HSCs is as a possible cure for monogenic disorders of erythrocytes, myeloid cells, and lymphocytes, including hereditary immunodeficiencies, thalassemias, and defects of phagocytes. These applications will be detailed in the following sections of this chapter.

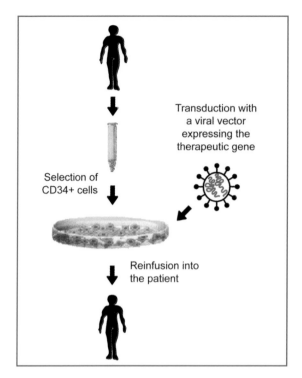

Fig. 4.2 Gene therapy of HSCs. CD34$^+$ HSCs are recovered from the bone marrow or the peripheral blood of a patient, maintained *ex vivo* to allow gene transfer, usually using a viral vector, and reinfused back into the patient

Table 4.3 Applications of gene therapy of hematopoietic stem cells

Gene therapy of monogenic hereditary disorders of red blood cells, granulocytes, monocytes-macrophages and lymphocytes
Modification of the immune response against cancer and virus antigens
Induction of resistance to HIV-1 infection
Bone marrow purging in autologous BMT
Increase of the therapeutic index of chemotherapy
Gene marking studies

(2) **Modification of the immune response against tumor or viral antigens**. Gene transfer into the HSC allows the modification of the specificity of T-lymphocytes. In particular, transfer of the T-cell receptor (TCR) genes offers the possibility to re-address these cells against specific antigens. This application is discussed further in the sections 'T-Cell Receptor (TCR) Subunits' and 'Gene Therapy of Cancer'.

(3) **Bone marrow purging**. Another interesting possibility is to exploit gene transfer into the HSC to interfere with the mechanisms leading to the development of leukemia and lymphoma. These disorders are commonly due to the occurrence of genetic defects at the HSC level, eventually leading to the pathological expansion of their progeny. A canonical example of this occurrence is chronic myeloid leukemia (CML), where malignant transformation of myeloid cells is sustained by a reciprocal translocation between the long arm

of chromosome 9 and the long arm of chromosome 22, leading to the formation of a characteristic abnormal chromosome, the Philadelphia (Ph') chromosome. This rearrangement, which occurs in the HSC, determines the transposition of the c-*abl* proto-oncogene, normally located on chromosome 9, into the *bcr* gene on chromosome 22 and leads to the formation of an abnormal gene consisting in the fusion of the first *bcr* exons with the *abl* 3' region. Since *bcr* is transcriptionally active in both the HSCs and their myeloid progeny, the pro-proliferative activity of *abl* drives these cells towards an uncontrolled proliferation. Autologous BMT is not effective for this disease, since most of the patients relapse due to the presence of residual tumor cells in the transplant. To eliminate these residual cells, the HSCs in the transplant can be treated with antisense oligonucleotides, ribozymes or siRNAs targeted against the nucleotide sequence corresponding to the *bcr-abl* fusion region, thus "purging" the bone marrow from the malignant cells before transplantation (cf. section on 'Non-Coding Nucleic Acids' and Figure 2.9).

(4) **Induction of resistance to HIV-1 infection**. HIV-1 infects CD4$^+$ lymphocytes and macrophages, which both derive from the HSC. Should this cell be genetically modified with genes conferring resistance to infection by either preventing entry of the virus or interfering with its replication, their resulting progeny would be protected against viral infection. This strategy is currently at the basis of different gene therapy clinical trials, which will be discussed in the section on 'Gene Therapy of HIV-1 Infection'.

(5) **Increase of the therapeutic index of chemotherapy**. Over recent years, SCT has been used as an adjuvant for the therapy of solid cancers, mostly breast, ovary, and testicle cancers, neuroblastoma, and non-small-cell lung carcinoma. These cancers can often be successfully treated with high-dose chemotherapy (the so-called "supramaximal chemotherapy"). The required doses, however, are very toxic for the bone marrow and the risk of aplasia constitutes a major limitation. As already reported above, autologous SCT can be used as an adjuvant to tolerate supramaximal chemotherapy; gene therapy of these cells can be of further utility in increasing the therapeutic index of chemotherapy (that is, the maximum dose of antiblastics tolerated by the patients). This can be accomplished by transferring, into the HSC, different genes conferring resistance to a few commonly used antiblastic drugs. Several of these genes (Table 4.4) were originally identified in studies aimed at understanding the mechanisms of cancer cell resistance to chemotherapy. The most investigated one is the *mdr-1* (multidrug resistance-1) gene, coding for a 120-kDa membrane glycoprotein (glycoprotein P). This protein belongs to the large family of ABC (ATP-binding cassette) membrane transporters, which selectively pump a vast series of hydrophobic (lipophilic) compounds out of the cells. Clinical trials conducted by retrovirus-mediated transfer of the *mdr-1* gene into the HSCs have shown that this procedure not only increases HSC resistance, but also determines expansion of the transduced cell pool *in vivo* under the selective pressure imposed by the drug (cf. also below). As alternatives to ABC proteins, other genes potentially useful for the same purpose are those coding for antioxidant enzymes, such as manganese superoxide dismutase (MnSOD), catalase, peroxidase, or those increasing the levels of glutathione. Another potentially useful gene codes for the enzyme O^6-alkylguanine-DNA alkyltransferase (or methylguanine methyltransferase, MGMT), which removes the alkyl groups of guanines. This enzyme can repair damage induced in DNA by alkylating drugs of the nitrosourea family, including BCNU, and of the nitrogen mustard family, including cyclophosphamide.

Table 4.4 Genes conferring resistance to HSCs against chemotherapy and/or allowing pharmacological selection of HSCs

Gene	Drug	Mechanism of action
Mdr-1/ABCB1 (glycoprotein P)	Doxorubicin, daunorubicin, vincristine, vinblastine, actinomycin D, paclitaxel, docetaxel, etoposide, teniposide, bisantrene	Membrane transporters; promote exit of several toxic compounds from the cell
Mrp-1/ABCC1	Doxorubicin, epirubicin, etoposide, vincristine, metotrexate	
Dihydrofolate reductase (DHFR)	Metotrexate, trimetrexate	Drugs are structural analogs of folic acid that inhibit endogenous DHFR
O^6-alkylguanine-DNA alkyltransferase (MGMT)	Alkylating agents, e.g., nitrosourea (BCNU) and nitrogen mustard derivatives, such as cyclophosphamide	Transfers an alkyl group from DNA to a cysteine in the enzyme active site
Glutathione S-transferase (GST)	Alkylating agents, e.g., nitrosourea (BCNU) and nitrogen mustard derivatives, such as cyclophosphamide; anthracyclines and cisplatin	Conjugates various molecules with glutathione, thus decreasing their toxicity and favoring their elimination

(6) **Gene marking studies**. The purpose of gene marking studies is to genetically label a target cell in order to follow its fate or that of its progeny once reinjected *in vivo*. This approach, which has no therapeutic intention, has so far provided answers to a number of outstanding questions concerning the biology of the HSCs. In particular, genetic labeling of HSCs using a retroviral vector has allowed the understanding of whether the relapse that is common after autologous BMT for some leukemias is due to the presence of residual cancer cells in the transplant or to the only partial success of myeloablation prior to transplantation. For example, it is known that autologous BMT is not very poorly successful in patients with CML or with other types of leukemia, who very frequently relapse. To understand the source of the cells causing the relapse, bone-marrow-derived HSCs were transduced with a retroviral vector before reinfusion into the patients; since the vector permanently integrated into the transduced cells' genomes, it became a permanent marker of these cells and of their progeny. After the patients relapsed, it was found that the tumor cells had the retrovirus integrated: it was thus possible to conclude that the cause of the relapse was the residual presence of tumor cells in the transplant.

Another field of application of gene marking studies is to comparatively assess the efficacy of different procedures for the mobilization and culture of HSCs by competitive repopulation experiments (Figure 4.3). Cells derived from the bone marrow are divided into two aliquots; each aliquot is maintained *ex vivo* using two different cytokine cocktails

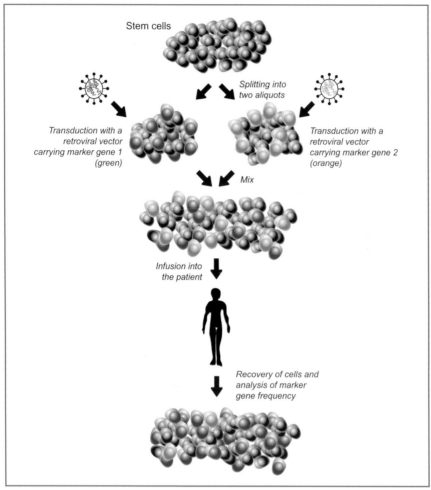

Fig. 4.3 Competitive repopulation experiments. Bone marrow cells or other progenitor cell types are recovered from an individual and divided into two (or more) aliquots, which are maintained in culture using different conditions or selected for different properties. Each fraction is then transduced with a retroviral vector carrying a different marker gene (marker genes green and orange in the figure). After transduction, the fractions are mixed and reinfused into the same individual. At different times from reinfusion, the cells are recovered and the frequency of each marker gene is evaluated, usually by quantitative PCR amplification. The gene marker that is found more frequently identifies the culture condition or the cell population showing the higher engraftment capacity

and then transduced with two retroviral vectors differing in the sequence of a reporter gene. The two cell fractions are then mixed together and reinjected into the patients. At different times after inoculation, the frequency of the cells containing each of the two reporter retroviruses is evaluated: the vector that is found more frequently indicates which of the two cytokine cocktails was more efficacious in preserving the capacity of HSCs to repopulate the hematopoietic system after transplantation.

4

Gene marking studies are not limited to HSC biology and can be extended to other *ex vivo* cultured cells. A very interesting application was implemented in the field of CTLs, in particular to understand whether the *ex vivo*-expanded lymphocytes recognizing a specific antigen might find their target once reinjected *in vivo*. The first gene therapy clinical trial, performed in 1989 in Bethesda, MD, USA was indeed a gene marking study of this kind. In this trial, the lymphocytes infiltrating the tumors (*tumor-infiltrating lymphocytes*, TILs) of five patients with metastatic melanoma were purified, transduced with a retroviral vector, expanded *ex vivo*, and then reinfused into the patients. At different times after inoculation, a considerable number of genetically modified cells was again found in the tumors, thus indicating that these cells are able to maintain their anti-tumor specificity despite having been expanded in the laboratory. This and similar observations have paved the way for the possibility of developing innovative therapeutic strategies based on adoptive immunotherapy of cancer (cf. section on 'Gene Therapy of Cancer').

4.2.4
Procedures for Gene Transfer into Hematopoietic Stem Cells

The true $CD34^+/CD38^-$ HSC is a small cell in the G0 phase of the cell cycle, with a condensed chromatin and an intact nuclear membrane. Thus, this cell cannot be transduced by gammaretroviral vectors unless it is stimulated to replicate (see section on 'Viral Vectors for Gene Therapy'). Entry of the HSCs into the cell cycle is promoted by culturing these cells using various combinations of stimulatory cytokines, including G-CSF, granulocyte-monocyte colony stimulating factor (GM-CSF), SCF, Flt-3 receptor ligand (Flt-3L), interleukin-3 (IL-3), IL-6, megakaryocyte growth and development factor (MDGF) and thrombopoietin (TPO) (Figure 4.4).

To favor their adhesion, $CD34^+$ cells are usually cultured on plastic surfaces covered by extracellular matrix proteins, such as fibronectin or peptides derived from this protein. Transduction is carried out by the addition, for 2 or 3 consecutive cycles, of cell culture supernatants from the packaging cells producing the retroviral vector (see section on 'Viral Vectors'). At the end of the transduction procedure, the HSCs are reinjected into the patient.

By means of this protocol, the efficiency of gene transfer is extremely high (98–100% of cells are transduced by retroviral vectors when analyzed *in vitro*). Unfortunately, this does not necessarily result in the transduction of the HSCs with truly stem cell properties, able to repopulate the bone marrow once reinjected (cf. below).

4.2.5
Gene Therapy Clinical Trials Entailing Gene Transfer into Hematopoietic Stem Cells: General Considerations

Using the above-described procedures, gene therapy has proven successful in a series of animal models of human disease, especially in mice knockout for the same genes that, in humans, are responsible for different immunodeficiencies, chronic granulomatous disease,

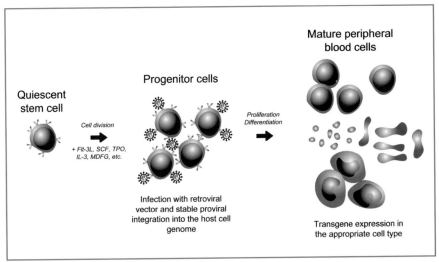

Fig. 4.4 Gene transfer into hematopoietic stem cells. HSCs are quiescent cells that need to be stimulated with a cytokine cocktail to re-enter the cell cycle. After stimulation, the cells are transduced with a retroviral vector and reinjected into the patient, where they mature into their final progeny in which they express the retrovirus transgene

lysosomal storage disorders, and others. Based on the exciting results obtained in these animal models, clinical experimentation began. The first hereditary disorder for which a gene therapy clinical trial was conducted, in 1990, was adenosine deaminase (ADA) deficiency, an autosomal recessive disorder mainly due to a T-lymphocyte defect (cf. below). This trial was conducted in the United States using a gammaretroviral vector to deliver the correct ADA cDNA into the peripheral blood T lymphocytes of two children with the disease. Subsequently, other gene therapy studies were performed based on gene delivery, usually using retroviral vectors, to both lymphocytes and CD34$^+$ precursors, obtained from either bone marrow or cord blood. To date, over 40 clinical studies have been conducted entailing retrovirus-mediated HSC gene transfer using the above-described protocols with the aim of treating different monogenic hereditary disorders (Table 4.5).

In contrast to the expectations raised by the mouse models, the overall results of these trials have been rather discouraging. *Ex vivo* HSC transduction was also found to be extremely efficient in patients, similar to the mouse models, with 80–100% of CD34$^+$ precursors being routinely transduced by gammaretroviral vectors. However, after reinfusion, the transduced cells did not persist inside the patients. With only some exceptions (cf. below), only a small percentage (0.01–1%) of provirus-containing cells was still detectable a few years after the transplant: too few, therefore, to confer a therapeutic benefic. This disappointing result is due to a peculiar characteristic of primate HSCs, namely that they lose their stem properties once cultured *ex vivo* to permit retroviral transduction. In particular, the cytokine cocktails that are commonly used to stimulate HSC proliferation at the same time also induce their differentiation: the transduced cells that are reinfused into the patients, therefore, are no longer stem cells but committed progenitors, having a much shorter life. Cytokine stimulation of HSCs is however required, because gammaretroviral

4

Table 4.5 Monogenic hereditary disorders for which gene therapy clinical trials were conducted by gene transfer into HSCs

Disease group	Disease	Defective gene
Severe combined immunodeficiency syndromes (SCID)	SCID-X1	Gamma common (γc) chain of interleukin receptors
	ADA-SCID	Adenosine deaminase
		JAK-3
	PNP-SCID	Purine-nucleoside phosphorylase (PNP)
Lysosomal storage disorders	Hurler's disease (MPS I)	α-L-iduronidase
	Hunter's disease (MPS II)	Iduronate-2-sulfatase
	Gaucher's disease	Glucocerebrosidase (β-glucosidase)
	Fabry's disease	α-Galactosidase A
	Sly syndrome (MPS VII)	β-Glucuronidase
Defects of phagocytes	Chronic granulomatous disease (CGD)	gp91^{phox}, p47^{phox}
	Leukocyte adhesion disorder	CD18 (β2-integrin)
Other diseases	Fanconi anemia, group C	FANCC

MPS, mucopolysaccharidosis

vectors necessitate, in order to integrate, that the cells undergo at least one mitosis, since their pre-integration complex has no access to the nucleus when the nuclear membrane is intact. In primates (including humans) only a small fraction of HSCs replicate in normal conditions. In contrast, a higher proportion of rodent HSCs physiologically remains inside the cell cycle and thus maintains a true HSC state also during the *ex vivo* culture. This explains the discrepant results obtained by gene therapy of murine knockout models in contrast to those of the human clinical trials.

An additional limitation of gammaretroviral vectors is the progressive silencing of their transgenes in the transduced cells. This process is sustained by cytosine methylation in the DNA of the vector LTR promoter, an event associated with chromatin remodeling towards a compact state (see section on 'Viral Vectors'). In a few HSC clinical trials in which the levels of expression of the therapeutic gene were analyzed a few years after treatment, it was indeed observed that several of the transduced cells were positive for the presence of proviral DNA, however they did not express the transgene, suggesting that silencing had occurred.

Loss of stem cell properties and silencing of therapeutic gene expression have been the main reasons for the poor success of most of the clinical trials so far conducted. To overcome these obstacles, several solutions have been proposed, which can also be implemented in combination. Besides the use of more efficient *ex vivo* HSC culture conditions, allowing the expansion of these cells without losing their stem properties, these include the use of lentiviral vectors to transduce quiescent HSCs, the treatment of patients with myeloablative drugs, and the genetic modification of HSCs by the inclusion of genes conferring drug resistance and providing a selective advantage to these cells.

Transduction of HSCs by lentiviral vectors. As discussed in the section on 'Viral Vectors', similar to wild-type HIV-1, lentiviral vectors have the capacity to enter the intact nuclear membrane thanks to the property of some of their proteins to cross the nuclear pores. Indeed, transduction of unstimulated CD34$^+$ human HSCs with these vectors, followed by their implantation in NOD/SCID mice (which can tolerate the presence of allogeneic cells due to their severe immunodeficiency) has shown that the transduced cells could completely regenerate the hematopoietic system. The utilization of gammaretroviral vector in the same experimental system led to completely negative results. In this respect, however, it should be observed that, while HIV-1 and the vectors derived from this virus are undoubtedly capable of infecting non-replicating cells, they still need the target cells to be metabolically active to efficiently integrate into the host genome. Unfortunately, due to their nature, HSCs fall into the category of metabolically inactive cells and therefore they appear to need a certain level of stimulation to be efficiently transduced (for example, by stimulation with TPO, which at least induces transition from the G0 to the G1 phase of the cell cycle). The first clinical trials with bone-marrow-derived CD34$^+$ HSCs transduced with lentiviral vectors for gene therapy of different disorders are currently ongoing and will certainly provide further information on the real efficiency of these vectors in a clinical setting.

Partial myeloablation. The success of HSC transplantation depends on the injected HSC dose and on the outcome of the competition between the injected HSCs and those of the host. For ethical reasons, all the initial HSC gene therapy protocols entailed the use of the same procedures as autologous BMT, with the exception that the patients were not treated with myeloablative drugs or radiation before transplantation. Once injected, the transduced HSCs had therefore to compete with the intact bone marrow of the recipient for homing and proliferation, a condition of clear disadvantage due to their relatively limited number and the lack of anatomical space for their expansion and differentiation. To overcome this problem, and also obviate the limited number of HSCs that retain their stem properties despite *ex vivo* stimulation and transduction, it is possible to treat the receiving patients with sub-lethal doses of radiations (a procedure commonly used in hematology for the so-called allogeneic "minitransplants", in which transplantation is performed still in the presence of a residual fraction of host bone marrow) or with antiblastic drugs such as busulfan or cyclophosphamide, in which the dose can be carefully titrated.

Transfer of selectable genes. An interesting possibility to increase the number of *ex vivo* transduced cells is to introduce into these cells, together with the therapeutic gene, an additional gene providing a selective advantage *in vivo*. The genes that are mostly considered for this purpose are those coding for the P-glycoprotein (*mdr-1*) or MGMT, the properties of which are described above. Experiments performed by HSC transduction with gammaretroviral vectors carrying *mdr-1* followed by their reinfusion into patients treated by multiple cycles of chemotherapy have indeed shown that more than 6% of white blood cells carried the provirus one year after transplantation. However, this approach should be considered with caution, since it does not entail the real expansion of transduced HSCs but the selective elimination of normal cells; in addition, it exposes the patients to chronic treatment with drugs having significant side effects.

4.2.6
Gene Therapy of ADA-SCID

The term severe combined immunodeficiency (SCID) refers to a series of rare monogenic hereditary disorders characterized by defects in the development and function of T lymphocytes and, either directly or indirectly, often of B-lymphocytes and natural killer (NK) cells, which depend on T lymphocytes for their activity. The clinical features are characterized by the occurrence of multiple episodes of infection from the first years of life, which are usually very severe and include interstitial pneumonitis, meningitis, and sepsis. Children with SCID are also commonly known as "bubble boys", since some of them were able to survive for several years in plastic bubbles in which the air coming inside was filtered and food and other objects were sterilized before introduction.

The overall frequency of SCID is between 1:50,000 and 1:100,000 children. More than 10 genetic defects causing SCID have been characterized, the most important of which are listed in Table 4.6 with their mode of inheritance and shown in Figure 4.5 along with the developmental pathway of T, B, and NK cells.

About 15–20% of SCID are caused by the deficiency of ADA, an enzyme involved in purine metabolism, which converts adenosine into inosine. Despite this protein being expressed ubiquitously, its defect is particularly relevant for the development of cells of the immune system and causes a SCID with autosomal recessive inheritance. The deficiency of ADA is responsible for increased levels of adenosine and 2'-deoxyadenosine in plasma and of nucleotides (in particular, dATP) in lymphoid tissues, red blood cells, kidney, and liver. The increase in intracellular purine metabolites determines a delay in T-, B-, and NK-cell proliferation, differentiation, and function, with a degree of severity that is

Table 4.6 Most frequent molecular defects leading to SCID, with the indication of their mode of inheritance and affected cell types

Disease	Mutated gene	Inheritance	Affected cells	Mechanism
Adenosine deaminase deficiency (ADA-SCID)	Adenosine deaminase (ADA)	AR	T, B, NK	Premature death of cells
SCID-X1	Common gamma chain (γc)	XL	T, NK	Survival defect due to lack of cytokine stimulation
	Janus kinase-3 (JAK-3)	AR	T, NK	
	IL7RA	AR	T	
Omenn syndrome	RAG1 or RAG2	AR	T, B	Defect in V(D)J recombination
	Artemis	AR	T, B	
	CD3 δ, ζ, ε	AR	T	Defect in pre-TCR or TCR signaling
	CD45	AR	T	

AR, autosomal recessive; *XL*, X-linked

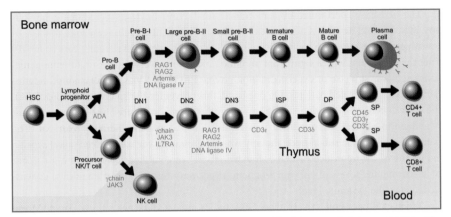

Fig. 4.5 Defects leading to the development of severe combined immunodeficiency (SCID). The defective genes are shown in *red* along the developmental pathways for B, T, and NK cells

inversely correlated with the residual levels of ADA. In addition to the immune cell defects, the accumulation of toxic metabolites can lead to alterations of non-lymphoid organs, such as kidney, liver, and brain, consistent with the features of a systemic disease.

Allogeneic BMT from HLA-identical twins usually leads to a complete cure of ADA deficiency; however, this is available only to a minority of patients. Instead, allogeneic BMT from unrelated, HLA-matched donors has high mortality and morbidity due to the frequent occurrence of infections and of GvHD, with an overall survival rate lower than 30%. The intramuscular injection, once a week, of the bovine ADA enzyme conjugated with poly(ethylene glycol) (PEG) – PEG increases the half-life of the injected protein by preventing its rapid degradation – available since the 1990s, corrects the metabolic defect, reduces the number of infections, and allows growth of the affected children. However, PEG-ADA is very expensive, 20% of the patients do not respond to treatment, and some patients develop neutralizing antibodies or show signs of autoimmunity after treatment.

ADA deficiency was the first human monogenic disorder to be treated by gene therapy. Despite being infrequent, this disease indeed constitutes an excellent model for gene therapy, since the ADA gene does not require regulation and an enzymatic activity as low as 10% of normal still allows reconstitution of the immune function. In addition, cells with a normal ADA gene possess a sight proliferative advantage over cells carrying the mutation, leading to a possible selective expansion once introduced in the patients.

Since the prevalent disease manifestations of ADA deficiency are in T lymphocytes, several of which are long-living cells, the first gene therapy clinical trial was performed by gene transfer into the peripheral blood T-lymphocytes of two children with the disease. The cells were recovered by apheresis and cultured *ex vivo* in the presence of a mitogen and IL-2 (to stimulate and maintain their proliferation, respectively); during culture, the lymphocytes were transduced with a gammaretroviral vector expressing the enzyme cDNA and then reinfused. Since 1990, several other gene therapy Phase I/II studies have been conducted for ADA deficiency in at least 6 different clinical centers; these studies have so far involved over 40 patients.

The first trials, performed by gene transfer in T lymphocytes, showed that gene therapy was safe and devoid of adverse effects. In some patients, a relatively high number of transduced lymphocytes (reportedly up to 20%) could be detected several years (over 10) after the transplant, with a significant immune function recovery. However, for ethical reasons, in this first phase of experimentation bovine PEG-ADA was simultaneously administered to all the treated patients. On one hand, this prevented the spontaneous selection of the transduced cells *in vivo*, while, on the other hand, it blurred the understating of whether the clinical improvement was really due to the gene therapy treatment. The patients in which the substitution therapy was progressively reduced indeed underwent an expansion of the transduced T cells. However, these were still too few and were incapable of sustaining the levels of enzyme production necessary for the correction of the metabolic defect at the systemic level.

When evaluated collectively, these first studies indicated the necessity to develop gene transfer protocols for the delivery of the ADA cDNA into HSCs rather then peripheral blood T-lymphocytes, to both increase the number of transduced cells and extend enzyme production to non-T -cells. However, the first clinical trials entailing gene transfer into HSCs, which were run in parallel with those on T lymphocytes, indicated that the efficiency of HSC transduction was very unsatisfactory, similar to all other HSC gene therapy studies. This problem was eventually overcome by using the myelotoxic drug busulfan to obtain partial myelosuppression before reinjecting the *ex vivo* transduced HSCs.

The first clinical study entailing non-myeloablative conditioning with busulfan was conducted in Italy in 2000 in two children not amenable to allogeneic BMT and not having access to PEG-ADA therapy. Thanks to partial myeloablation and despite the use of standard *ex vivo* HSC culture conditions using HSC prestimulation with Flt-3L, SCF, TPO, and IL-3 followed by gammaretroviral transduction, the efficiency of bone marrow repopulation was high (up to 10%), thus obtaining the production of ADA by all the HSC progeny cells and the subsequent restoration of the immune function. This result represents an outstanding achievement, the importance of which extends beyond ADA gene therapy since it clearly indicates that partial myelosuppression is an essential requisite for the success of all clinical protocols entailing *ex vivo* gene transfer into the HSCs. Long-term follow-up of these 2 patients, together with that of another 8 patients treated according to the same protocol, has now confirmed immune reconstitution with increase in T-cell counts and normalization of T-cell function in 9 out of these 10 children, in the absence of major adverse events. These results ultimately confirm that gene therapy represents a safe and effective treatment for ADA-SCID.

4.2.7
Gene Therapy of SCID-X1

The most frequent SCID (about 40–60%) is due to defects in a gene positioned on the X chromosome, which is thus transmitted with X-linked inheritance (SCID-X1). This gene codes for the common gamma chain (γc), a membrane protein common to several interleukin receptors. In particular, the receptors for IL-2, -4, -7, -9, -15, and -21 consist of specific α and/or β chains and of a γ chain, which is identical for all of them (Figure 4.6). The

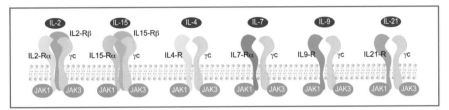

Fig. 4.6 Composition of the interleukin receptors. The γ chain (γc) gene was originally cloned as a component of the IL-2 receptor, where it interacts with the IL2-Rβ chain. It was eventually recognized that the same protein binds the IL15Rβ chain to form the IL-15 receptor. The γ chain also interacts with other chains to form receptors for IL-4, IL-7, IL-19, and IL-21. These receptors signal through the Janus-activated kinase (JAK)/signal transducer and activator of transcription (STAT) pathway. Cytokine binding determines JAK1 and JAK3 activation, which, in turn, allows STAT dimerization, nuclear translocation, and, ultimately, gene transcription

γc is an essential component of these receptors, since it is required for the activation of the JAK-3 (Janus kinase-3) tyrosine kinase, which initiates signal transduction downstream of the receptors. Mutation of the γc results in the lack of T lymphocytes, since these cells depend on the activity of IL-7 for their survival, proliferation, and maturation in the thymus, and of NK cells, which require IL-15 for their growth and differentiation. The number of B cells is usually normal, however these cells do not produce antibodies due to the absence of helper T cells. Children with SCID-X1 usually do not survive beyond the first or second year of life due to severe infections. The only possible conventional therapy is allogeneic BMT.

The first gene therapy clinical trial for SCID-X1 was conducted at the end of the 1990s in Paris, France. CD34$^+$ cells from the bone marrow of 10 children with the disease were pre-stimulated *ex vivo* for 24 h with SCF, Flt-3L, MDGF, TPO, and IL-3 on a fibronectin substrate and then transduced over the following 3 days with an amphotropic gammaretroviral vector based on the Mo-MLV virus, carrying the γc cDNA. The transduced cells (10–50 million/kg body weight) were then reinfused into the patients, in the absence of conditioning myelosuppression. As can be observed, the protocol for this study did not significantly differ from that used for ADA-SCID or, more generally, for all other gene therapy clinical trials entailing HSC gene transfer. However, in sharp contrast to these trials, in which the overall outcome was very disappointing in terms of bone marrow repopulation by the transduced cells, gene therapy of SCID-X1 resulted in a striking success: 18 months after transplant, 9 out of 10 patients showed complete reconstitution of their immune system, left their protected, sterile environment, and came back to normal life, in the absence of any relevant infection. Virtually all T and NK cells of the treated patients expressed the γc protein delivered by the retroviral vector. Analogous results were in parallel obtained in another series of children treated for the same disease in London, UK, using a very similar transduction protocol, with the exception that the gammaretroviral vector was slightly different and it was pseudotyped with the GaLV envelope (cf. section on 'Viral Vectors') in order to increase transduction efficiency.

The success of SCID-X1 gene therapy in 2000 was universally saluted as the first important clinical achievement accomplished by gene therapy. What were the conditions that rendered it possible and how was it different to the many other gene therapy clinical trials that showed very limited levels of transduced cells month- or year-long periods after transplantation? The answers to these questions relate to the peculiar biology of X-SCID. In this disease, lymphocyte precursors are continually produced from the HSC, however, due to the missing γc, they rapidly undergo apoptosis. After correction of the defect, although the efficiency of gene transfer was initially as low as in the other clinical studies, the presence of γc conferred an enormous selective advantage to the lymphocytes that expressed it. These lymphocytes therefore continued to proliferate and eventually reconstituted the whole immune system over a few months, thus completely curing the disease.

Despite the apparent success of SCID-X1 gene therapy in the short term, the two trials have later dramatically highlighted the insertional mutagenesis problem that accompanies gammaretroviral gene transfer. Two years after transplantation, two of the patients treated in Paris developed a T-cell acute lymphoblastic leukemia (T-ALL), which was later shown to be due to the insertion of the retroviral vector inside the LMO2 (LIM domain only 2) proto-oncogene, thus causing its activation. Subsequently, another two children from the Paris cohort and one out of the ten treated in London also developed T-ALL. In all five patients, the LMO2 was found to be overexpressed due to retroviral insertional mutagenesis.

The LMO2 gene codes for a protein binding to different transcription factors (including SCL/AL1, GATA1, and GATA2) that regulate cell differentiation during hematopoiesis. In normal conditions, the LMO2 gene is active during the first phases of thymopoiesis and is then rapidly switched off during T-cell differentiation. In functional terms, the LMO2 protein can be considered as an oncogene, since its deregulation, due to chromosomal translocations, was associated to the development of T-ALL in some patients. Additionally, its deregulated expression in the mouse determines the formation of T-cell lymphomas. In the case of the two patients who developed T-ALL in the French gene therapy trial, the retroviral vector was found to be integrated in correspondence with the LMO2 promoter in one case and within the gene first exon in the second case; in both cases, retroviral integration determined increase and disregulation of LMO2 gene expression (Figure 4.7). Thus, these appear to be canonical insertional mutagenesis events, in which the viral LTR activates the promoters of the cellular gene close to its integration site (cf. also section on 'Viral Vectors').

It is still controversial whether the transcriptional activation, although inappropriate, of the LMO2 gene might alone cause the development of leukemia. Experiments performed in mice by the generation of transgenic animals or by transplantation of HSCs transduced with retroviral vectors expressing the LMO2 cDNA have shown that the acti-

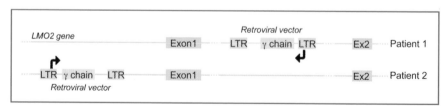

Fig. 4.7 Insertional mutagenesis in two children treated by gene therapy of SCID-X1. In Patient 1, the retroviral vector expressing the γ chain was found inserted in the first intron of the LMO2 gene, and in Patient 2 in the promoter region of the same gene. The *arrows* indicate the direction of transcription from the vector LTR

vation of this gene is responsible for a marked hyperproliferation of the cells in the T-lymphocyte compartment, however without determining true malignant transformation. Other experiments have shown that the therapeutic gene, namely the γc itself, when expressed at non-physiological levels, might act as a potential oncogene for the cells. Thus, it is likely that leukemia in the treated children was the consequence of an unfortunate functional cooperation between γc and LMO2, the former expressed at high levels from the vector and the latter activated through insertional mutagenesis. This cooperation might also explain why LMO-2-activated clones were only selected in the SCID-X1 trials and not in the more than 40 clinical trials for other disorders entailing gene transfer into HSCs, in which at least 250 patients have been treated so far, including those for ADA-SCID.

The results obtained in these SCID-X1 trials and those obtained in the CGD trial (see following section) raise essential safety concerns concerning the use of gammaretroviral (and probably lentiviral) vectors for gene transfer, which are discussed in detail in the section on 'Viral Vectors'.

4.2.8
Gene Therapy of Chronic Granulomatous Disease (CGD)

Chronic granulomatous disease (CGD) is an inherited disorder of phagocytes characterized by recurrent bacterial and fungal infections of several organs. The disease is due to the incapacity of phagocytes (neutrophil granulocytes and macrophages) to undergo, after phagocytosis, the oxidative burst, that is the very rapid and intense process that, by using oxygen, leads to the formation of different chemical species with high oxidative power, which are released into the phagocytosis vesicles. This process is usually triggered by the production of superoxide anion (O_2^-); this is then converted into a series of oxidizing compounds (reactive oxygen species, ROS) having high bactericidal activity, including hydroxyl radical ($OH^·$), hydrogen peroxide (H_2O_2), peroxynitrite anion ($ONOO^-$), hypochlorous acid ($HOCl$), singlet oxygen ($^1O_2$), and oxyhalides (HOX, in which X is usually chloride).

Superoxide anion is generated by a membrane NAPDH oxidase, also called phagocyte oxidase (*phox*), which is defective in CGD. In patients with this disease, phagocytosis occurs normally, however the microorganisms persist and multiply inside the phagocytosis vacuoles, leading to persistent and recurrent infection by bacteria and fungi, with the formation of large granulomatous masses, hence the name.

NADPH is a multi-component enzyme, localized on the phagocyte membranes, which accepts electrons from NADPH on the cytosolic side of the membrane and donates them to molecular oxygen on the other side, that is either in the extracellular space or inside the phagosomes. The enzyme is quiescent in resting phagocytes and becomes active upon binding of opsonized microorganisms to cell surface receptors (Figure 4.8). The enzyme consists of at least 5 subunits; 2 of these are integral membrane proteins named, according to their apparent mass, p22phox and gp91phox, of which the latter is heavily glycosylated; these two proteins are also known as the α and β subunits of flavocytochrome b558, since gp91phox associates with a FAD molecule. The other 3 subunits (p47phox, p67phox, and p40phox) are localized inside the cytosol in resting conditions and translocate to the mem-

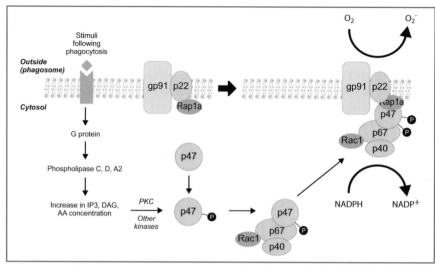

Fig. 4.8 Activation of phagocyte NAPDH oxidase. The phagocyte NAPDH oxidase is an enzyme complex consisting of different subunits, which receive electrons from NADPH from the cytoplasmic side of the membrane and donate them to oxygen on the opposite side. In inactive phagocytic cells, two subunits (p22^{phox} and gp91^{phox}) are present on the plasma membrane and bind a heme. These two subunits form cytochrome b$_{558}$. Two other subunits (p67^{phox} and p47^{phox}) are present in the cytoplasm. Phagocyte cell activation is triggered by a series of stimuli following phagocytosis (for example, opsonized microorganisms, the C5a component of complement, formylated peptides deriving from the lysis of dead microorganisms, bioactive lipids, or proteins secreted by activated cells). The time between phagocyte stimulation and superoxide production is less than 5 seconds. All these stimuli act through specific cell surface receptors, which eventually lead to the activation of a G protein. This in turn activates three phospholipases (C, D, and A2), which lead to the synthesis of a series of second messengers. In particular, phospholipase C produces 1,2-diacylglycerol (DAG) and 1,4,5, inositol-triphosphate (IP3) starting from phosphatidylinositol-4,5-diphosphate. Phospholipase D hydrolyzes phosphatidylcholine to phosphatidic acid, which in turn is degraded to DAG. Finally, phospholipase A2 is responsible for the production of fatty acids and, in particular, of arachidonic acid (AA). DAG, AA, and IP3 activate protein-kinase C (PKC) and other less defined intracellular kinases. The net result of this activation is the phosphorylation of p47^{phox}, which then binds p67-^{phox} and p40-^{phox} and translocates to the cell membrane, where it binds and activates the gp91^{phox}/p22^{phox} complex. The function of the activated complex also depends on additional proteins. In particular, the regulation of the complex is obtained through the interaction of GTP-binding, low-molecular-weight G proteins. Of these, Rac-1 specifically binds p67^{phox} while Rap1-A translocates to the membrane independent of the other cytosplamic proteins and specifically binds the cytochrome

brane to associate with the flavocytochrome b$_{558}$ only upon activation. The translocation of these subunits to the membrane is essential to confer enzymatic activity to the cytochrome by inducing a conformational change in the complex. Two other GTP-binding, low-molecular-weight proteins participate in the formation of the oxidase: one in the cytoplasm – which alternatively can be Rac-1 in macrophages or Rac-2 in neutrophil granulocytes – and one associated to the membrane, Rap-1A.

Defects in the genes coding for gp91^{phox}, p22^{phox}, p47^{phox}, or p67^{phox} impair or abolish NADPH oxidase activity and are thus responsible for CGD. Since the gp91^{phox} gene is located on the X chromosome, its defects are transmitted with an X-linked pattern of inheritance (X-CGD); over 350 different mutations are known, accounting for 50–70% of the CGD cases. The other defects are instead transmitted as autosomal recessive disorders. The estimated incidence of CGD is 1–4:250,000 individuals. The diagnosis is usually made early in life due to recurrent infections; long-term prophylaxis with antibiotic and antifungal drugs and, in some cases, use of interferon γ reduce the frequency of life-threatening infections. Morbidity and mortality however remain high in these patients. When a compatible donor is available, allogeneic BMT is curative, although its beneficial effects are counterbalanced by the risk associated with immunosuppressive therapy, since CGD patients are severely compromised by the presence of chronic infections.

CGD represents an ideal candidate for gene therapy, since the disease has recessive inheritance, its phenotype is only apparent in phagocytic cells, and a partial correction should probably be sufficient, since in some female carriers of the gp91^{phox} defect, due to skewed X-chromosome inactivation, the number of normal cells is less than 10% and nevertheless they are asymptomatic. Since phagocytic cells have a short half-life (a few hours for neutrophil granulocytes), the gene therapy target is necessarily the HSC. Experimental evidence indeed indicates that gene transfer of the corrected cDNA into CD34$^+$ HSCs using gammaretroviral vectors completely restores NADPH function in myeloid cells differentiated *ex vivo* from HSCs derived from CGD patients (Figure 4.9).

Different clinical trials have addressed the possibility to correct both X-CGD (gp91^{phox} mutations) and the autosomal form due to p47^{phox} defects by gene therapy. The first series of Phase I/II studies, conducted in the United States, have exploited gammaretroviral vectors and transduction protocols similar to those used by most other HSC trials. The results of these studies have indeed shown that it is possible to permanently correct the CGD defect, however the number of transduced cells expressing a functional NADPH oxidase has been disappointingly low (from 0.004% to 0.6% of cells), since the corrected myeloid progenitors, in contrast to SCID-X1 and, to a lesser extent, ADA-SCID lymphocytes, do not gain any selective advantage from gene transfer.

For these reasons, in 2002 and 2004 two additional trials were organized in Europe, in which patients were treated with mild (in the former trial) or more aggressive (in the latter) partial myelosuppression using busulfan, similar to the successful ADA-SCID gene therapy protocol. In the two X-CGD patients treated in the 2004 trial, a significant portion (12% and 31%) of phagocytes carrying a functional NADPH oxidase was detected after infusion, with the eradication of the pre-existing bacterial and fungal infections. Unexpectedly, these percentages progressively increased over time, to reach 50–60% of all peripheral blood granulocytes. It was later found that this increment was due to the insertion of the vector in close proximity to three genes physiologically involved in the control of cell proliferation, with their consequent transcriptional activation. Two of these genes code for the zinc finger transcription factors MDS1-EVI1 and PRDM16 and one for a factor (SETBP1) originally identified for its property to bind the SET protein, which is involved in the development of acute leukemia with undifferentiated cells. These observations again highlighted the mutagenic potential of gammaretroviral insertion. In contrast to SCID-X1, however, in which LMO2 activation is the main leukemogenic event,

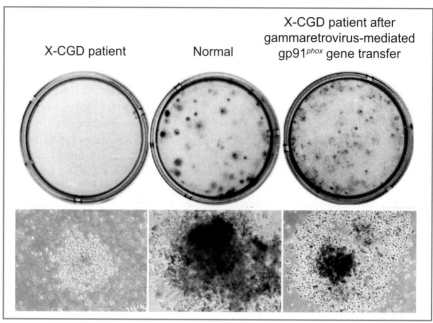

Fig. 4.9 Functional correction of NAPDH activity in myeloid colonies from an X-CGD patient after gene transfer of the gp91phox cDNA into CD34+ hematopoietic stem cells. The function of the phagocyte NADPH oxidase can be assessed by the nitroblue tetrazolium (NBT) assay, which focuses on the ability of phagocytes to produce oxygen radicals and subsequently reduce a soluble nitroblue tetrazolium dye to insoluble formazan, which has a brownish color. The upper part of the picture shows methylcellulose cell culture dishes in which myeloid colonies were obtained starting from bone marrow CD34+ cells of an X-CGD patient (*left*), a normal individual (*middle*), and an X-CGD patient after transduction with a gammaretroviral vector expressing the normal gp91phox cDNA, after staining by the NBT assay. The lower part of the figure shows magnifications of individual colonies

the activation of the three genes observed in the X-CGD trial confers a more benign hyperproliferative potential to the myeloid compartment. Whether this might constitute a preleukemogenic event (by increasing the probability of occurrence of other genetic changes) or, on the contrary, might extinguish over time still remains to be understood. Four other CGD patients who were later treated using a similar myeloablative protocol showed significantly lower levels of enrichment of the transduced cells after transplantation.

It remains to be understood why the insertional mutagenesis events leading to the expansion of some of the transduced cell clones occurring in the SCID-X1 and X-CGD trials were not observed in several other patients treated with analogous protocols for other disorders. It might be possible that subtle variations in the structure of the gammaretroviral vectors used might be responsible for these differences. In particular, the vector used for the X-CGD trial shows a transcriptional activity that is significantly higher in HSCs compared to the vectors used by other studies. While this appears beneficial for the full reconstitution of NADPH activity, it might have increased the probability of activation of genes neighboring the insertion sites. Furthermore, the CGD trial entailed transduction of peripheral blood CD34$^+$ cells after mobilization with G-CSF: it is well known that these HSCs are more enriched in myeloid precursors compared to bone marrow HSCs, utilized by most of the other studies.

Finally, one of the two patients with X-CGD originally treated in the apparently successful trial eventually died two and a half years after treatment due to a severe sepsis, potentially indicative of a relapse of an immunodeficient condition. In this patient, the levels of gp91^{phox} expression were very low, almost below the threshold for detection, a result compatible with the possibility of silencing of retroviral gene expression over time despite persistence of provirus-containing cells.

When these results are collectively considered, it appears clear that the future of CGD gene therapy will strictly depend on the development of novel vectors with an increased safety profile, in order to avoid or minimize the problem of insertional mutagenesis (in particular, the SIN vectors; cf. section on 'Viral Vectors').

4.2.9
Gene Therapy of Lysosomal Storage Disorders

The so-called "inborn errors of metabolism", a definition originally introduced by the English physician Archibald Garrod at the beginning of the last century, and now better named "hereditary diseases of metabolism", include a heterogeneous series of disorders due to mutations of single proteins, usually enzymes, required for the correct function of a metabolic pathway. Traditionally, these disorders were classified according to the affected metabolic pathway. They included *disorders of carbohydrate metabolism* (for example, galactosemia, glycogenosis or glycogen storage disease, hereditary fructose intolerance, etc.), *disorders of metabolism and transport of amino acids* (for example, phenylketonuria, homocystinuria, most organic acidurias, cystinuria, urea cycle disorders), *disorders of organic acid metabolism* (for example, methylmalonyl-CoA mutase deficiency), and *lysosomal storage disorders* (see below). Currently, this classification is considered incomplete, since, over the last 20 years, mutations of hundreds of genes involved in various other metabolic pathways have been discovered. These include genes taking part in the *metabolism of porphyrins* (acute intermittent porphyria), *steroids* (congenital adrenal hyperplasia), *purines and pyrimidines* (Lesch-Nyhan syndrome), *lipids* (familial hypercholesterolemia), *peroxysomal function* (Zellweger's syndrome, adrenoleukodystrophy), and others.

The inborn errors of metabolism are mainly inherited as autosomal recessive traits, since, in most cases, the defective proteins are enzymes and thus the mutation of a single gene is compensated by the wild-type allele in heterozygotes. As a consequence, these disorders are very amenable to treatment by gene therapy. In particular, the lysosomal storage disorders have been, and continue to be, the focus of several gene therapy clinical trials.

The lysosomal storage disorders (LSDs) are a group of hereditary diseases of metabolism characterized by the defect of one or more proteins involved in the activity of lysosomes. More than 40 different defects are well characterized, all showing a recessive pattern of inheritance, with an overall incidence of 1:7500 newborns. The LSDs can be classified into different groups according to the substance accumulating inside the lysosomes.

4

The most prevalent disorders are the mucopolysaccharidoses (MPS), due to a defect in the degradation of mucopolysaccharides (high-molecular-weight proteins exerting essential roles in the extracellular environment), the sphingolipidoses, due to a block in the lysosomal degradation of sphingolipids, and the oligosaccharidoses, due to defects in the degradation of oligosaccharides and glycoproteins.

Most LSDs are due to the total absence or significant decrease (<10%) of the activity of a soluble lysosomal enzyme. A subclass of LSDs can also be caused by defects of non-enzyme proteins, for example the so-called sphingolipid activator proteins (SAPs), a family of glycoproteins required for the catabolism of sphingolipids by specific acid hydrolases; absence of these proteins determines the accumulation of non-metabolized sphingolipids inside the lysosomes. Mutation of integral membrane proteins of the lysosomes can also cause an LSD. For example, this is the case of the Niemann-Pick disease type C, in which the mutated proteins are the two cholesterol transporters NPC1 and NPC2, or of neuronal ceroid lipofuscinosis (NCL), characterized by the deposit of indigested brown or autofluorescent material (lipofuscin) in neurons, followed by the degeneration of these cells. Finally, LSDs can also be caused by mutations of proteins required for the intracellular trafficking of lysosomal enzymes. For example, mucoliposis types II and III are caused by defects in a phosphotransferase of the Golgi apparatus, which is necessary for the addition of mannose-6-phosphate to the lysosomal enzymes. In the absence of this modification, these enzymes are directed toward the secretory pathway rather than being transported to the lysosomes.

The clinical characteristics of LSDs are a consequence of the intra-lysosomal accumulation of non-metabolized macromolecules, which determines a progressive increase in the volume of the relevant organ and its malfunction. The signs of the disease include hepatosplenomegaly, cardiac abnormalities, and skeletal defects, with variable involvement of the kidney, immune system, and central nervous system.

The lysosomal enzymes are normally synthesized in the endoplasmic reticulum and then processed post-translationally. In particular, they are glycosylated in the endoplasmic reticulum and, subsequently, transported into the Golgi apparatus, where they are phosphorylated at position 6 of the terminal mannose residue (M6P). The phosphorylated enzyme then binds the mannose-6-phosphate/IGF-II receptor (M6PR/IGFIIr), present in the Golgi apparatus, and the enzyme/receptor complex is thus addressed toward the lysosomes through the endosomal vesicle sorting system. Once in the mature lysosomes, the enzyme/receptor complex disassembles due to the low pH of these organelles. The receptor then moves back to the Golgi apparatus or is transported, again using the intracellular vesicle trafficking system, to the plasma membrane. A small quantity of enzyme leaves the transport route towards the lysosome and is secreted outside the cells. This extracellular fraction can thus bind either the M6PR receptor present on the surface of all cells (in case the enzyme retains M6P) or the mannose receptor normally present on the plasma membrane of cells of the reticuloendothelial system (in case the enzyme exposes non-modified mannose residues). Both receptors can mediate endocytosis of the enzyme and transport it to the lysosomes (Figure 4.10). This internalization route constitutes the basis upon which the current enzyme substitution therapies for some LSD work: a purified or recombinant lysosomal enzyme is administered systemically to the patient, and, thanks to these receptors, the protein is transferred from the circulation into the cells.

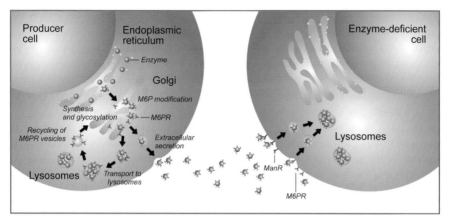

Fig. 4.10 Lysosomal enzyme trafficking. Nascent lysosomal enzymes (*blue spheres*) are glycosylated in the endoplasmic reticulum and then acquire the mannose 6-phosphate modification (*red*) in the Golgi apparatus, where they bind the mannose 6-phosphate receptor (M6PR; *red*). The majority of the enzymes are then trafficked to the mature lysosomes. A minority of the lysosomal enzymes are secreted from the cell. Extracellular phosphorylated or non-phosphorylated enzyme can bind the plasma membrane-localized M6PR (*red*) or the mannose receptor (ManR; *blue*), respectively. Both receptors mediate the endocytosis and subsequent lysosomal targeting of the exogenous enzymes. This process can occur in both genetically modified and enzyme-deficient cells. It is important to note that the M6PR is ubiquitously expressed, whereas expression of the ManR is limited to cells of the reticuloendothelial system

Ample preclinical evidence of efficacy in animal models shows that transfer of the enzyme cDNAs into the relevant cell types can cure various forms of LSD. Usually, tight regulation of expression of these enzymes is not required, and levels as low as 10% of normal can be sufficient to achieve a normal phenotype. On the other hand, however, most lysosomal enzymes are expressed ubiquitously, and their absence causes a vast range of pathological manifestations, including involvement of the central nervous system. This would in principle prevent the possibility of developing a gene therapy approach for these disorders. However, in the LSD specifically due to defects of lysosomal enzymes, the cells are both able to secrete these enzymes and to internalize them once present in the extracellular environment. Therefore, it is also possible to cross-correct the enzymatic activity in cells in which gene transfer has not occurred directly.

The overall purpose of gene therapy of LSD is thus to produce systemic levels of a missing enzyme able to cross-correct the defect in all the relevant cells in the body. To achieve this goal, the strategy followed so far has consisted in the modification of the patients' cells by *ex vivo* gene transfer, followed by reinfusion of the corrected cells. In most trials, the target cells for gene transfer were the HSCs, especially because these cells are also the precursors of cells of the reticuloendothelial system, in particular macrophages, which are among the cell types most affected by LSD.

The first clinical trials were in patients having the non-neuropathic form of Gaucher's disease and, later, type I (Hurler's disease) and type II (Hunter's disease) mucopolysaccharidosis (Table 4.5). The initial protocols entailed transplantation of CD34$^+$ HSCs transduced *in vitro* with gammaretroviral vectors carrying the correct cDNA. Transplantation was carried out under non-myeloablative conditions and the success of these studies has therefore been minimal. To increase efficiency, various clinical trials are now exploiting different *ex vivo* HSC culture and transduction conditions, inclusion of myeloablative conditioning, or use of lentiviral vectors.

Several LSD involve the central nervous system; in these cases, cross-correction is not possible, since the lysosomal enzymes are not able to pass through the blood–brain barrier. As a consequence, any gene therapy approach aimed at correcting the genetic defect in the brain has to be based on the direct intracerebral injection of the therapeutic genes. In this context, lentiviral and AAV vectors can be considered as vehicles for the intracerebral delivery of the enzyme cDNAs, due to their exquisite property to transduce neurons at high efficiency *in vivo*.

4.3
Gene Therapy of Cystic Fibrosis

From the early years of gene therapy, it was immediately evident that cystic fibrosis (CF) represented a sort of ideal disease for this technology. Indeed, this is a monogenic disorder, with a recessive phenotype, which is mainly manifested in an organ, the lung, that can be easily accessed through a non-invasive, natural route: all features, in principle, ideal for gene therapy. Gene therapy of CF, however, has proved much more difficult than initially anticipated.

CF is transmitted as an autosomal recessive disorder with a frequency of 1:1800 children. The disease is due to mutations of the CFTR (cystic fibrosis transmembrane conductance regulator) gene, coding for a 250-kDa transmembrane protein belonging to the ABC family of transporters. The CFTR protein is activated by cAMP and acts as a chloride ion channel, mediating secretion of these ions outside the cells. Over 1400 different CFTR mutations are now known. The most frequent mutation (60% of the CF cases in the Caucasian population) is ΔF508, caused by the deletion of a nucleotide triplet coding for phenylalanine at position 508 of the protein. Presence of this mutation impairs the correct folding of CFTR and causes its accumulation and consequent degradation in the endoplasmic reticulum.

The absence or malfunction of CFTR mostly affects the respiratory and gastrointestinal tracts. In particular, about 90% of patients with CF die because of a respiratory disease, since the absence of the protein determines the production of very thick mucus, which is readily colonized by pathogen microorganisms causing chronic airway inflammation. The currently available treatments have the objectives of keeping the airways patent and suppressing infection. However, notwithstanding the enormous effort made over the last decades in this respect, CF inevitably leads to respiratory insufficiency, with a median survival age less than 30 years.

The CFTR gene was cloned in 1989, the same year the first gene therapy clinical trial was conducted. Since then, the disease was the object of several clinical studies in the first part of the 1990s. In particular, during those years, 25 Phase I/II clinical studies were conducted, which involved over 400 patients. Most of these initial trials aimed to show that transfer of the CFTR cDNA into the nasal epithelium of CF patients, as a surrogate model for lung, was safe and effective. In these studies, efficacy was evaluated by measuring the extent of correction of the abnormal transepithelial voltage that is usually observed in CF patients due to defective chloride ion transport. Once efficacy of treatment was verified, administration of the normal CFTR cDNA continued by direct instillation, into the respiratory tree, using a bronchoscope or by aerosol.

Most of these initial trials used first-generation adenoviral vectors as carriers for the CFTR cDNA. As described in the section on 'Viral Vectors', these vectors are excellent tools for gene delivery, however they are fraught with strong immunogenicity and induction of robust inflammation. As a paradigm of this condition, it is worth mentioning the results of one of the first CF gene therapy trials, conducted in New York in the early 1990s. In this trial, four patients who were treated with a first-generation adenoviral vector, carrying deletions of the E1 and E3 genes, developed an acute inflammatory reaction, which was particularly severe in one child, concomitant with the production of very high levels of IL-6 starting a few hours after instillation of the vector.

In addition to the problems connected with the inflammatory response and with their immunogenicity, one of the reasons that have additionally restricted the use of adenoviral vectors for gene therapy of the lung is the relatively low expression of the CAR receptor on the apical membrane of the airway epithelial cells. On the other hand, the immunogenicity of these vectors means that this problem cannot be solved by repeated administrations.

For all the above-described reasons, the initial optimism concerning the relative simplicity of CF gene therapy progressively dampened. Adenoviral vectors were subsequently substituted by AAV vectors as CFTR cDNA carriers. In particular, after several years of preclinical development and validation, a few Phase I/II clinical trials were conducted using an AAV2 vector, which, however, generated negative results. The project, which was supported by a biotechnology company in the United States, was abandoned in 2005. The inefficacy of AAVs for gene transfer into the lung is however not completely surprising, since the epithelial cells of the respiratory tract are not among the post-mitotic cell types for which AAV shows natural tropism. It might however be possible that the use of alternative serotypes (in particular, AAV5) might show better efficacy.

Parallel to the overall lack of success of gene transfer using viral vectors, more than 10 different clinical trials have exploited non-viral methods for CFTR gene delivery to the respiratory epithelium. In general, most of these studies showed, as a proof-of-principle, that the non-viral methods can lead to a restoration of CFTR function as high as 25% of normal. In contrast with viral vectors, however, these non-viral formulations, mainly based on liposomes or cationic lipids, can be administered repeatedly, thus leading to an incremental increase of functional improvement. A few vast clinical studies using these delivery methods are currently ongoing in the United States and United Kingdom, especially thanks to the economic support of charities and patients' associations.

4.4
Gene Therapy of Muscular Dystrophies

One of the most important goals of gene therapy is the development of safe, efficient and long-lasting procedures for gene transfer into the skeletal muscle. On one hand, this tissue is where a series of hereditary disorders, often dramatically, are manifested; on the other hand, thanks to its accessibility and mass, skeletal muscle also represents a possible source for the release of therapeutic proteins locally or into the circulation, thus allowing the development of innovative, gene therapy-based strategies to cure the disorders of the soluble components of blood (for example, coagulation defects) or produce factors acting locally (for example, factors inducing therapeutic angiogenesis in peripheral artery obstructive disorders).

4.4.1
Dystrophin and Dystrophin-Associated Proteins

The congenital muscular dystrophies (MDs) include a heterogeneous series of severe neuromuscular degenerative disorders, genetically determined, causing progressive atrophy of skeletal muscles and having a broad and usually severe phenotypic spectrum. Nine main different types of defects are commonly classified as canonical MDs (Table 4.7), however over 100 other different diseases have some symptoms or signs proper of MD. In addition, most MDs are systemic disorders, in which, besides the skeletal muscles, the heart, gastrointestinal tract, nervous system, endocrine glands, skin, eyes, and other organs are also affected.

Most genes causing MD code for proteins having the function to connect the cell cytoskeleton to the extracellular matrix. Dystrophin (427 kDa) is a very long filamentous protein located on the cytosolic side of the plasma membrane of striated muscle fibers (sarcolemma), particularly concentrated in correspondence with the neuromuscular junctions. The protein is composed of 4 structural domains: an N-terminal region (actin-binding domain, ABD); a central rod domain consisting of a series of repeats each composed of a three α-helix bundle, similar to that of spectrin; a cysteine-rich domain (CR) and a C-terminal domain (CT) (Figure 4.11A). The N-terminal region binds the actin filaments, while the CR domain is essential to bind and localize to the sarcolemma a series of proteins called the dystrophin-associated glycoprotein complex (DGC). The DGC connects the internal cytoskeleton to the extracellular matrix, thus stabilizing the sarcolemma while the muscle fiber alternatively stretches or shortens. In addition to this structural role, dystrophin and the DGC participate in a variegate series of intracellular signaling processes. Both in humans and in animal models, the lack of dystrophin also determines a secondary absence of the DGC proteins from the muscle fiber sarcolemma.

The DGC consists of more than ten different proteins (Figure 4.11B). The central component of the complex is dystroglycan, which is initially synthesized as a single protein and then cleaved to generate β-dystroglycan (43 kDa), a transmembrane protein, and α-dystroglycan (156 kDa), which associates to the former on the outer side of the sar-

Table 4.7 Muscular dystrophies

Disease	Mutated gene
Duchenne muscular dystrophy (DMD)	Dystrophin
Becker muscular dystrophy (BMD)	Dystrophin
Emery-Dreifuss muscular dystrophy	Emerin, lamin A or lamin C
Limb-girdle muscular dystrophy (LGMD)	Over 15 different genes
	Autosomal dominant: LGMD 1A: myotilin LGMD 1B: lamin A/C LGMD 1C: caveolin 3 and others
	Autosomal recessive: LGMD 2A: calpain-3 LGMD 2B: dysferlin LGMD 2C: γ-sarcoglycan LGMD 2D: α-sarcoglycan LGMD 2E: β-sarcoglycan LGMD 2F: δ-sarcoglycan and others
Facioscapulohumeral muscular dystrophy (FSHD) or Landouzy-Dejerine muscular dystrophy	Not known
Myotonic dystrophy (MD) or Steinert's disease	DMPK (DM1) and ZNF9 (DM2)
Oculopharyngeal muscular dystrophy (OPMD)	Poly(A)-binding protein nuclear 1 (PABPN1)
Distal muscular dystrophy (DD)	Different genes (dysferlin, titin, desmin, and others)
Congenital muscular dystrophy (CMD)	Different genes (laminin α2-merosin, fukutin, type VI collagen, integrin α7, and others)

colemma. The intracellular portion of β-dystroglycan directly binds the CR domain of dystrophin, which in turn binds the actin cytoskeleton. The extracellular α-dystroglycan instead binds different components of the extracellular matrix, in particular with laminin α2 (the α chain component of laminin 2, also called merosin), anchored onto the basal membrane. Another essential protein sub-complex of the DGC is that of the sarcoglycans, which bind laterally to the dystroglycans. These consist of four transmembrane proteins (α-, β-, γ-, and δ-sarcoglycan, of 50, 43, 35, and 35 kDa respectively), which associate stoichiometrically to form a heterotetramer, the SG sub-complex. Other proteins taking part in DGC formation are dystrobrevin and the syntrophins, a complex of five proteins that bind the dystrophin CT domain, and sarcospan, a transmembrane protein. Finally, a more

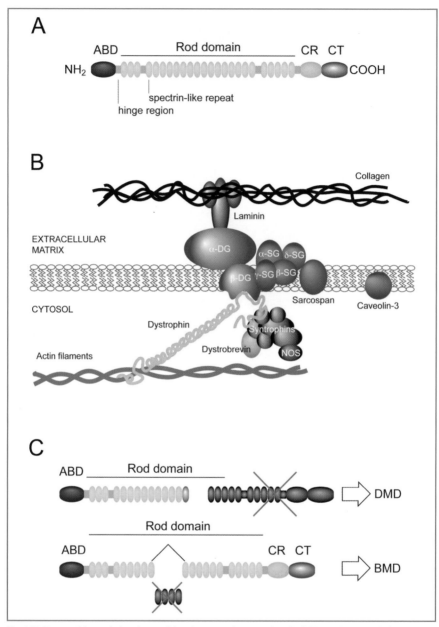

Fig. 4.11 Dystrophin and the dystrophin-glycoprotein complex (DGC). **A** Structural domains of dystrophin. *ABD*, actin-binding domain; *CR*, cysteine-rich domain; *CT*, C-terminal domain. **B** Schematic representation of the major proteins forming the DGC, connecting the internal cytoskeleton to the extracellular matrix (see text for description). **C** Dystrophin mutations. Mutations leading to premature protein truncation usually leads to severe Duchenne muscular dystrophy (DMD). Mutations causing deletions in the central rod domain but leaving the N- and C-terminal regions intact generate the milder phenotype of Becker muscular dystrophy (BMD)

relaxed association is found with caveolin-3, a muscle-specific isoform of caveolin, the main constituent of caveolae (plasma membrane microdomains involved in a specific route of endocytosis, see section on 'Endocytosis').

4.4.2
Duchenne and Becker Muscular Dystrophies

Duchenne muscular dystrophy (DMD) is the most frequent of the MDs, with an incidence of about 1:3500 males. The disease, which is inherited as a recessive X-linked trait, is caused by a defect in the dystrophin gene, located on the X-chromosome. Defects of the same gene are also responsible for Becker muscular dystrophy (BMD), which is less frequent (about 1:20,000) and has a milder clinical course.

The dystrophin gene is the longest gene in the human genome, having a length of 2.4 Mbp, with 79 exons and 8 tissue-specific promoters. The gene requires 16 h to be entirely transcribed and, after splicing, its mRNA has 14 kb, of which 11 kb correspond to the coding portion. The dystrophin protein has a mass of 427 kDa. The gene is mainly expressed in skeletal muscle, heart, and cortical neurons.

Several mutations of the dystrophin gene have been identified; about 2/3 of these are deletions concerning two major areas of the gene, one corresponding to the first 20 exons and the other one centered around exons 45–53. The remaining 1/3 of the patients carry point mutations causing the introduction of Stop codons, frame shifts, or modification of splicing signals, or are located in the promoter region. In patients with DMD, dystrophin is quantitatively very reduced or the CT domain is absent; in BMD patients, the levels of the protein are only slightly reduced and the ABD (N-terminus) and CT (C-terminus) domains are usually intact, while the protein contains deletions in its central rod domain (Figure 4.11C).

DMD is a disease with a progressive and devastating course. At birth, affected males are apparently normal, and the first symptoms develop between 3 and 5 years of age as a mild muscular weakness, evidenced by difficulty in walking up stairs and sitting upright, or by frequent stumbling. The musculature progressively weakens, with an inexorable worsening of symptoms. By the age of 10 the affected children are usually in a wheelchair and most die before age 20. There are no specific therapies available for the disease. Finally, it is important to observe that, besides skeletal muscle, patients with DMD and BMD also show a more or less pronounced involvement of the myocardium, which becomes progressively more important the longer patients live, by evolving into a frank dilative cardiomyopathy. Any gene therapy protocol aimed at curing the disease, therefore, will need to take into account the necessity to also correct the dystrophin defect in the heart.

4.4.3
Gene Therapy of Duchenne and Becker Muscular Dystrophies

Mutations of dystrophin or the DGC proteins increase fragility of the plasma membrane and determine progressive loss of muscle fibers. These are initially substituted by satellite cells, which however carry the same genetic defect. The progressive exhaustion of the muscle regeneration potential is responsible for muscle degeneration over time, and for the substitution of muscle fibers with fibro-adipose tissue, which is progressively invalidating and eventually fatal to the patient due to respiratory failure.

Gene therapy of muscle dystrophies poses important conceptual and technological problems. Skeletal muscle constitutes about 40% of body mass, and thus requires the development of very efficacious and diffuse gene transfer methodologies. Additionally, skeletal muscle consists of multinucleated fibers, incapable of division, which are maintained by the replication and fusion of specialized stem cells, the satellite cells. It is indeed conceivable to stably transduce satellite cells *ex vivo* to exploit their regenerative potential; however, these cells are only capable of a limited number of replications, before entering senescence. These characteristics currently prevent the possibility of their utilization.

Several preclinical studies have taken advantage of two popular animal models of DMD: the *mdx* mouse, in which the dystrophin gene has a single point mutation causing a premature termination of translation, and the *xmd* dog, a golden retriever in which dystrophin carries a different point mutation causing exclusion of exon 7 of the protein. While young *mdx* mice, in contrast with DMD children, show a minimal clinical phenotype, with little or no muscle fibrosis, neonatal *xmd* dogs show a severe phenotype, similar to the human disease. Starting from the *mdx* mouse, other mouse models were developed, showing different degrees of pathology, including the *u-dko* mice, in which neither dystrophin nor utrophin (cf. below) are present, and the *m-dko* mice, in which, in addition to dystrophin, the muscle-specific MyoD transcription factor is also absent. Both these knockout mice develop a MD resembling most of the characteristics of the human disease, including involvement of the heart.

These animal models have been exploited to develop a vast series of preclinical gene transfer protocols. A first approach consisted in the direct injection of the whole dystrophin cDNA, in the form of a plasmid, into the skeletal muscle, in order to take advantage of the ability of muscle fibers to internalize naked DNA when present in the extracellular environment. The first clinical trial of muscular dystrophy gene therapy, which was closed in 2004, indeed consisted in the inoculation of a plasmid containing the entire dystrophin cDNA under the control of the CMV IE promoter into the radialis muscle of nine patients with DMD or BMD. The levels of transduction and transgene expression were too low to confer a therapeutic benefit, and the distribution of dystrophin was non-homogeneous. To increase the efficiency of transfection after direct injection of a plasmid, different strategies can be considered, including the use of amphiphilic block co-polymers, ultrasound, ultrasound plus microbubbles, or electroporation (see Chapter on 'Methods for Gene Delivery').

The gene transfer procedures based on viral vectors have so far shown much more efficacy in the animal models. The entire dystrophin cDNA (14 kDa) or even its coding portion (11 kb) are too long to be cloned in AAV, lentiviral, and first-generation adenoviral vectors. However, first-generation adenoviruses can accommodate the cDNA of utrophin, a protein encoded by a different gene located on chromosome 6, which is very similar both structurally and functionally to dystrophin. Utrophin is normally present at the mature muscle fiber neuromuscular junctions, and is also expressed on the sarcolemma of fetal muscle fibers and regenerating muscle. Utrophin is also expressed on the sarcolemma in DMD patients, however its levels are too low to compensate for the missing function of dystrophin.

The overexpression of exogenous utrophin, using first-generation adenoviral vectors, which are very effective in infecting non-replicating muscle fibers, was found to significantly improve the disease phenotype in the DMD animal models, since utrophin protein can functionally associate with the DGC complex, thus increasing its level on the sarcolemma. However, first-generation adenoviral vectors cannot find clinical use due to the induction of inflammation and, in particular, immune response, which leads to the destruction of the transduced cells within a few weeks of inoculation and prevents the possibility of inoculating the vector again. Gutless (or helper-dependent) adenoviral vectors not only are devoid of the immune problems elicited by first-generation adenoviral vectors, but also offer the possibility to clone up to 30 kb of DNA and can thus accommodate the entire dystrophin cDNA (14 kb). Unfortunately, these vectors are still difficult to produce and are commonly contaminated by clinically unacceptably high levels of helper adenoviral vectors (cf. section on 'Viral Vectors').

In terms of efficiency of gene transfer to the skeletal muscle, AAV vectors are certainly the tool of choice at the moment, since they show a marked tropism for muscle fibers and, most importantly, persist in these fibers for very prolonged periods of time, in the absence of inflammatory or immune response, or transgene silencing. Unfortunately, these vectors can accommodate DNA sequences no longer than 4.5 kb, including the promoter driving expression of the therapeutic gene. This limit can however be respected thanks to the possibility of using dystrophin variants that are significantly shorter than the wild-type protein. Some BMD patients with mild forms of the disease indeed produce shorter dystrophin proteins characterized by common deletions of the central rod and the CT domains, however possessing the other domains intact, in particular the CR domain (Figure 4.12). When the cDNAs coding for these minidystrophins (~6–7 kb) or microdystrophins (~4 kb) are transferred into the skeletal muscle of the DMD animal models, they compensate for the lack of the full-length wild type protein with acceptable efficiency. The minidystrophins can be transferred using adenoviral vectors and the minidystrophins also

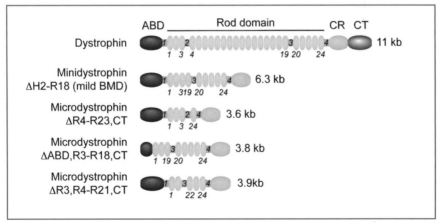

Fig. 4.12 Mini- and microdystrophins. The figure shows the structure of one minidystrophin and three microdystrophins found to compensate for lack of dystrophin in DMD animal models. The spectrin-like repeat composition of these proteins is indicated compared to the wild-type protein

4

by AAV vectors. Based on the encouraging results obtained in the animal models, a Phase I/II clinical trial is currently ongoing involving the injection of an AAV5 vector transferring a minidystrophin into the biceps muscle of a cohort of DMD patients.

As discussed above, clinical success of gene therapy in patients will strictly depend on the possibility of injecting the therapeutic cDNA systemically to reach as many muscle fibers as possible. This requirement, however, demands that the vectors, once present in the blood, are able to pass the junctions between the endothelial cells of the vessel walls, or cross the cells themselves through transcytosis, in order to make contact with the muscle fiber sarcolemma. AAV2 is inefficient in this process, unless vascular permeability is altered by locally increasing blood pressure, which can be achieved by injecting the vectors by hydrodynamic pressure (see Chapter on 'Methods for Gene Delivery'). Alternatively, in the *mdx* mouse, it was possible to transduce more than 90% of the muscles by using an AAV6 vector, injected together with recombinant vascular endothelial growth factor (VEGF), a powerful inducer of vascular permeability. Much more effectively, it is now possible to use AAV vectors pseudotyped with the most recent AAV serotypes, namely AAV8 and AAV9, which are spontaneously capable of transducing muscle and heart with high efficiency once injected intravenously, in the absence of any permeabilizing agent. These vectors now represent the system of choice for gene transfer into the skeletal muscle and heart.

As discussed above, a few dystrophin mutations that cause premature termination of the protein or removal of the central domain but avoid frame shifts in the C-terminal half generate proteins that are only mildly affected in their function, and thus cause minor BMD symptoms. In contrast, more than 75% of DMD patients have point mutations of the protein causing a frame shift with the consequent production of a protein that is severely altered or prematurely truncated (Figure 4.11C). These patients could therefore be treated by inducing the exclusion of the intron containing the point mutation from the final mRNA, a process known as *exon skipping*. This strategy can be pursued by treating the cells with antisense oligonucleotides (ASOs) pairing and thus masking the normal splicing signals on the dystrophin pre-mRNA, thus determining exclusion of the pathologic exon from the mature mRNA. An example, taken from a recent clinical trial aimed at inducing dystrophin exon 51 skipping, is shown in Figure 4.13.

Non-modified ASOs are rapidly degraded both in the cells and in the extracellular environment; however, a series of chemically modified ASOs are available, including phosphorothioates, morpholinos, LNAs, PNAs, and ENAs (cf. section on 'Modified Oligonucleotides'). These chemical modifications on one hand provide the ASO resistance to nucleases while, on the other hand, increasing their affinity to the target nucleic acid. The efficiency of these molecules in inducing exon exclusion in cultured cells is variable; only a few of these molecules have effectively been tested *in vivo* in experimental models consisting of knock-in mice engineered to express the human dystrophin gene. Currently, two clinical trials aimed at inducing exon skipping are ongoing. The first aims to verify the effects of 2'-*O*-methyl-modified ribose molecules with a full-length phosphorothioate backbone (2OMePS), and the second phosphorodiamidate morpholino oligomers, both injected first intramuscularly and later systemically. The first results of the former trial, aimed at inducing exon 51 skipping as outlined in Figure 4.13, are already available. The trial was run by injecting the ASO in the tibialis anterior muscle, followed by muscle biopsy 28 days later. The results showed restoration of dys-

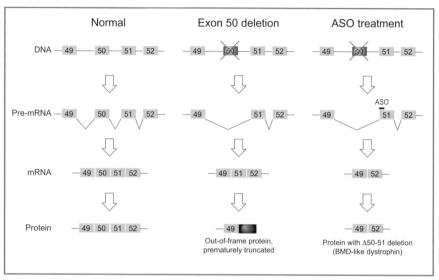

Fig. 4.13 Induction of exon skipping by antisense oligonucleotides. The *leftmost panel* shows the exon–intron composition of the normal dystrophin gene in the region encompassing exons 49–52. In a patient with a deletion of exon 50, an out-of-frame transcript is generated in which exon 49 is spliced to exon 51. As a result, a stop codon is generated in exon 51, which prematurely aborts dystrophin synthesis (*middle panel*). When cells are treated with a sequence-specific antisense oligonucleotide (ASO) binding the exon 51-internal sequences required for the correct inclusion of this exon during splicing, exon 51 is skipped. This restores the open reading frame of the mature mRNA and allows the synthesis of a still functional dystrophin protein (*right panel*)

trophin expression at the sarcolemma of the majority of patients and total levels of protein equal to 3–12% of normal, thus encouraging further extension of the study.

Finally, it should be remembered that treatment with ASO is by its own nature transient and would anyhow require repeated administration of the compound. An interesting alternative to obtain permanent exclusion of an exon carrying a pathological mutation is to express, inside the cells, small nuclear RNAs (snRNAs) containing antisense sequences using AAV vectors. A clinical trial based on the use of an AAV1 vector expressing an antisense sequence within the U7 snRNA is ongoing.

4.4.4
Gene Therapy of Limb-Girdle Muscular Dystrophy

Although less frequent and with a lower social and sanitary impact than DMD, some of the diseases caused by defects of the DGC proteins can be of high interest to gene therapy, especially since the genes encoding for these proteins are significantly shorter than dystrophin and their cDNAs can thus be delivered using AAV vectors. This is the case of limb-girdle muscular dystrophy (LGMD), a collective denomination including a group of clinically and genetically heterogeneous neuromuscular disorders characterized by weak-

4

ness of the limbs (hip and shoulder) and proximal muscles (the limb-girdle muscles) of variable severity and progressiveness. At least five autosomal dominant (LGMD1, from A to E) and at least ten autosomal recessive (LGMD2, from A to J) forms of LGMD are known (Table 4.7). Among the genes responsible for the former group are those coding for myotilin, lamin A/C, and caveolin-3; among those mutated in the recessive forms, which are responsible for over 90% of cases, are calpain-3 (the most frequent defect among all LGMD), dysferlin, and the sarcoglycans. The forms that are most severe and have early onset are usually caused by mutations of α-sarcoglycan (LGMD 2D), β-sarcoglycan (LGMD 2E), γ-sarcoglycan (LGMD 2C), and δ-sarcoglycan (LGMD 2F). The absence of each of these proteins determines the disappearance of the whole complex from the sarcolemma, thus causing both MD and cardiomyopathy.

The first available LGMD animal mode was the Bio4/6 Syrian hamster, which bears a large deletion of the δ-sarcoglycan gene. Since this animal develops both muscular and cardiac defects, it represents an excellent model to test the efficacy of gene therapy approaches having as a target both organs.

Recently, Phase I/IIa clinical trials were initiated in France and the United States aimed at evaluating safety and preliminary efficacy of the intramuscular injection in a single muscle of the forearm of AAV1 vectors carrying the γ-sarcoglycan and α-sarcoglycan cDNAs in patients with type 2C and type 2D LGMD, respectively. The first results obtained by the latter trial showed safety of the administration, and, most relevant, complete restoration of the full sarcoglycan complex in all the treated subjects, thus warranting more extensive experimentation.

4.5
Gene Therapy of Hemophilia

Hemophilia A and B, due to genetic defects of Factor VIII and Factor IX of the coagulation cascade respectively, are among the main disorders addressed by preclinical gene therapy studies, and among those for which clinical experimentation is most advanced. The interest in gene therapy for these diseases is due to multiple reasons, including: (i) the availability of animal models mimicking human disease, in both small (knockout mice) and large (hemophilia B dog) animals; (ii) the possibility to easily measure the efficacy of treatment biochemically, using standard assays that measure coagulation; (iii) the need to reach a relatively low level of correction to provide full therapeutic benefit, since 1–5% of the physiological levels of Factors VIII or IX are sufficient to confer improvement and 30% to reconstitute a normal phenotype; (iv) the possibility to produce the missing proteins in organs different from the liver, which is the natural source, since the coagulation factors are secreted in the circulation; and (v) finally, the necessity to find alternatives to the extremely high-cost substitutive therapy, which is prohibitive to the vast majority of affected patients, considering that the overall worldwide prevalence of hemophilia is 1:5000 males.

When all the above issues are collectively considered, it does not appear surprising that, since the beginning of gene therapy, hemophilia has constantly been considered a very promising and appealing candidate for correction by gene transfer.

4.5.1
Blood Coagulation

Bleeding due to loss of vessel wall integrity is a dramatic event for an organism and evolution has thus selected very efficient mechanisms to combat this condition. In mammals,
the process leading to cessation of bleeding is named hemostasis. This consists of two
essential components, known as primary and secondary hemostasis. During primary
hemostasis, occurring immediately after vessel damage, platelets form a hemostatic plug
in the damaged region; subsequently, during secondary hemostasis, the cascade activation
of a series of soluble factors circulating in plasma generates a network formed by the protein fibrin, which stabilizes and strengthens the platelet plug. Fibrin is normally found in
the circulation in the form of fibrinogen, which is not able to aggregate; the process ultimately determining the conversion of fibrinogen into fibrin is named coagulation.

The coagulation cascade follows two activation modalities, the intrinsic (by contact-
activation) and the extrinsic (by tissue factor) activation pathways. These two pathways
converge into a final common pathway, which leads to the formation of the covalently
cross-linked fibrin network. The extrinsic pathway is the major activator of blood coagulation in physiologic conditions (Figure 4.14).

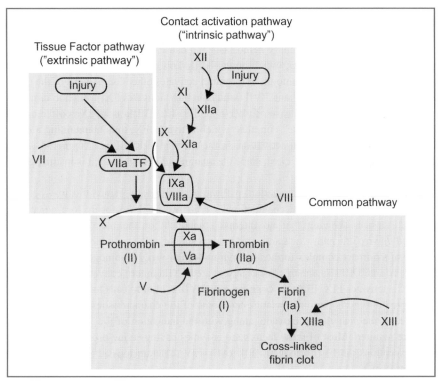

Fig. 4.14 Schematic representation of the coagulation cascade. The three major pathways (intrinsic,
extrinsic, and common pathways) are shown. See text for description

Each of the three pathways of the coagulation cascade is composed of a series of sequential biochemical reactions, which progressively amplify the process. In each reaction, a zymogen (that is, an inactive precursor of an enzyme), becomes active and thus triggers the subsequent reaction in the cascade, sometimes interacting with a glycoprotein co-factor. The extrinsic and intrinsic pathways converge at the level of activation of Factor X (FX) into Factor Xa (FXa) and then both proceed along the common pathway.

Both the extrinsic and intrinsic pathways become activated following damage to the vascular endothelium. In the extrinsic pathway, the process is initiated by a transmembrane glycoprotein, named tissue factor (TF) or tissue thromboplastin. This protein is constitutively expressed in functional form on the surface of various cells outside the endothelium, however not of endothelial or circulating cells. Endothelial damage thus exposes TF to circulating FVII, which thus becomes activated to FVIIa. The TF-FVIIa complex in turn activates FIX and FX; binding of FXa (which can also be generated by the intrinsic pathway) with FVa forms the prothrombinase complex, which activates prothrombin into thrombin. Finally, thrombin converts fibrinogen into fibrin, and FXIIIa allows fibrin precipitation and formation of covalent bonds between different fibrin molecules one on top of the other, thus leading to the formation of a clot stabilizing the platelet plug.

Activation of the intrinsic pathway thus depends on exposure to TF, the availability of which is limited, and is further inhibited by the presence of an inhibitory pathway (TF inhibitory pathway, TFPI), which blocks TF/FVIIa activity. However, the process is maintained and amplified thanks to a complex formed by FIXa (enzyme) and FVIIIa (enzyme co-factor). The exposed extracellular matrix in the damaged region binds the platelets that had induced primary hemostasis and causes the aggregation of circulating von Willebrandt (vWF) factor, which binds FVIII. This event generates a phospholipid platform onto which a small amount of thrombin activates other components of the coagulation cascade, including FV and FVII (which, in turn, activates FXI, which in turn activates FIX), and stimulates release of vWF-bound FVIII. FVIIIa is a FIXa co-factor and together they form the "X-ase" complex, which activates FX, thus maintaining a cycle leading to sustained thrombin activation. In the absence of FVIII or FIX, therefore, the coagulation process is not efficient, since the quantity of FXa generated is insufficient to maintain hemostasis.

The intrinsic (or contact-activation) pathway was originally named this way because it is possible to trigger blood coagulation *in vitro* by simple exposure to a negatively charged surface, showing that no internal component of the blood was apparently required. *In vivo*, the pathway is activated by exposure of collagen, which triggers the formation of a primary complex formed by high-molecular-weight kininogen (HMWK), prekallikrein, and FXII (Hageman's Factor). Activated FXIIa in turn converts FXI into FXIa, and FXIa activates FIX. FIXa, in complex with FVIIIa, in turn leads to FX activation.

Table 4.8 reports the nomenclature and some of the characteristics of the 13 factors involved in the coagulation cascade, along with the indication of the pathway in which they are required. Most of these factors are enzymes having serine-protease activity; the only exceptions are FI (fibrinogen), FIII (TF), FIV (calcium ions), FV/FVI, and FVIII (glycoproteins).

All the factors required for the coagulation are produced by the liver, with the exception of vWF, which is synthesized by megakaryocytes and platelets.

Table 4.8 Coagulation factors

Factor	Name	Activated form	Enzymatic characteristics	Activation pathway	Activity
FI	Fibrinogen	Fibrin	Serine-protease	Common	Forms the fibrin clot
FII	Prothrombin	Thrombin		Common	Activates FI, FV, FVII, FXII, protein C, platelets
FIII	Tissue factor (TF, tissue thromboplastin, CD142)			Extrinsic	Co-factor of FVIIa
FIV	Calcium			Common	Required for binding of coagulation factors to phospholipids
FV	Proaccelerin			Common	Co-factor of FX; together with FXa forms the prothrombinase complex
FVI	Accelerin – corresponds to activated Factor V (FVa)	FVa			
FVII	Proconvertin	Convertin	Serine-protease	Extrinsic	Activates FIX
FVIII	Antihemophilic factor A	FVIIIa		Intrinsic	Co-factor of FIXa
FIX	Antihemophilic factor B – Christmas factor	FIXa	Serine-protease	Intrinsic	Activates FX
FX	Stuart-Prower factor	FXa	Serine-protease	Common	Activates FII; forms the prothrombinase complex with FVa
FXI	Plasma thromboplastin antecedent	FXIa	Serine-protease	Intrinsic	Activates FXII, FIX, and prekallikrein
FXII	Hageman factor	FXIIa	Serine-protease	Intrinsic	Activates precallicrein and fibrinolysis
FXIII	Transglutaminase	FXIIIa	Transglutaminase	Common	Determines the formation of covalent bonds in fibrin

4.5.2
Hemophilias

The term hemophilia refers to some hereditary disorders due to loss or malfunction of some of the proteins involved in coagulation. The most common form, hemophilia A, is due to mutations of the gene coding for FVIII; hemophilia B is due to a defect of FIX. Both diseases have an X-linked inheritance and show a global prevalence of 1:5000 males; hemophilia B is 5 times less frequent than hemophilia A. Together with von Willebrandt's disease – due to deficiency of vWF, which causes a defect of primary hemostasis – hemophilia A and B include from 95% to 98% of all the hereditary coagulation disorders. The remaining diseases, which are usually transmitted as autosomal recessive traits, are rare. They include FVII deficiency (total prevalence in the general population of 1:500,000), prothrombin (FII) and FXIII deficiencies (1:2 million prevalence), FXI deficiency (once defined hemophilia C, causing a relatively less important disorder which is common among Ashkenazi Jews), and hypofibrinogenemia, caused by a defect of fibrinogen.

FVIII is a glycoprotein of 2351 amino acids that circulates in plasma in a complex with vWF, which protects it from proteolytic degradation and concentrates it to the sites of vascular damage. FIX is a serine-protease of 415 amino acids, representing the largest of the vitamin K-dependent proteins – vitamin K plays an important role by allowing carboxylation of a series of glutamic acid residues that are essential for the normal function of the protein. The plasma concentration of FIX is about 50 times higher than that of FVIII.

From a clinical perspective, hemophilia A and hemophilia B are indistinguishable. The severe forms are characterized by repeated bleedings in the joints, which are particularly painful since blood irritates the synovial membranes. The only possible therapeutic approach is substitutive therapy, consisting in the administration of the missing factor in the form of recombinant protein (available since the 1990s) or hemoconcentrates (available since the 1970s). The cost of hemoconcentrates for therapy of an adult individual with hemophilia in the United States is estimated at between 50,000 and 100,000 USD per year.

Both FVIII and FIX are encoded by two genes located on the X chromosome. As far as the FIX gene (33,500 bp) is concerned, more than 2000 different mutations are known, most of which are single nucleotide mutations, two thirds of which determine a frame shift. The FVIII gene is much longer (186,000 bp) and consists of 26 exons; inside exon 22 two additional genes are located, F8A and F8B, having unknown function; two additional copies of F8A are present outside the FVIII gene, located about 500 kb in telomeric position. In about 45% of the hemophilia A patients, the disease is due to a large inversion and translocation of exons 1–22, together with their corresponding introns, away from exons 23–26, due to a genetic rearrangement triggered by the homology between the F8A gene located inside intron 22 and one of the two F8A copies positioned outside the FVIII gene. This recombination event almost exclusively occurs in male germ cells. The vast majority of the other defects are point mutations. The characterization of the molecular defects of patients with hemophilia is particularly relevant, since patients carrying mutations leading to the production of a truncated protein or not producing a protein at all (due to large deletions, non-sense mutations or, in the case of hemophilia A, the inversion of intron 22) have a much higher probability of developing antibodies that inactivate the factor administered for replacement therapy compared to patients having missense mutations or small deletions.

4.5.3
Gene Therapy of Hemophilia A and B

The length of the FVIII cDNA (>8 kb) poses cloning problems into the most common conventional vectors. Some strategies were devised to possibly circumvent this problem, including deletion of the non-essential domain B of the protein or simultaneous use of two vectors to deliver different parts of the gene, one containing a 5' donor splice site and the other a 3' acceptor splice site, in order to reconstitute mRNA continuity by a trans-splicing mechanism. In contrast to FVIII, the FIX cDNA has a length of 1.4 kb and can thus be accommodated into virtually all the currently available vectors, also allowing insertion of genetic elements driving tissue-specific or otherwise regulated expression.

The first clinical trial for hemophilia A was conducted at the end of the 1990s by transduction of the liver with a Mo-MLV vector carrying a FVIII cDNA missing the B domain. The results of this trial were largely unsatisfactory, still underlining the strict requirement of gammaretroviral vectors for replicating cells. Indeed, liver cells can be induced to replicate after partial hepatectomy, which activates a process of liver regeneration, or are still replicating in the newborn. Gene therapy under these conditions is however still limited to preclinical experimentation.

In contrast to gammaretroviruses, high-efficiency liver transduction and even over-physiological FVIII and FIX expression can be achieved by using first-generation adenoviral vectors in both knockout mice and hemophilic dogs. However, as already discussed, the use of these vectors is accompanied by a strong inflammatory response and immune activation causes a rapid drop in therapeutic factor production a couple of weeks after injection. In particular, in the hemophilia B dog injected with an adenoviral vector expressing FIX, factor production was found to be above normal levels immediately after inoculation, then became <1% of normal in only three weeks and 0.1% at a few months. Similar results were also observed by the inoculation of an adenovirus-FIX in the liver of non-human primates.

The gutless adenoviral vector generation allows delivery of the whole FVIII cDNA. Experiments of liver transduction in the mouse using these vectors have effectively demonstrated transgene expression to persist in the absence of signs of liver toxicity. However, translation of these results to the clinics does not appear devoid of risk. The first patient who was recently enrolled in a trial entailing injection of a gutless vector to the liver at low dosages showed signs of inflammation, myalgia, and fever immediately after vector injection; these events were however expected, due to the immediate reaction to the vector capsid (see section on 'Viral Vectors'). More worryingly, 7 days after inoculation, transitory signs of liver damage were observed, suggesting that activation of the immune response can still occur even using the gutless adenoviral vectors. These findings have discouraged further enrollment in this trial.

The clinical experimentation using AAV vectors is much more advanced, both because of the safety of these vectors and their ability to transduced post-mitotic cells. Due to the relatively limited cloning capacity of AAV, the clinical studies have addressed gene therapy of hemophilia B. Both in knockout mice and the hemophilic dog these vectors proved very effective in correcting the defect and showed permanent restoration of normal coagulation activity. Based on these encouraging results, a first Phase I/II trial was conducted in 1999 based on the intramuscular injection of an AAV2 vector. Patient follow-up effectively confirmed the safety of AAV vectors and revealed factor expression for at least a few years after injection. However, the levels of circulating FIX were too low (<1–2% of normal) to allow therapeutic benefit to the patients.

The results of subsequent studies entailing AAV vector transduction of the liver by direct injection or portal vein inoculation generated much more encouraging results. After these treatments, therapeutic or even higher than normal levels of FIX were detected in the circulation in mice, hemophilic dog, and non-human primates. These results prompted the organization of clinical trials entailing portal vein inoculation of an AAV2 vector in which FIX expression was under the control of a liver-specific promoter. At the highest of the doses used, one of the patients showed levels of circulating FIX higher than 10% of normal, peaking at 2 weeks after injection and persisting for at least 4 weeks. However, in contrast to what was observed in the animal models, the production of the factor progressively decreased afterwards until it became undetectable 14 weeks after treatment. Apparently, this unexpected occurrence was not due to the presence of anti-FIX antibodies, however to the development of an immune response against the AAV vector capsid proteins, by which the transduced hepatocytes were eliminated by the patient's $CD8^+$ lymphocytes. How to explain, on one hand, that such an occurrence has never been observed in the animal models and, on the other hand, that it occurred very late after injection? A possible interpretation is that AAV2 is a common infectious agent for humans and not for other animal species, and thus transduction might have reactivated preexisting immune recognition of the surface protein of the virus. The delayed kinetics by which the immune response developed could be due to the prolonged persistence of the AAV capsid proteins in the transduced cells (the AAV gene coding for the capsid proteins is not present in the vectors and thus *de novo* synthesis of these proteins cannot occur). Should this be the correct interpretation of the observed events, the problem might be solved by inducing a transient immunosuppression in the patients receiving the vectors, lasting for a period of time sufficient for the internalized capsid proteins to be completely degraded, or by using the novel AAV serotypes to which man is not naturally exposed. In particular, AAV8 seems to show hepatic tropism, which is 10–100 times higher than AAV2, at least in the mouse, using internalization and intracellular processing pathways different from AAV2.

Finally, a single gene therapy clinical study for hemophilia A needs to be reported. This was conducted by the *ex vivo* transfection of fibroblasts derived from the derma of 9 patients using a plasmid coding for FVIII under the control of the fibronectin promoter, followed by the selection of the transfected cells and their implantation in the omentum. The therapeutic efficacy in this trial was modest (0.5–4% of FVIII activity compared to normal) and transitory. However, this trial is an important reminder that hemophilia gene therapy might be performed by *ex vivo* gene transfer into cell types different from the hepatocytes, followed by the selection of clones producing high levels of the protein. This approach could be of particular interest to overcome the problem of the large size of the FVIII cDNA.

4.6
Gene Therapy of Cancer

Cancer is the second leading cause of death after heart disease; yearly, over 12 million new cancer cases and 7 million cancer deaths are estimated worldwide. Despite the tremendous progress made over the last decades, the application of current treatment techniques (surgery, radiation therapy, chemotherapy, and biological therapy) results in the cure of nearly two out of three patients diagnosed with cancer. Thus, the need for innovative therapeutic strategies is burning.

In a very simplified view, cancer arises as the consequence of, on one hand, the accumulation of genetic modifications leading to uncontrolled cell replication and, on the other hand, the incapacity of the immune system to counteract this occurrence. Cancer gene therapy can thus follow two alternative approaches, namely either to target the cancer cells themselves or to improve the efficacy of the immune system in recognizing and destroying them. The strategies followed by the several gene therapy cancer trials so far conducted – which, as reported in the previous Chapters, represent the vast majority of the gene therapy clinical studies – are shown in Table 4.9, divided according to the different strategies followed.

The approaches targeting cancer cells can essentially have one of three major objectives: (i) to inhibit tumor cell proliferation, by inducing restoration of cell cycle control or blocking proteins essential for replication; (ii) to induce cancer-cell specific cytotoxicity, by introducing suicide genes into cancer cells; and (iii) to exploit the property of virus mutants to selectively replicate and lyse cancer cells.

Alternatively, cancer gene therapy can stimulate tumor cell destruction by the immune system by: (i) increasing antigenic stimulation of cancer cells (anti-cancer vaccination); (ii) increasing the cytotoxic response against cancer cells; or (iii) redirecting the immune system against cancer cells through the genetic modification of $CD8^+$ T lymphocytes. There is a conceptually important difference between the approaches directly targeting cancer cells and those having as a target the immune system, since the former are more demanding in requiring treatment of all tumor cells from both primary tumor and metastases while the latter rely on the efficiency of the immune system to recognize and destroy all cancer cells.

A final application of gene therapy in the cancer field is to exploit gene transfer to increase the therapeutic index of chemotherapy, by transferring, into the HSCs, genes conferring resistance of these cells to high-dose chemotherapy, such as *mdr-1*. This application has already been discussed in the section on 'Gene Therapy of Hematopoietic Stem Cells'.

4.6.1
Inhibition of Cancer Cell Proliferation or Survival

The objective of one of the strategies used for cancer gene therapy is to transfer, into the cancer cells, genes or non-coding nucleic acids able to inhibit their proliferation or to induce apoptosis. From the molecular point of view, tumors are characterized by an alteration of the molecular mechanisms physiologically controlling cell proliferation. The alteration is essentially due to several mutations affecting tumor suppressor genes, such as p53, Rb, and BRCA1. Thus, transfer of the normal alleles of these genes into the cancer cells could reconstitute normal cell cycle control. Other possible therapeutic cDNAs are those coding for the mutated and non-functional forms of proteins transducing proliferative signals, such as c-Jun or H-Ras.

Alternatively, different clinical studies have exploited various therapeutic genes (mod-

4

Table 4.9 Strategies for gene therapy of cancer

Target cell	Strategy	Goal	Therapeutic gene
Cancer cells	Inhibition of cancer cell proliferation	Restoration of cell cycle control	Tumor suppressors (p53, Rb, BRCA1)
			Antisense oligonucleotides, ribozymes, siRNAs or intracellular antibodies against oncogenes, cdc2, cyclins, PCNA, tyrosine kinase receptors, signal transducers, etc.
	Transfer of suicide genes into cancer cells	Specific induction of cytotoxicity in the suicide gene-expressing cells	Gene activating a cytotoxic pro-drug, for example HSV-TK (cf. Table 4.10)
	Oncolytic viruses	Selective lysis of cancer cells by viral replication	
Cells of the immune system	Immunotherapy	Increase of antigenic stimulation by cancer cells (active immunization, cancer vaccination)	Tumor-specific antigens (TSAs and TAAs; cf. Table 4.11)
			Genes coding for cytokines increasing antigen stimulation (IL-2, IL-12, IFN-γ, GM-CSF)
		Increase of the cytotoxic T-cell response against cancer cells	Genes coding for immunoregulatory cytokines (IL-2, IL-12, IL-7, GM-CSF, IFN-γ, IL-6, TNF-α)
			Genes coding for co-stimulatory proteins (B7, ICAM-1, LFA-3)
			Genes coding for immunogenic proteins (MHC I and II alloantigens)
		Genetic modification of effector T cells to redirect them towards cancer cells (adoptive immunotherapy)	TCR genes
Hematopoietic stem cells (HSCs)	Increase of the therapeutic index of cancer chemotherapy	Transfer of genes preventing toxicity of chemotherapy into HSCs	Mdr-1

ified antisense oligonucleotides, ribozymes, siRNAs, intracellular antibodies) able to inhibit expression or function of cellular proteins essential for cell proliferation or survival. These approaches have had multiple targets, including activated oncogenes (*c-myc*, *c-fos*, *c-myb*), cell cycle kinases (*cdc2*) or cyclins (*cyclin A*, *cyclin E*), DNA polymerases and their accessory factors (such as, for example, *proliferating cell nuclear antigen*, PCNA, a processivity factor for DNA polymerase δ) or, finally, anti-apoptotic genes, such as *bcl-2* or *survivin*, with the objective of turning off expression or function of these genes.

Antisense oligonucleotides, ribozymes, or siRNAs can be either administered systemically as short nucleic acids or directly expressed inside the cancer cells using retroviral or, more efficiently, adenoviral vectors. As already mentioned above, the success of all the trials having the inhibition of cell proliferation or survival as their objective requires that all the cancer cells, including those from the metastases, are reached by the therapeutic gene. This ambitious objective appears more realistic when administration of small nucleic acids occurs systemically rather than pursuing their expression using viral vectors. On the other hand, the latter options has the advantage of a more persistent expression over time, without requiring continuous administration.

The strategies used by the major cancer gene therapy clinical trials using antisense oligonucleotides are reported by Table 2.2 in the section on 'Non-Coding Nucleic Acids'.

4.6.2
Gene Therapy of Cancer Using Suicide Genes

Another approach for cancer gene therapy is based on the delivery, into the cancer cells, of genes inducing cell death in a pharmacologically controllable manner. The patient is treated with an otherwise inactive drug (prodrug) which becomes exclusively activated in the cells in which the therapeutic gene is expressed, usually thanks to the enzymatic activity of the protein encoded by this gene. This approach is also known as *prodrug gene therapy*, i.e., therapy using prodrug-activating genes. A list of the developed prodrugs and of the respective suicide genes activating them is reported in Table 4.10.

A commonly used strategy is the intratumoral injection of viral vectors coding for the thymidine kinase (TK) gene of the herpes simplex virus type 1 (HSV-1). Intratumoral expression of HSV-TK is harmless by itself, however it becomes toxic when the patient is treated with drugs such as gancyclovir or acyclovir, originally developed for the treatment of HSV infection. These are nucleoside analogs that cannot be phosphorylated by human TK and are thus not incorporated into DNA under normal conditions. HSV-TK, however, is capable of phosphorylating these prodrugs, which thus become precursors for DNA polymerization during the S-phase of the cell cycle. However, once incorporated into DNA, these modified nucleotides block further DNA elongation, and cells eventually die by apoptosis (Figure 4.15).

This approach has additional interest, since the cells transduced with HSV-TK, once treated with a prodrug, also transfer the TK-activated drug and other secondary toxic compounds to neighboring, non-transduced cells (the so called *bystander effect*). The bystander effect determines diffusion of the killing effect beyond the transduced cells and is thus useful to increase the therapeutic efficacy of gene transfer. The bystander effect

Table 4.10 Prodrug gene therapy

Suicide gene	Prodrug	Mechanism of action
Herpes simplex type 1 thymidine kinase (HSV-TK)	Gancyclovir (GCV), acyclovir (ACV), valacyclovir	Inhibition of DNA synthesis
E. coli cytosine deaminase (CD)	5-Fluorocytosine (5-FC)	Inhibition of DNA and RNA synthesis
CYP2B and CYP3A enzymes of human cytochrome P450	Cyclophosphamide and isophosphamide	Alkylating agents
E. coli xanthine-guanine phosphoribosyltransferase (XGPRT)	6-Tioxantine (6-TX)	Inhibition of DNA synthesis
E. coli purine-deoxynucleoside phosphorylase (PNP; gene *deoD*)	6-Methylpurine-2'-deoxyribonucleoside (MeP)	Inhibition of DNA synthesis
E. coli nitroreductase	5-Aziridine-1-il-2,4-dinitrobenzamide (CB1954)	Alkylating agent

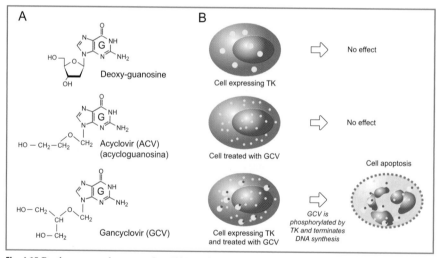

Fig. 4.15 Prodrug gene therapy using HSV-1 thymidine kinase. **A** Chemical structure of deoxy-guanosine and its structural analogue drugs acyclovir (ACV) and gancyclovir (GCV). **B** Suicide gene therapy. In a cell expressing thymidine kinase (TK), the pro-drug is activated by the enzyme and blocks DNA synthesis, followed by cell apoptosis

requires direct cell-to-cell contact, since local transfer of the toxic compounds occurs through the gap junctions that form between the cells in various tissues.

The main gene therapy clinical trial based on HSV-TK was a large Phase III study for brain tumors (glioblastoma, astrocytoma), which have very poor prognosis despite any conventional therapy. This trial involved over 40 clinical centers in North America and

Europe and enrolled over 200 patients. The experimental approach was based on the intracerebral injection, at the end of neurosurgery to remove the tumors, of packaging cells producing gammaretroviral vectors expressing HSV-TK. The rationale of this approach was twofold: first, the vector should transduce only replicating cells (i.e., cancer cells and not neurons); second, only replicating cells should be sensitive to prodrug activation by the suicide gene. In addition, packaging cells, which are tolerated despite being of murine origin since the brain is a site of immune privilege, should continue to produce retroviral vectors *in situ*.

Despite the outstanding preclinical results obtained in animals following a similar approach, however, the clinical results were discouraging in terms of extending the life of the treated patients. Post mortem analysis revealed that these disappointing results were probably due the relatively low efficiency of suicide gene transduction in a sufficiently large number of cancer cells and to the limited extension of the bystander effect, which was significantly less pronounced than in the animal experiments. The former problem could be addressed by using vectors different from retroviruses, in particular adenovirus-es; extension of the bystander effect could instead be stimulated by different means, including the co-transduction of cellular proteins involved in gap junction formation (for example, connexin-43 or -26) or modification of HSV-TK by inclusion of a peptide from the Tat protein of HIV-1, which confers, to heterologous proteins containing it, the prop-erty of being secreted by the expressing cells and taken up by neighboring cells (see Chapter on 'Methods for Gene Delivery').

More encouraging results were indeed obtained using adenoviral vectors (both repli-cation-defective and oncolytic) in Phase I and II clinical trials entailing transduction of HSV-TK either alone or in combination with cytosine deaminase (a bacterial enzyme acti-vating the 5-fluorocytosine prodrug; cf. below) in patients with prostate carcinoma. Experimental evidence indicates that adenoviral vectors are capable of transducing their genes in a much larger number of cells *in vivo*; in addition, the inflammatory and immune responses usually accompanying the use of these vectors might represent an efficient adju-vant in anticancer therapy. The use of an adenoviral vector expressing HSV-TK has now been extended to the treatment of glioblastoma in a large Phase III clinical trial in Europe.

Another interesting application of HSV-TK in the gene therapy field is as a tool to phar-macologically control the expansion of the cells injected into the patients during cell ther-apy, for example for adoptive immunotherapy using allogeneic cells (cf. below). A specif-ic example of this application is the control of CTL proliferation in leukemia or lymphoma patients treated with allogeneic BMT. In these patients, the donor T lymphocytes contained in the transplant have a prominent therapeutic role, since they recognize and destroy the residual cancer cells (GvL). However, the same lymphocytes also react against the normal cells of the recipient, and are thus responsible for GvHD. The use of these cells in trans-plantation, therefore, needs to be accurately titrated, in particular to prevent their uncon-trolled proliferation once reinfused into patients. For this purpose, encouraging results were obtained by a few clinical trials entailing *ex vivo* transduction of the donor T-lymphocytes using gammaretroviral vectors before their reinfusion into the recipients. These trials revealed that the transduced cells survived for several months and exerted efficient anti-tumoral activity. In the patients who developed GvHD, treatment with gancyclovir effi-ciently controlled the severity of the disease (see also below in the section on

'Immunotherapy of Cancer'). The use of HSV-TK gene transfer in allogenic transplantation is now being evaluated in a large Phase III clinical trial.

Other prodrug/activating enzyme pairs are shown in Table 4.10. Among these, *E. coli* cytosine deaminase (CD) has particular relevance. This enzyme converts 5-fluorocytosine (5-FC), which is non-toxic by itself, into 5-fluorouracil (5-FU), which is used as an antiblastic drug since it inhibits the enzyme thymidylate synthase, converting deoxyuridine monophosphate (dUMP) into thymidine monophosphate (dTMP), which is subsequently phosphorylated to thymidine triphosphate (dTTP) for use in DNA replication and repair. In this manner, synthesis of DNA is inhibited; in addition, the 5-fluoro-dUTP that is formed also becomes incorporated into RNA, with the overall consequence of determining death of both resting and proliferating cells. Furthermore, 5-FU sensitizes cells to ionizing radiation and can thus be used synergistically with this modality of anticancer therapy.

4.6.3
Oncolytic Viruses

Another innovative approach for cancer therapy is based on oncolytic viruses. The idea to cure human cancer by infecting patients with viruses that selectively replicate in, and thus lyse, cancer cells while leaving normal cells unaffected stems from some anecdotal observations, the first dating back more than 100 years, that patients with advanced cancers surprisingly recovered after infection with a virus that was no better characterized. Despite these remarkable reports, the search for natural viruses with truly oncolytic properties (that is, viruses that selectively destroy cancer cells) has been disappointing so far. However, it should be considered that, for their own intracellular replication, most viruses need to neutralize the same cellular proteins that mutate or are anyhow inactivated during the process of neoplastic transformation, including those encoded by the genes involved in cell cycle checkpoints or regulating apoptosis. Based on these observations, the use of an adenovirus mutant able to selectively replicate in cells in which the tumor suppressor p53 was inactivated was first proposed a few years ago.

Wild-type adenovirus contains two genes encoded by its E1 region, E1A and E1B; as detailed in the section on 'Viral Vectors', both genes are deleted in first-generation adenoviral vectors, which are thus incapable of replicating autonomously. The E1A protein interacts with a series of cellular proteins that regulate the cell cycle; among these, it binds and inactivates the retinoblastoma (Rb) protein, thus allowing S-phase entry and consequent viral replication. On the other hand, however, E1A also activates p53, which would block cell proliferation and induce apoptosis, thus impairing viral replication. To counteract this effect, during infection with the wild-type virus, the E1B protein binds and inactivates p53, thus allowing viral replication to proceed and preventing cell apoptosis. Due to this function, E1B plays an essential role during adenoviral replication: cell infection with an E1B mutant virus is rapidly extinguished, since the infected cells undergo apoptosis before significant production of viral progeny. However, if an E1B-mutant adenovirus infects cells in which p53 is mutated, replication is as efficient as with the wild-type virus and culminates with cell lysis and release of new viral particles that infect neighboring cells (Figure 4.16). Considering that mutation of p53 is one of the most frequently observed mutations in human cancers, a virus having specific tropism for p53-minus cells allows selective destruction of these cells *in vivo*, while sparing non-transformed cells.

An E1B-mutated adenovirus, called dl1520 or better known with the commercial name of ONYX-015, was originally used in Phase I and II clinical trials based on the intra-tumoral injection of the virus in patients with recurrent head and neck cancer. Following the original trials, over 15 additional clinical studies, involving over 100 patients, are currently ongoing in the United States. A variant of the virus, called H101, is also used in China. The results of the first trials with ONYX-015 revealed that injection of this virus was usually well tolerated even when doses as high as 1×10^{13} viral particles were injected, with side effects limited to moderate fever and malaise. However, the effect of virus injection alone was limited and probably independent of the presence of p53 mutations. Indeed, it now appears that mutant adenovirus injection exerts most of its beneficial effects through the activation of the immune system, the inhibition of tumoral angiogenesis, or the sensitization of cancer cells to simultaneously administered antiblastic drugs. As a matter of fact, injection of the mutant virus currently finds its best application in conjunction with conventional chemotherapy protocols.

Other clinical trials are currently ongoing or planned using a new generation of oncolytic adenoviruses, in which E1B is deleted and E1A is under the transcriptional control of a tissue-specific promoter (in particular, the promoter of the prostate antigen PSA), in order to achieve selective viral replication in specific cell types, or in which E1A carries mutations improving viral replication in cancer cells.

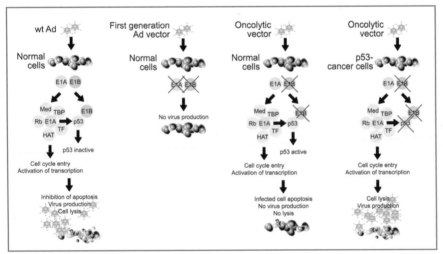

Fig. 4.16 Oncolytic adenoviral vectors. Wild-type adenovirus expresses two genes, E1A and E1B. The former is required to induce cell cycle entry and activation of transcription through its interaction with a number of cellular proteins (in *yellow*); the latter inhibits p53 activation and the consequent induction of apoptosis. Both genes are deleted in first-generation adenoviral vectors, which are thus incapable of replication. In the oncolytic adenoviral vectors, E1A is wild type, thus allowing viral replication, however E1B is mutated. As a consequence, viral replication only occurs in cancer cells in which p53 is inactive, since in normal cells viral replication rapidly induces cell apoptosis

Less advanced, however very encouraging, are a few clinical trials using other defective, oncolytic viruses. These include studies with HSV-1 mutants in gliomas and other solid cancers such as melanoma and carcinomas, or vacciniavirus mutants in melanoma. In particular, the use of replication-competent, attenuated HSV-1 variants carrying mutations in the gene coding for the neurovirulence factor ICP34.5 is generating very interesting, albeit preliminary results. The same viruses have shown excellent safety and efficacy profiles in preclinical animal experimentation; the main features of these viruses are discussed in the section on 'Viral Vectors').

Finally, it should be remembered that an additional possibility offered by the oncolytic virus approach is to include, within the viral genome, genes coding for immunomodulatory proteins, in particular the same cytokines that are used by the more traditional, non-replicating vectors (see below), with the purpose of simultaneously stimulating activation of the immune system against cancer cells.

4.6.4
Immunotherapy of Cancer

The vast topic of cancer immunotherapy essentially includes three different approaches, based on: (i) the administration of molecules (usually antibodies) recognizing cancer cells (*passive immunotherapy*); (ii) the direct stimulation of the patient's immune system to recognize specific antigens expressed by cancer cells (*active immunization* or *anti-cancer vaccination*); and (iii) the infusion of immune cells activated *ex vivo* against cancer cells (*adoptive immunotherapy*). Of note, the same concepts of passive immunotherapy, active immunization, and adoptive transfer of activated immune cells can be applied to the immunotherapy of viral infections.

The major strategies used in anticancer active and adoptive immunotherapy are schematically shown in Figure 4.17.

Passive immunotherapy aims to treat cancer patients with molecules recognizing specific targets expressed by cancer cells, followed by inhibition of their function or destruction of the cells expressing them. The most powerful application of passive immunotherapy entails the use of monoclonal antibodies against specific cancer antigens. These include, among others, cetuximab (a monoclonal antibody against the epidermal growth factor receptor (EGFR) – expressed by colorectal cancer cells; trastuzumab (against the HER2/neu receptor encoded by the c-erbB2 gene in different breast cancers); rituximab (against the CD20 antigen expressed by low-grade B-cell non-Hodgkin lymphomas); and bevacizumab (against VEGF, an essential cytokine for tumor angiogenesis in colorectal cancer and other tumors). Collectively, over 15 monoclonal antibodies with anti-cancer activity are now used clinically, while several hundred are in advanced experimentation.

In principle, gene therapy could be used to express the genes coding for these antibodies in the form of single-chain antibodies (see section on 'Antibodies and Intracellular Antibodies'), with the possible advantage of a sustained and continuous production of the therapeutic molecule, despite an obvious loss in manageability of treatment. No application of this kind, however, has so far reached clinical experimentation.

Active immunization (or *cancer vaccination*) instead consists in the direct stimulation

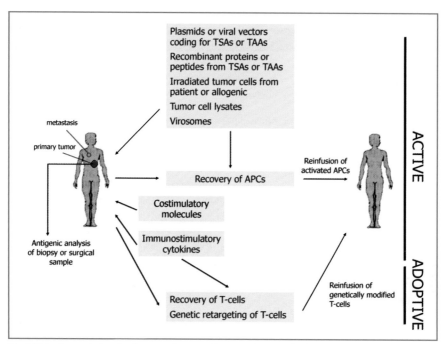

Fig. 4.17 Active and adoptive immunotherapy of cancer. See text for description. APC, antigen-presenting cells

of the patient's immune system and, in particular, of its CTLs to recognize and destroy the cancer cells. Gene therapy can play a fundamental role in reaching these targets in at least two ways, namely by transferring the tumor-specific antigens against which an immune response is to be elicited and by enhancing the immune response itself (Table 4.9).

4.6.4.1
Tumor-Specific Antigens

Most of the gene therapy clinical trials so far have aimed at cancer vaccination exploiting the antigenic differences between normal and cancer cells. In particular, the antigens recognized by T cells can be divided into two main categories: the *tumor-specific antigens* (TSAs), which are specifically expressed by cancer cells and not by normal cells, and the *tumor-associated antigens* (TAAs), which are normal proteins however expressed at abnormally high levels in cancer cells, while produced at low levels or absent in normal cells (Table 4.11).

TSAs include the variable portions of antibodies and TCRs expressed by the tumor cells in B-cell or T-cell lymphomas respectively. In both B and T cells, rearrangement of the immunoglobulin or TCR genes generates proteins in which the antigen recognition site (composed of the V_H and V_L regions in the case of antibodies and the $V\alpha$ and $V\beta$ domains in the case of T lymphocytes) is a unique antigenic determinant. Since lymphomas arise

Table 4.11 Cancer cell antigens

		Antigen	Cancer
Tumor-specific antigens (TSAs)	Antigens specific to cancer cells	Antibody idiotype	B-cell lymphoma
		T-cell receptor (TCR) idiotype	T-cell lymphoma
	Mutated proteins participating in cellular transformation	Mutated p21ras protein	~10% of cancers
		p210$^{bcr-abl}$ fusion protein	Chronic myelogenous leukemia (CML)
		Mutated p53 protein	>50% of cancers
	Viral proteins expressed by cancer cells	Human papilloma virus (HPV) E6 and E7 proteins	Cervical cancer
		Epstein-Barr virus (EBV) EBNA-1 protein	Hodgkin's disease (HD), EBV-positive non-Hodgkin's lymphomas (NHL)
Tumor-associated antigens (TAAs)	Normal proteins expressed at abnormally high levels	PSA, HER2/neu, MUC-1	Various carcinomas
	Oncofetal antigens	CEA, AFP	Various carcinomas
	Differentiation antigens	Melan-A/MART-1, tyrosinase, gp100	>50% of melanomas
	Cancer-testis antigens (CTA)	Proteins of the MAGE, BAGE, GAGE, LAGE, PRAME, NY1-ESO-1, etc. families	Melanoma, bladder carcinoma, non-small-cell lung carcinoma, and other cancers

as a consequence of the uncontrolled proliferation of a single cell clone, in both cases the antibody or TCR idiotypes exclusively characterize the tumor cells. The genetic regions coding for the idiotypic portions can be cloned, starting from the lymphoma cells' DNA, using PCR and then expressed either as recombinant proteins or in the context of plasmid or viral vectors (cf. section on 'Antibodies and Intracellular Antibodies').

Other TSAs consist in cellular proteins bearing mutations that create novel antigenic determinants. Among these are some tumor suppressors, such as p53, which mutate along the process of neoplastic transformation, or the fusion proteins generated by tumor-specific translocations. A paradigmatic example of the latter category is the fusion protein between the cellular *bcr* and *abl* genes, arising as a consequence of the translocation between chromosomes 9 and 22 occurring in chronic myelogenous leukemia (CML). Finally, a few cancers are associated with viral infections, and some of the viral genes are

continuously present and expressed in the cancer cells. This is the case of human papillo-
ma virus (HPV), the causative agent of virtually all cervical cancers, in which the viral E6
and E7 proteins are expressed in the cancer cells, or of Epstein-Barr virus (EBV)-positive
lymphomas, which express the viral EBNA-1 antigen.

A second category of tumor antigens are the TAAs, consisting of normal proteins
expressed in an inappropriate manner in tumors. In some cases, these are non-mutated proteins that are expressed at low levels in normal conditions but at abnormally high levels in
tumors. These include the prostate-specific antigen (PSA) in prostate carcinoma, or the MUC-
1 protein in breast cancer and other carcinomas. Other proteins are only expressed during
embryonic development, before the immune system becomes immunocompetent; once these
antigens (known as oncofetal antigens) become expressed in cancer cells, they might be
exploited to elicit an immunological response. This is the case of α-fetoprotein (AFP) and the
carcino-embryonic antigen (CEA). AFP is expressed at very high levels (milligrams per milliliter) in fetal serum and at much lower levels (nanograms per milliliter) in normal adult
serum. Very high levels, however, are found in patients with hepatic carcinoma. CEA is a
membrane glycoprotein expressed by gastro-intestinal and hepatic cells in the fetus between
the second and sixth month of pregnancy. The protein becomes expressed again in about 90%
of patients with advanced colorectal carcinoma and in about 50% of patients in the initial
stages of colorectal cancer and in other carcinomas. The CEA serum levels are routinely
checked to monitor disease evolution after surgical removal of colorectal cancer.

An additional TAA class consists of the so-called "differentiation antigens", namely proteins which, under normal conditions, are only expressed during differentiation of some cell
types (in particular, of melanocytes) but are switched off in differentiated cells. Some of
these proteins become expressed again in the transformed cells derived from the same tissues and can thus be used both as markers and as antigens for immunological stimulation.
Examples of these proteins are tyrosinase, gp100, and Melan-A/MART-1 in melanomas.

Finally, another class of TAAs are the so-called *cancer-testis antigens* (CTAs), proteins that in normal conditions are exclusively expressed in the germ cells of the testis and
in no other cell type. During neoplastic transformation, expression of several of the genes
coding for these proteins is aberrantly activated. At least 44 different CTA gene families
are now known; most of these genes are located on the X chromosome in clusters containing several homologous genes (for example, the MAGE-A, GAGE, and SSX gene families). The function of most CTAs is not known; some of them might be involved in the
control of chromosome structure during spermatogenesis. Bladder carcinoma and non-
small-cell lung carcinoma express high CTA levels; expression is less pronounced in
breast and prostate cancer and lower in kidney and colorectal carcinoma.

4.6.4.2
Antigen Presentation

Immunization against TSAs or TAAs can be obtained by different modalities, the choice
of which essentially determines the type of immune response that it elicited. It is thus
important to briefly summarize the molecular mechanisms by which antigens are presented to the immune system (Figure 4.18).

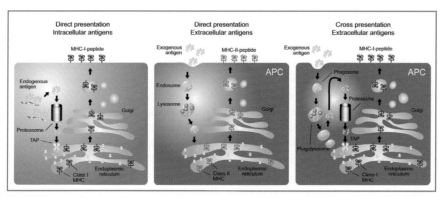

Fig. 4.18 Presentation of antigens to the immune system. Endogenous antigens are presented in the context of the MHC Class I proteins by all cell types (*left panel*). Exogenous antigens are internalized and then expressed within MHC Class II molecules (*middle panel*) or, thanks to the cross-presentation pathway, within MHC Class I molecules (*right panel*); professional APCs are able to present antigens according to the last two modalities.

Cell-mediated, cytotoxic immune response, sustained by CD8+ CTLs, is directed against antigens presented on the cell surface in the context of the major histocompatibility complex (MHC) Class I molecules. Intracellular antigens, usually consisting of normal or mutated cellular proteins, or viral proteins when the cell is infected by a virus, are processed by the proteasome and the digested peptides are transported by the TAP transporter into the endoplasmic reticulum, where they become complexed with MHC Class I molecules and exposed onto the cell surface. The MHC Class I:peptide complexes are recognized by CTLs (the CD8 receptor directly binds MHC Class I) and, following this interaction, lysis of the presenting cells occurs. With the exception of a few cell types such as erythrocytes, MHC Class I is expressed by all cells; as a consequence, T-cell-mediated cytotoxicity is the main immune response against transformed cells or cells infected by a virus.

In contrast, extracellular antigens need to be presented to the immune system by specialized cells, named professional APCs, consisting of dendritic cells (DCs), macrophages, and B lymphocytes. These cells phagocytose the antigen (in the form of extracellular proteins, bacteria, parasites) and digest it into peptides through lysosomal degradation. The antigenic peptide-containing vesicles then fuse with MHC Class II-containing vesicles derived from the endoplasmic reticulum; the peptides assemble with MHC Class II molecules and the complexes are transported towards the cell surface by the intracellular vesicle transport system. Within the lymph node, the MHC Class II:peptide complexes are then recognized by CD4+ helper T lymphocytes (the CD4 receptor directly binds MHC Class II). These cells indeed play an essential co-stimulatory role for both the cytotoxic (mediated by CTLs) and humoral (mediated by B-cell-produced antibodies) responses.

Besides presenting extracellular antigens in the context of MHC Class II molecules, APCs and, in particular, DCs can also do so in the context of Class I molecules, thus generating a cytotoxic response. In fact, these cells are able to phagocytose virus-infected dead cells or transformed cells, or endocytose their proteins. Thanks to the vesicular transport mechanisms, the internalized molecules are transferred to the lysosomes, where protein degradation generates polypeptides that, in turn, reach the cell cytosol and are processed by the proteasome. With analogy to the above-described MHC Class I presentation, the TAP transporter mediates translocation of the generated peptides into the endoplasmic reticulum, where these are complexed with the MHC Class I and exposed onto the cell surface. The process by which proteins deriving from other cell types become associated with the APC MHC Class I molecules is known as *cross-presentation*. This process is essential to generate an efficient cytotoxic response against tumor antigens as well as against viral antigens, since most viruses do not directly infect the APCs. In fact, antigenic presentation within the MHC Class I on the surface of most cells is sufficient to render these cells targets for the cytotoxic activity, but insufficient to prime the immune response. Only professional APCs, indeed, express the co-stimulatory molecules required for the activation of the immune response, in particular the members of the B7 family, which interact with the T-cell CD28 receptor. The process of cross-presentation is also essential to generate a cytotoxic T response against tumor or viral antigens (vaccination) when these are administered as recombinant proteins.

4.6.4.3
Cancer Vaccination

Stimulation of the immune system to react against a tumor antigen can be obtained by various modalities, both *ex vivo* or *in vivo* (Figure 4.17). The antigen can be directly administered to the patient in the form of: (i) radiation-inactivated cancer cells, either derived from the primitive tumor or consisting of allogenic tumor cell lines; (ii) cell lysates obtained from these cells; (iii) recombinant proteins corresponding to the TAAs or TSAs; (iv) synthetic peptides; (v) viral capsids, derived from different viruses, conveying the antigen of interest (virosomes); or alternatively, by exploiting gene therapy, in the form of: (vi) naked plasmid DNA; or (vii) viral vectors expressing the DNA of interest. The last two strategies bear the collective name of *DNA vaccination* or *genetic vaccination* or *DNA immunization*.

The above-listed strategies for antigen administration can also be utilized *ex vivo* on purified APCs, usually DCs. The main sources of DCs for clinical experimentation are bone marrow or peripheral blood CD34$^+$ cells, circulating DCs, or monocytes. In particular, DCs can be obtained from the patients' monocytes by treating these cells with IL-4 and GM-CSF, with the possible addition of other factors (for example, the cytokines IL-1β, IL-6, and TNFα, and prostaglandin E2 (PGE2)). *In vitro* cultured DCs can be exposed (or "loaded" or "pulsed") with a desired antigen in the form of synthetic peptide, tumor lysates, or RNA derived from the tumor cells, and reinjected into the patients, usually intradermally or subcutaneously. From the injection site, the activated DCs migrate into the secondary lymphoid organs, in particular the lymph nodes, where expression of their co-stimulatory molecules increases, followed by activation of CD8$^+$ T cells to differentiate.

Efficient anti-tumor (or anti-viral) vaccination also requires, besides CD8$^+$ T-cell stimulation, activation of CD4$^+$ T cells, which provide an essential helper function. When protein antigens are used, activation of CD4$^+$ T cells is directly stimulated by the DCs thanks to the mechanisms of antigen presentation on MHC Class II. In the case of genetic vaccination, in which the antigens are produced intracellularly and thus expressed in the context of MHC Class I, CD4$^+$ helper activity is only generated when the antigens are released from the transduced DCs or when these cells undergo lysis. In both conditions, the antigens are phagocytosed by other DCs and presented in the context of their MHC Class II molecules, as reported above. In addition, in the case of immunization using plasmid DNA, the CD4$^+$ helper function is also directly stimulated by the non-methylated CpG sequences present in the plasmid DNA itself. These sequences directly interact with members of the Toll-like receptor (TLR) family, in particular TLR9, which activate the innate immunity and, in the APCs, the secretion of inflammatory cytokines that, in turn, promote CD4$^+$ T-cell activation.

Based on the above-described properties, vaccination using antigens in the form of peptides or recombinant proteins mainly stimulates an antibody response, since the antigens are presented in the context of MHC Class II, which activates CD4$^+$ T cells with helper function. In contrast, DNA vaccination, which is based on intracellular antigen production, also results, besides CD4$^+$ T-cell stimulation and antibody production, in the activation of CD8$^+$ CTLs. Furthermore, the antigenic proteins are expressed endogenously in their native conformations, without prior denaturation or modification as often occurs for recombinant antigens. Therefore, the elicited immune response is against an antigen identical to the natural one. Finally, in the case of DNA vaccination, antigenic stimulation is prolonged, and thus able to generate a significant immunological memory. Plasmid DNA is injected *in vivo* intramuscularly or intradermically. In both cases, the DCs internalize the plasmid and present the encoded protein to the immune system; in contrast, muscle fibers express low levels of MHC Class II molecules and lack the co-stimulatory molecules that are required for efficient antigenic stimulation.

When the genetic vaccines based on plasmid DNA are compared to those based on viral vectors (typically, adenovirus or vacciniavirus; see below), the former immediately show some intuitive advantages. Plasmid vaccines can be obtained and purified in large quantities, and manipulated and stored without particular requirements, due to the relative stability of circular, double-stranded DNA. Furthermore, the vaccines based on viral vectors can generate an unwanted, dominant immune response against viral antigens, which can diminish the desired response against the tumor antigen. Finally, the production of neutralizing antibodies against viral proteins after the first administration usually prevents re-inoculation of the same vaccine as a booster. On the other side, antigenic presentation after viral transduction lasts significantly longer than using plasmid DNA. The two modalities of antigen delivery are however not exclusive, and can be successfully alternated for primary and booster immunizations.

As a final note, it is important to remember that most of the above-reported considerations on tumor antigens and on the properties of DNA vaccination also apply to immunization against viral antigens.

4.6.4.4
Cancer Vaccine Trials Exploiting Gene Therapy

In different animal models of adoptive immunotherapy, in which cancer cells are implanted in syngeneic mice followed by the evaluation of tumor growth, prophylactic vaccination against various TAAs shows a very strong protective effect. However, in contrast to their prophylactic efficacy, TAA-based vaccines are much less efficient in eradicating an already implanted tumor. In agreement with these animal data, little success was also observed in a series of gene therapy clinical trials based, as a sole therapeutic strategy, on the transfer of plasmids encoding different TAA genes, including CEA, Melan-A/MART-1, and gp100. Notwithstanding the use of large quantities of plasmid DNA (5–10 mg) injected intramuscularly, the observed immune response was limited and the clinical impact modest.

To obtain more vigorous activation of the immune system, ameliorations in both the levels of antigen expression and its presentation can be introduced. Antigen gene transfer is more effective when the encoding plasmids are delivered by the *gene gun* technology or by electroporation (see Chapter on 'Methods for Gene Delivery'), or by using viral vectors. As far as viral gene delivery is concerned, the clinical trials so far conducted have been based on two different vectors: recombinant poxviruses, used by most trials, and adenoviruses.

The first clinical trials using poxviruses were based on replication-competent vaccinia strains, subsequently substituted by non-replicative strains in which some virulence genes had been removed (virus NYVAC) or by strains adapted to replicate in avian but not in human cells (modified vacciniavirus Ankara, MVA). Finally, more recently, to overcome the potential pathogenicity concerns about the use of human poxviruses, vaccines were developed based on avian poxviruses, such as canarypox (attenuated strain ALVAC) or fowlpox, which are incapable of replicating in mammalian cells. The most utilized tumor antigens have been MUC-1 for lung and prostate cancer, the E6 and E7 proteins for cervix cancer, the oncofetal protein 5T4 for different carcinomas, and the MAGE-1 and MAGE-3 TAAs for melanoma.

As for adenoviral vectors, the clinical trials so far conducted have revealed that these vectors are less efficient than poxviruses, probably because of the more widespread levels of anti-adenovirus immunity in the general population.

To improve antigenic presentation, different clinical trials are now based on the association of the tumor antigen gene with immunostimulatory cytokine-coding genes, including, typically, IL-2, IL-12, interferon-γ (IFN-γ), and GM-CSF (Table 4.9).

Another problem encountered by the clinical trials based on cancer vaccination is related to the intrinsically low immunogenicity of different tumors, despite TSA and TAA expression, and the development of immune escape mechanisms. During tumor growth, the selective pressure exerted by the immune system and the intrinsic genetic instability of cancer cells determine selection of tumor cell clones in which the levels of expression of MHC Class I (required for CTL activation) are reduced, or the co-stimulatory molecules necessary for proper antigenic presentation are lacking, or the TAA or TSA amounts are reduced. Other escape mechanisms involve the reduction or total inhibition of expression of the Fas receptors, which trigger apoptosis upon binding to the FasL cytokine produced by CTLs, or, conversely, the expression of FasL itself, which determines apoptosis of CTLs. Finally, immune escape of tumor cells can also be mediated by the production of cytokines such as TGF-β or IL-10, which exert inhibitory effects on CTLs.

To overcome some of the above-reported issues, three different approaches have been devised by the current gene therapy trials, based on the transfer of: (i) genes coding for immunostimulatory cytokines able to activate CTLs, such as IL-2, IL-12, IL-7, GM-CSF, IFN-γ, IL-6, and TNF-α); (ii) genes coding for co-stimulatory cytokines such as B7-1, ICAM-1, and LFA-3, able to improve antigenic presentation to CTLs; and (iii) genes coding for allogeneic proteins (for example, the HLA-B7 protein), thus able to change the antigenic identify of the cancer cells and force their recognition as foreign cells by the immune system (Table 4.9). In particular, different trials are now using a vaccine based on a recombinant vacciniavirus or on the fowlpox virus, named TRICOM (TRIad of CO-stimulatory Molecules), entailing expression of three co-stimulatory molecules (B7-1, ICAM-1, and LFA-3). This vaccine, delivered subcutaneously, is combined with the administration of different TAAs, such as MUC-1 or CEA, for the treatment of different carcinomas or as PSA for metastatic prostate carcinoma.

The clinical trials that so far exploited the above-summarized approaches were mainly directed against various carcinomas or advanced melanoma, usually metastatic. The initial results of several studies revealed a significant reduction of the primitive tumor mass, and often of metastases, following vaccination. However, significant results on the increase of survival of the treated patients are still missing and are currently the focus of larger, still ongoing Phase II and III trials.

Quite surprisingly, some of the patients enrolled in a few of the cancer vaccination trials completely recovered from their diseases, including the disappearance of both the primary tumor and metastases following vaccination. For example, this is the case of a single patient in a group of 16 with metastatic melanoma, in whom a recombinant adenovirus carrying the Melan-A/MART-1 antigen was administered. In this respect, it is however important to note that similar results were only obtained in an anecdotic, and thus not statistically significant, manner. However, these findings probably indicate that the success of cancer vaccination depends on immunological and experimental variables that are still not completely understood, and are probably different from patient to patient. Although anecdotic, these results nevertheless strengthen the enthusiasm for a vaccination approach for cancer therapy.

4.6.4.5
Adoptive Immunotherapy

In addition to passive immunotherapy and anti-cancer vaccination, a third form of immunotherapy is the adoptive transfer of cellular immunity, or *adoptive immunotherapy*. This consists in the *ex vivo* expansion of allogenic CTLs directed against specific antigens, followed by their infusion into the patients (Figure 4.17). Therapy with antigen-specific CTLs is now progressively finding applications in modern clinical medicine, for the treatment of different viral diseases and for cancer immunotherapy. The applications in virology mainly include the control of EBV and CMV infections in immunocompromised patients, typically after BMT or in individuals with HIV/AIDS. In these patients, infusion of CTLs recognizing and destroying the virus-infected cells significantly decreases the extent of the infection.

In the field of anticancer immunotherapy, adoptive transfer of allogenic T cells represents the only effective strategy having so far shown curative capacity against cancers of non-viral origin. Already at the end of the 1970s, it was first observed that the percentage of leukemia or lymphoma relapse after BMT was significantly lower in patients receiving allogenic compared to autologous BMT. This effect was mediated by the reactivity of the donor's T lymphocytes against the recipient's tumor cells (GvL). Infusion of donor lymphocytes is currently used in the context of BMT procedures. As already discussed above, a collateral effect is the development of reactivity against the normal cells of the recipient (GvHD), which can be controlled by CTL transduction with a retroviral vector expressing a suicide gene, should proliferation of the infused CTLs became uncontrolled or otherwise undesired.

The efficacy of allogeneic CTL infusion in the context of BMT indeed indicates that a potentially efficacious strategy for cancer immunotherapy could be based on the *ex vivo* expansion and activation of autologous T lymphocytes from the patients, selected for high affinity and avidity against tumor antigens. Of notice, active cancer vaccination holds the same purpose *in vivo*, however the inhibitory environment generated by the tumor might often render this process less effective. Autologous CD8$^+$ T lymphocytes directed against tumor antigens can be directly obtained from the patient, for example by recovering them from the tumor itself in the form of TILs (cf. section on 'Genes as Drugs'), followed by their *ex vivo* expansion. This procedure, however, usually generates low amounts of CTLs able to recognize the antigens of interest with high affinity.

A more efficacious way to obtain large quantities of CTLs having a desired target specificity is to expand large amounts of primary CD8$^+$ T lymphocytes from the patient, irrespective of their specificity, and then modify their target recognition specificity by transferring the genes coding for a given TCR. As discussed in the Chapter on 'Therapeutic Nucleic Acids', the simplest modality to modify TCR specificity is to transfer the genes coding for the TCR α and β chains specific for an antigen of interest. Currently, the α and β chains for different TAAs are available, including those recognizing the Melan-A/MART-1 and gp100 antigens of melanoma, the NY-ESO-1 CTA antigen, and one epitope of the mutated p53 protein. When these genes are transferred into T lymphocytes, usually by retroviral vectors in which the two chains are separated by an internal ribosome entry site (IRES), they confer these cells the capacity to recognize and destroy the cells presenting the respective antigens. A recent clinical trial consisted in the treatment of 15 patients with metastatic melanoma resistant to conventional therapy, with autologous lymphocytes engineered to express a TCR against Melan-A/MART-1. In two of the treated patients, the transduced lymphocytes persisted in very high numbers for at least two months after infusion, determining a marked regression of the metastatic lesions.

Finally, it should be mentioned that lentiviral vectors might be preferable to gammaretroviral vectors to achieve expression of the desired genes in CD8$^+$ T lymphocytes. The former class of vectors, indeed, should permit longer expression of their transgenes without silencing over time and allow transduction of HSCs rather than already differentiated CD8$^+$ T lymphocytes.

4.7
Gene Therapy of Neurodegenerative Disorders

Neurons in the central nervous system are highly differentiated, post-mitotic cells whose proliferative potential completely ceases after birth. Additionally, a vast number of neurons are physiologically lost during life. A few studies indicate that each individual is born with about 19–22 billion neocortical neurons, of which over 80,000 are lost every day, thus determining an overall loss of more than 10% of the total neurons by 80 years of age. Despite this continuous neuronal death, outside a few areas where neuronal stem cells able to proliferate and differentiate into new neurons might exist, most of the brain regions do not possess any regenerative capacity.

In addition to this physiologic neuronal loss, several specific diseases accelerate neuron deprivation, thus determining the occurrence of neurodegenerative disorders. These are invariably characterized by more or less diffuse neuronal depletion, which involves specific types of neurons (for example, cholinergic neurons in Alzheimer's disease or motoneurons in amyotrophic lateral sclerosis) or defined brain regions (for example, dopaminergic neurons in the substantia nigra in Parkinson's disease).

The progressive aging of the population is currently determining a very significant increase in the prevalence of neurodegenerative disorders, which, together with neurological disorders of vascular origin (ictus), are now among the leading causes of death and morbidity in Western countries. For example, from 6 to 7 million people are estimated to be affected by Alzheimer's diseases (the main cause of dementia) in Western Europe, while the prevalence of this disorder doubles every 5 years in individuals over 65 years of age, to reach 1 person out of 3 at 80 years.

4.7.1
Neurotrophic Factors

The real cause of accelerated neuronal depletion is still elusive in most neurodegenerative disorders, despite some recent insights into the pathogenetic mechanisms of disease development. Independent of the cause, however, a therapeutic goal common to most neurodegenerations is to preserve the viability and function of the residual neurons as long as possible. The discovery of a series of cytokines exerting a trophic and protective effect on neurons now renders this possible.

Already between the 1920s and 1940s, it was realized that neuron survival strictly depends on the presence of soluble factors, produced in limited quantity by the tissues targeted by their axonal projections. These factors are collectively known as "neurotrophic factors". A paradigmatic example of the function of these factors is observable for the motoneurons in the anterior horn of the spinal cord. Survival of these cells strictly depends on the production of neurotrophic factors by the innervated muscle: only the neurons reaching the target muscle survive during development, while the others undergo apoptosis. In addition to this pro-survival function, neurotrophic factors also stimulate neuronal proliferation during development and, in adult organisms, regulate a series of metabolic functions in neurons, including protein synthesis and the ability to synthesize specific neurotransmitters.

The list of soluble factors exerting a trophic effect on neurons is now very long (Table 4.12). It includes a series of cytokines selectively active on neurons, but also factors exerting activities in other districts, such as insulin-like growth factor-1 (IGF-1), having a trophic and anti-apoptotic activity in different districts including heart and skeletal muscle, or VEGF, the main factor regulating blood vessel formation both during development and in adults.

The neurotrophic factors of specific neurological interest can be grouped into three main families: the neurotrophins, the GDNF family, and the CNTF/LIF family.

Neurotrophins. The prototype member of the neurotrophin family is nerve growth factor (NGF), which was originally discovered in 1940. The other members of the family, which are structurally closely related to NGF, are brain-derived neurotrophic factor

Table 4.12 Major factors with neurotrophic activity

Family	Main factors
Neurotrophins	Nerve growth factor (NGF)
	Brain-derived neurotrophic factor (BDNF)
	Neurotrophin (NT)-3, -4/5, -6, -7
GDNF	Glial cell line-derived neurotrophic factor (GDNF)
	Neurturin (NTN)
	Persefin (PSP)
	Artemin (ART)
CNTF/LIF	Ciliary neurotrophic factor (CNTF)
	Leukemia inhibitory factor (LIF)
	Cardiotrophin-1 (CT-1)
	Cardiotrophin-1-like cytokine (CLC)
Hepatocyte growth factor (HGF) or scatter factor (SF)	
IGF	Insulin
	Insulin-like growth factor (IGF)-1 and -2
FGF	Acidic fibroblast growth factor (aFGF, FGF-1)
	Basic fibroblast growth factor (bFGF, FGF-2)
Epidermal growth factor (EGF)	
Neuregulins	Glial growth factor (GGF)-1, -2, -3
Interleukins	Interleukin-6 (IL-6)
Neuroimmunophilins	Cyclophilin
	FK506-binding protein (FKBP)
Platelet-derived growth factor (PDGF)	
Protein S-100	
Transforming growth factor-β (TGF-β)	
Vascular endothelial growth factor (VEGF)	

(BDNF), neurotrophin-3 (NT-3) and NT-4/5. These factors have been shown to enhance the survival and differentiation of several classes of neurons *in vitro*, including sensory neurons, dopaminergic neurons in the substantia nigra, basal forebrain cholinergic neurons, hippocampal neurons, and retinal ganglial cells.

The neurotrophins interact with two different types of receptors. The high-affinity receptors consist of a family of three integral plasma membrane tyrosine kinases: TrkA, TrkB, and TrkC. NGF directly interacts with TrkA, and BDNF and NT-4/5 with TrkB. Similar to the other tyrosine kinase receptors, activation of the Trk receptors depends on their ligand-induced dimerization. This event leads to enzymatic activation of the receptor, which becomes phosphorylated in its cytoplasmic domain, and to the subsequent activation of a series of intracellular transducers, including PI3K, *ras*, and PLCγ-1, which transduce the activation signal inside the neuron; several of the activation pathways are shared with other tyrosine kinase receptors. Neurotrophins also bind a low-affinity receptor known as p75^{NTR}; in contrast to the Trks, all neurotrophins bind this receptor with equal affinity. The p75^{NTR} receptor facilitates interaction between the Trks and their ligands, besides activating autonomous signal transduction pathways. In the absence of Trk receptor activation, isolated p75^{NTR} stimulation determines neuron apoptosis instead of protection from apoptosis.

GDNF family. A second family of neurotrophic factors consists of a group of at least four proteins: glial cell line-derived neurotrophic factor (GDNF), neurturin (NTN), artemin (ART), and persefin (PSP), showing structural homology to transforming growth factor-β (TGF-β). These factors act on different cell populations in the central, peripheral, and autonomous nervous system, including dopaminergic neurons in the substantia nigra. Outside the nervous system, these factors are essential for the development of kidney and the differentiation of spermatogonia.

The GDNF family members transduce their activation signal through a common receptor subunit, the c-Ret receptor, originally identified as a proto-oncogene. Binding specificity is ensured by a series of co-receptors (GFRα1 for GDNF, GFRα2 for NTN, GFRα3 for ART, and GFRα4 for PSP), which are attached to the plasma membrane through a glycosyl-phosphatidyl-inositol (GPI) anchor.

CNTF/LIF family. Finally, a third family of neurotrophic factors includes a few cytokines that are structurally related to IL-6, including the ciliary neurotrophic factor (CNTF), which was first identified as a survival factor for neurons from the ciliary ganglion of chicken embryos, and leukemia inhibitory factor (LIF), which maintains embryonic stem cells in an undifferentiated, pluripotent state. Besides their neurotrophic activity, these factors modulate the kind of neurotransmitter used by neurons and, in particular, modulate the conversion from noradrenergic to cholinergic phenotype.

CNTF exerts its actions through the activation of the high-affinity CNTF receptor complex, which contains the ligand-binding α subunit (CNTF Rα) and two signal-transducing β subunits (LIF Rβ and gp130). Upon activation, these receptor complexes activate cytoplasmic JAK tyrosine kinases, which, among other substrates, phosphorylate the STAT transcription factors, leading to transcriptional activation of several cellular genes.

4.7.2
Gene Therapy of Neurodegenerative Disorders Using Neurotrophic Factors

Besides their fundamental role during nervous system development, several of the neurotrophic factors exert a powerful anti-apoptotic activity in adult life and are thus important for neuronal survival. In various animal models of neurodegenerative damage, administration of neurotrophic factors prevents neuronal loss and significantly improves performance. Based on these observations, various clinical trials have been conducted based on the administration of some of these factors in the form of soluble, recombinant proteins. Neurotrophic factors do not cross the blood–brain barrier and thus need to be administered by infusion in the liquor. However, when administered through this route, some of them (for example, NGF for the treatment of Alzheimer's disease) exert significant collateral effects, which were found to limit or completely prevent the possibility of clinical use.

In light of these observations, it is thus evident that the possibility to obtain site-specific production of these factors *in vivo* using gene therapy is a very appealing possibility. This possibility is further reinforced by the observation that some of the most efficient viral vectors that are currently available, namely AAVs and lentiviruses, are very efficient at transducing neurons *in vivo* and driving therapeutic gene expression for very long periods of time, essential characteristics for prolonged treatment.

The following sections report the current advances of gene therapy clinical trials for the most relevant neurodegenerative disorders (Alzheimer's disease, Parkinson's disease, Huntington's disease, and the motoneuron degenerative disorders – amyotrophic lateral sclerosis and spinal muscular atrophy). In addition to their general ability to respond to neurotrophic factor administration, some of these disorders are also the focus of alternative gene therapy attempts, aimed at interfering with the pathogenetic mechanisms specifically underlying disease development. Table 4.13 reports the current gene therapy approaches for the major neurodegenerative disorders.

4.7.3
Gene Therapy of Alzheimer's Disease

The term "senile dementia" refers to a clinical syndrome, typical of aging individuals, characterized by memory loss and alteration of cognitive functions (including speaking, capacity to solve problems, ability to calculate, attention, manual ability, etc.) serious enough to compromise the social and occupational activities of the patients. Alzheimer's disease (AD) is the most common form of dementia in the elderly: about 7% of subjects over 65 years and more than 30% over 80 years are affected by AD. Due to the progressive aging of the population, AD thus represents one of the most pressing current social and health problems. The second most frequent cause of dementia is the brain disorders of vascular origin; next are dementia with Lewy bodies (DLB), frontotemporal dementia associated with parkinsonism, and dementia due to alcoholism, drug addiction, HIV/AIDS, syphilis, brain tumors, and various metabolic disorders.

AD is a neurodegenerative disorder characterized by the selective alteration of the cholinergic neurons of the neocortex, entorhinal area, hippocampus, amygdala, basal

Table 4.13 Current approaches for gene therapy of neurodegenerative disorders

Disease	Therapeutic gene	Delivery system	Therapeutic strategy
Alzheimer's disease (AD)	NGF, BDNF, FGF-2	Patients' fibroblasts transduced *ex vivo* with a retroviral vector	Inoculation of transduced fibroblasts in the nucleus basalis
		AAV	Injection of vectors in the nucleus basalis
Parkinson's disease (PD)	GAD	AAV	Injection of vectors in the subthalamic nucleus
	NTN, GDNF, BDNF, artermin	AAV	Injection of vectors in the putamen
	AADC	AAV	Injection of vectors in the striatum
	Various neutrophic factors	Cells overexpressing neurotrophic factor encapsulated in semipermeable membranes	Implantation of capsules containing the cells in the basal ganglion region
Huntington's disease (HD)	CNTF, NGF, NTN	Cells overexpressing CNTF encapsulated in semipermeable membranes	Implant of capsules containing the cells in the lateral ventricle
Amyotrophic lateral sclerosis (ALS)	GDNF, CNTF, BDNF, IGF-1, VEGF-A and -B	AAV or lentiviral vectors	Injection of viral vectors into the spinal cord or skeletal muscle
		Cells overexpressing neurotrophic factor encapsulated in semipermeable membranes	Implant of capsules containing the cells in the intrathecal space
	Bcl-2, IAP EEAT2	AAV or lentiviral vectors	Injection of viral vectors into the spinal cord or skeletal muscle
	Anti-SOD1(G93A) siRNAs or shRNAs	Naked DNA or viral vectors	Intramuscular injection (for vectors)
Spinal muscular atrophy (SMA)	BDNF	AAV or lentiviral vectors	Injection of viral vectors into the spinal cord or skeletal muscle
	Normal SMN1 cDNA	AAV or lentiviral vectors	Injection of AAV-SMN1 or LV-SMN1 into the spinal cord or skeletal muscle
	Antisense oligonucleotides restoring inclusion of SMN2 exon 7	Antisense oligonucleotides	Injection of oligonucleotides to modify SMN2 splicing

nuclei, anterior portion of the thalamus, and various monoaminergic nuclei of the brain stem. In the areas involved, the neuronal cytoskeleton appears grossly altered. In particular, neurofibrillary tangles are detected inside the cells, consisting of filamentous inclusions in the neuronal soma and dendrite proximal regions. These abnormal inclusions are formed by hyperphosphorylated and insoluble forms of tau, a microtubuli-associated protein which is normally soluble. Eventually, the cells presenting these alterations die and the neurofibrillary tangles also became detectable in the intercellular space.

The AD-affected regions also contain the so-called senile plaques, composed of extracellular deposits of amyloid surrounded by dystrophic axons and astrocyte and microglia cell protrusion (inflammatory cells). The term *amyloid* refers to insoluble fibrous protein aggregates composed of polypeptide filaments with a β-sheet conformation, staining positive with Congo red. The major constituent of amyloid is a 4-kb peptide-defined Aβ amyloid. This derives from the hydrolysis of a larger protein, the amyloid precursor protein (APP), which is normally present in neuron dendrites, cell soma, and axons, where it exerts a function that is still not fully understood. APP is synthesized in the endoplasmic reticulum, glycosylated in the Golgi apparatus, and transported onto the surface of neurons as an integral membrane protein. Aβ is generated by sequential cleavage of APP by β- and γ-secretases and exists as short and long isoforms, of which the most common are Aβ1-40 and Aβ1-42 respectively. Aβ1-42 is especially prone to misfolding and builds up aggregates that are thought to be the primary neurotoxic species involved in AD pathogenesis.

AD is usually a sporadic disorder, however a fraction of patients show familial transmission, with autosomal dominant inheritance and almost complete penetrance. These cases are due to mutations in the APP gene, which increase the levels of secreted Aβ peptide, or in the presenilin-1 and presenilin-2 genes, two transmembrane protein influencing the levels of produced Aβ peptide. Other genetic variations predisposing to AD are some polymorphisms in the apolipoprotein E (ApoE) gene, a glycoprotein required for transport of cholesterol and other fatty acids in the blood (in particular, the ApoE4 allele would facilitate Aβ accumulation), or in the α2-macroglobulin gene, a protein contributing to the removal of Aβ deposits from the synaptic regions.

Currently, there is no specific therapy for AD. Those commonly administered are aimed at ameliorating some specific symptoms occurring over the course of the disease, including depression, sleep disturbances, hallucinations, and illusive perceptions. The drugs used act by potentiating the cholinergic system of the basal forebrain, which is usually one of the most affected regions.

A large series of preclinical studies in animals shows that the neurotrophic factor NGF both prevents neuronal death and stimulates cholinergic neuron function in rodents and primates; in addition, NGF administration is known to improve learning and memory. The factor, however, cannot be administered systemically, since it does not cross the blood–brain barrier, and its infusion in the ventricles causes a series of collateral effects, including pain (since it stimulates nociceptive neurons in the dorsal root ganglia) and hypophagia, besides promoting migration and proliferation of Schwann cells. As already discussed above, these problems could be overcome by the local delivery of the growth factor coding gene by gene therapy.

In 2001, the first gene therapy clinical trial for AD was conducted, entailing the administration of the NGF cDNA in 8 patients in the initial phase of the disease. Dermal fibroblasts

4

were obtained from skin biopsies and transduced *ex vivo* with a retroviral vector expressing the NGF cDNA. Following selection and expansion, the NGF-producing cells were implanted using a stereotactic apparatus close to the nucleus basalis of Meynert. This brain region contains the soma of cholinergic neurons projecting towards the whole brain cortex, which typically undergo degeneration in AD. Five years after cell implantation, no adverse reactions were observed in any of the treated patients. Of interest, analysis of brain function using positron emission tomography (PET) revealed a significant increase in the metabolic activity of the whole cortex, a result compatible with improved function of the nucleus basalis. These results were confirmed by evaluating the patients' cognitive performance.

Although the relatively small number of treated patients does not allow definitive conclusions to be drawn, collectively these results confirm the beneficial role that NGF production exerts in slowing down AD progression. After the full development of the AAV technology for gene delivery, the brutal *ex vivo* approach utilized by this trial is now leaving space for more refined NGF cDNA *in vivo* gene transfer trials using this class of vectors.

4.7.4
Gene Therapy of Parkinson's Disease

Parkinson's disease (PD) is a common disorder, having a prevalence of 1–3% after 65 years and thus representing the second most common form of neurological disorder. The disease is typically due to the progressive loss of dopaminergic neurons in the pars compacta of the substantia nigra, one of the basal nuclei playing a fundamental role in the extrapyramidal system of motor control.

The basal nuclei (or basal ganglia) include a series of subcortical regions, extensively interconnected, receiving input from the cortex and the thalamus and projecting towards the cortex through the thalamus and some of the nuclei of the brain stem. Thus, the basal nuclei are the main components of the sub-cortical re-entry circuits connecting the cortex and the thalamus. These nuclei are involved in the control of movement and have long been considered as the main components of a specific motor system, independent of the pyramidal (or corticospinal) motor system, known as the extrapyramidal motor system. While it is now clear that extensive interconnections exist between the pyramidal and extrapyramidal systems and that the two systems strictly cooperate in motor control, specific lesions of the two systems still generate completely different clinical conditions. While defects in the pyramidal tract are characterized by spasticity and muscle paralysis, those of the basal nuclei determine the so-called extrapyramidal syndrome, characterized by three typical disturbances in movement control: (i) presence of tremor and other involuntary movements; (ii) alteration of the posture and muscle stiffness; and (iii) diminished or slow movement (hypokinesia), but not accompanied by paralysis.

The basal ganglia include four different nuclei (Figure 4.19A):

(1) The *striatum*, composed of three essential parts: nucleus caudatus, putamen (which together form the dorsal striatum), and ventral striatum.

(2) The *globus pallidus*, or pallidus, subdivided into an external and an internal segment. The internal pallidus is functionally similar to the substantia nigra pars reticulata; neurons in both areas use γ-aminobutyric acid (GABA) as a neurotransmitter.

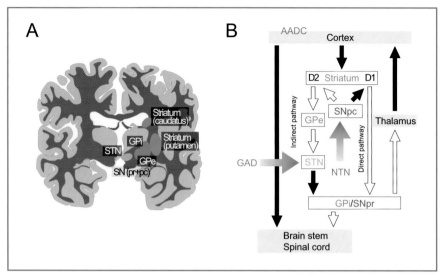

Fig. 4.19 Basal nuclei and major gene therapy approaches for Parkinson's disease. **A** Schematic representation of the anatomical localization of the basal ganglia (*GPi*, globus pallidus internal segment; *GPe*, globlus pallidus external segment; *STN*, subthalamic nucleus; *SN*, substantia nigra; *pr*, pars reticulata; *pc*, pars compacta). **B** Schematic representation of the activating (*black arrows*) and inhibitory (*white arrows*) circuits of the basal ganglia. D1 and D2 denote dopaminergic neurons of the striatum. The red arrows indicate the targets of the three current major gene therapy approaches for Parkinson's disease, based on the delivery of the AADC (aromatic L-amino acid decarboxylase), GAD, and NTN (neurturin) genes

(3) The *substantia nigra*, which is located in the mesencephalon, composed of a pars reticulata and a pars compacta, in which neurons use either GABA or dopamine as neurotransmitters, respectively. The substantia nigra is named after the presence, in the neurons of the pars compacta, of neuromelanin, a dark pigment derived from the oxidation of dopamine polymers.

(4) The *subthalamic nucleus*. This structure is anatomically strictly connected with both segments of the pallidus and the substantia nigra; its neurons are excitatory and use glutamate as a neurotransmitter.

Among the four basal ganglia, the striatum receives most of the neuronal inputs. In particular, this structure receives excitatory, glutamatergic projections from all areas of the cortex, in addition to excitatory signals from the thalamus, dopaminergic projections from the mesencephalon, and serotoninergic signals from the raphe nuclei. About 90% of the striatal neurons are GABAergic. In contrast, most of the output signals originate from the internal pallidus and the substantia nigra pars reticulata; the projections emanating from these structures exert a tonic inhibition of their target nuclei in the thalamus and brain stem.

The striatum (input nucleus) and the internal pallidus/substantia nigra pars reticulata (output nuclei) are connected by two pathways, having opposite functions on the efferent inhibitory signals (Figure 4.19B). The *direct pathway* consists of GABAergic neurons that directly connect the striatum with the two efferent nuclei; the *indirect pathway* initially connects the striatum with the external pallidus (GABAergic neurons), then the external pallidus with the subthalamic nucleus (GABAergic neurons), and, finally, the subthalamic nucleus with the two efferent nuclei (glutamatergic neurons). The last glutamatergic connection between the subthalamic nucleus and the two efferent nuclei is the only excitatory connection of the basal nuclei, while all the rest are GABAergic and inhibitory.

The direct and indirect pathways exert opposite effects: the direct pathway inhibits the two efferent nuclei, and thus determines activation of the thalamus and consequent increase in activity of the thalamus-cortical projections. In contrast, activation of the indirect pathway, which essentially consists in the excitation of an inhibitory effect, results, as its final outcome, in additional inhibition of the thalamus-cortical neurons. As a consequence, activation of the direct pathway facilitates movement while activation of the indirect pathway inhibits it.

The dopaminergic neurons of the substantia nigra pars compacta project towards the striatum and influence the direct and indirect pathways differently. The striatal neurons in the direct pathway express D1 dopamine receptors, which facilitate synaptic transmission, while the neurons in the indirect pathway express D2 dopamine receptors, which reduce synaptic transmission. However, although these synaptic activities are opposite, since the direct and indirect pathways also have opposite function, the net result of the dopaminergic afferences is in both cases the same, namely a reduction in the inhibition of the thalamus-cortical neurons and the consequent facilitation of the movements initiated by the cortex.

In patients with PD, loss of the dopaminergic neurons of the substantia nigra pars compacta increases activity of the internal pallidus/substantia nigra pars reticulata efferent nuclei, which, in turn, augments inhibition of the thalamus-cortical neurons that normally facilitate movement. In particular, loss of dopamine in the striatum increases inhibition of the thalamus ventral lateral nucleus, which sends excitatory signals to the motor cortex.

Symptoms start to develop when the levels of striatal dopamine are less than 40% of normal. These essentially consist of a triad: tremor at rest, usually of limbs, which ceases or reduces when movement becomes voluntary; stiffness of the musculature, which determines a characteristic gait and posture, and amimic expression of the face; and bradykinesia, causing difficulty in initiating and arresting movements and general slowness. Several of the patients also experience depression and other neuropsychiatric symptoms, possibly due to the degeneration of other dopaminergic circuits in the brain. The disease has a chronic and usually progressive course. Although cognitive decline is usually not present at disease onset, the development of dementia occurs in over 25% of patients in the end stages of the disease.

Patients are treated pharmacologically with L-3,4-dihydroxyphenylalanine (L-DOPA), a precursor of dopamine, or with other molecules that imitate its action (dopaminergic drugs, including bromocriptine and others); the latter are preferred to L-DOPA as a first therapy, in order to avoid dyskinesia, which is associated with long-term therapy with L-DOPA. Other useful treatments are based on anticholinergic drugs (for example, biperidene), which suppress tremor at rest, and drugs blocking dopamine degradation, including the inhibitors of monoaminooxidase-B (MAO-B), such as selegeline. In some patients amantadine, an antiviral drug, also shows efficacy, since it potentiates the effects of dopamine.

Substitutive therapy with L-DOPA has the potential to substantially modify the natural history of the disease. Treatment with this drug is usually initiated in the advanced stages of the disease and is very effective in improving symptoms. The drug is commonly administered together with an inhibitor of dopa-carboxylase (carbidopa), which prevents L-DOPA degradation in blood and peripheral tissues, thus increasing the amounts able to cross the blood–brain barrier.

After 5–10 years of L-DOPA therapy, however, patients become less tolerant and their pharmacological treatment becomes problematic. In patients with advanced disease and uncontrollable symptoms, surgical pallidoctomy is available, consisting in the induction of a small lesion in the internal pallidus under stereotactic guidance to inhibit function of this area.

More recently, a therapeutic option alternative to surgery is deep brain stimulation (DBS), consisting in the surgical implantation of an electrode, usually at the level of the subthalamic nucleus or the internal pallidus, aimed at electrically stimulating the region. This procedure reduces the extent of symptoms by 50–60%. Since DBS does not increase the levels of dopamine, it has no effect on the neuropsychiatric symptoms of PD (depression, anxiety and cognitive decline).

Neither L-DOPA nor surgical therapy prevent the progressive loss of dopaminergic neurons and the progressive worsening of symptoms. For this reason, the development of alternative therapies able to cure or at least control disease progression is highly desirable. In this context, PD has several very favorable characteristics for gene therapy. The disease is due to a defect in a precisely localized region of the brain, which can be targeted using standard stereotactic procedures; furthermore, although the actual cause of the disease is still unclear, several of the pathogenetic mechanisms leading to neuronal death are known, along with the protective role that different cytokines exert in this process.

Over the last 10 years, gene therapy of PD using viral vectors to deliver neurotrophic factors or cDNAs coding for enzymes involved in neurotransmitter metabolism was successful in various animal models of the disease. These models are usually based on the injection of toxic substances either in the brain (for example, 6-hydroxydopamine – 6-OHDA) or systemically (1-methyl-4-phenyl-1,2,3,6-tetrahydropyridine – MPTP), able to selectively induce dopaminergic neuron loss in the substantia nigra. Animals treated in this manner show the same symptoms as PD patients; indeed, the activity of MPTP was originally identified in 1983 when a few individuals who had inadvertently taken this drug as a contaminant of a synthetic heroin batch developed severe PD symptoms.

Three of the experimental approachess have now progressed towards Phase I and Phase II clinical trials. They are all based on the intracerebral injection, under stereotactic guidance, of AAV2 vectors, due to the ability of these vectors to transduce neurons with high efficiency and express the therapeutic gene for prolonged periods (Figure 4.19B).

(1) The first gene therapy clinical trial for PD was based on the delivery of the cDNA coding for the glutamic acid decarboxylase (GAD) enzyme in the subthalamic nucleus. GAD is the limiting enzyme in the metabolic pathway leading to the synthesis of GABA, the main inhibitory neurotransmitter in the brain. In PD, the main inhibitory efferent nuclei of the basal ganglia (external pallidus and substantia nigra pars reticulata) are continuously stimulated along the indirect pathway by the subthalamic nucleus, which is hyperactive and disinhibited and whose axonal projections release the excitatory neurotransmitter glutamate. Overproduction of GABA following GAD gene transfer into the subthalamic nucleus converts the same from excitatory to inhibitory, thus normalizing the overall efferent signals emanating from the basal ganglia circuit.

A first Phase I clinical trial consisted in the injection of an AAV2-GAD vector in the subthalamic nucleus of 12 patients with PD. The results of this trial showed an absence of collateral effects or immunological response, in addition to significant improvement of motor function. Based on this promising result, a large Phase II clinical trial was initiated.

(2) A second approach for PD gene therapy was aimed at preventing or slowing down neuronal loss in the substantia nigra. This was based on the delivery of an AAV vector expressing the NTN gene to the putamen neurons, which are the major targets of dopaminergic axons. NTN is a member of the GDNF family of neurotrophic factors, sharing the same receptors and mechanisms of action. A vast series of studies in rodents and non-human primates have shown that GDNF, NTN and other ligands of the same family have the potential to improve symptoms and prevent disease progression. A recent clinical trial based on the infusion of recombinant GDNF protein in the putamen of PD patients gave similarly positive results.

The Phase I gene therapy trial, based on the administration of AAV2-NTN – designated CERE-120 – to the putamen, proved safe and showed preliminary evidence of efficacy, upon which a Phase II trial was organized. This trial was a double-blind, controlled clinical trial that completed enrollment of 58 patients with advanced PD. Patients were enrolled across nine leading academic medical centers in the United States, with two thirds of patients receiving CERE-120 via stereotactic neurosurgery into the putamen, and one third enrolled in a control group. Unfortunately, the results of this trial, released at the end of 2008, unexpectedly did not demonstrate an appreciable difference between patients treated with CERE-120 vs. those in the control group. These results again highlight the substantial difference between experimentation in animal models (including non-human primates) and that in patients, and further stress the absolute necessity of human clinical trials.

(3) A third approach was based on the use of AAV vectors for the delivery of the aromatic L-amino acid decarboxylase (AADC) enzyme in the striatum, with the purpose of increasing conversion of L-DOPA into dopamine. During the natural progression of the disease, the levels of AADC in both the substantia nigra and the striatum dopaminergic neurons decrease, thus determining the progressive requirements of increasing L-DOPA dosage. However, excessive L-DOPA levels cause important collateral effects, mainly due to the hyperstimulation of the mesolimbic pathways, which maintain normal AADC levels. Transfer of the AADC gene in the striatum, therefore, can extend the therapeutic window when L-DOPA can be safely administered. A similar approach has shown outstanding success in a monkey model of PD, in which the therapeutic efficacy of the transferred gene lasted at least 3 years after vector inoculation. In a recently completed Phase I clinical trial, patients received low-dose and high-dose AAV2-AADC in the putamen (5 patients in each cohort). Six months after treatment, the functional disease scores were significantly improved in both groups, however the surgical vector injection procedure was found to be associated with an increased risk of intracranial hemorrhage.

Finally, alternative approaches are currently considering the possibility to genetically modify primary or established cell lines *ex vivo* to obtain expression of genes with potential therapeutic activity, such as growth factors or enzymes involved in dopamine biosynthesis. These modified cells are then implanted into the relevant basal ganglion area, encapsulated within semipermeable membranes allowing passage of oxygen and nutrients and release of therapeutic factors, but keeping the cells separate from the surrounding tissue and the immune system. This approach is showing success in various PD animal models and it is thus likely that it will rapidly proceed towards the clinics.

4.7.5
Gene Therapy of Huntington's Disease

Huntington's disease (HD) (or Huntington's chorea, from the Greek *choros*, dance, a term introduced by Paracelsus to describe the typical rotary and sinuous involuntary movements of some of the affected patients) is a form of hyperkinetic disorder associated with alterations of the basal nuclei. The disease has a prevalence of 5–10 in every 100,000 individuals and is genetically inherited. It becomes evident between the third and fifth decades of life and neurological deterioration inexorably progresses until death, usually occurring 15–20 years after the beginning of symptoms. Due to its invalidating clinical characteristics, inexorable progression, and social impact, HD is unanimously considered one of the most devastating neurological disorders affecting humans.

While PD is due to the loss of the dopaminergic activity in the striatum and the consequent hyperactivity of the efferent nuclei of the basal ganglia circuit, HD is caused by the prominent loss of the striatal neurons from which the indirect pathway originates. As a consequence, inhibition of the external pallidus is reduced and the neurons of this area become hyperactive in inhibiting the subthalamic nucleus. This causes excessive motor activity, resulting in the characteristic symptoms of the disease. These include slow torsions of the extremities (atetosis), rapid involuntary movements of the limbs and face (chorea), and abnormal posture associated with slower movements performed with the simultaneous contraction of agonist and antagonist muscles (dystonia). Finally, localized lesions of the subthalamic nucleus (commonly due to small ictuses) can determine appearance of flinging movements, often violent, of the upper and lower extremities of one side of the body, contralateral to the lesion. This condition is known as hemiballism, in which the term *ballism* is used to underline the similarity of these movements with the act of throwing (Greek *ballein*, to throw).

The disease has autosomal dominant inheritance with very high penetrance. The responsible gene (the *huntingtin* gene), localized on chromosome 4, was identified in 1993. The first intron of the gene contains, in normal individuals, a few repetitions (<40) of the trinucleotide repeat CAG, coding for the amino acid glutamine. HD patients show amplification of the number of these repeats: once above 40, these become unstable and their number tend to further increase from generation to generation, causing disease onset at a progressively younger age (a genetic characteristics know as *anticipation*). The huntingtin protein is localized in the cytoplasm; when the gene undergoes triplet amplification, the vast number of glutamines determines precipitation of the protein in the nucleus of neurons, and the subsequent degeneration and death of these cells. Although HD is characterized by diffuse neuronal degeneration affecting the whole brain, neuronal loss is more prominent in the striatal neurons from which the indirect pathway originates. The consequent inactivation of the subthalamic nucleus explains the chorea-like movement; rigidity and akinesia observed in the advanced phases of the disease are instead due to the loss of the striatal neurons projecting towards the internal pallidus. There are no conventional therapies for this disease.

HD is essentially due to neuronal loss secondary to mutated Huntingtin protein precipitation in the nucleus of the striatal neurons. The administration of neurotrophic factors, such as NGF, BDNF, and CNTF, has been shown, in animal models of the disease, to protect neurons from degeneration and thus prevent disease progression. Starting from these observations, a first clinical trial was conducted entailing transfer of the CNTF cDNA into a cultured cell line in the laboratory. Following selection of high-producer clones, the cells were implanted under stereotactic guidance into the lateral ventricle of 6 HD patients within capsules surrounded by a semipermeable synthetic polymer. The study lasted 2 years and entailed replacement of the cell-containing capsules every 6 months. The procedure turned out to be safe, however only 11 of the 24 implanted capsules were still capable of releasing CNTF after recovery.

More recently, following the initial success of AAV vectors in PD, and given the similarly efficient transduction of neurons using lentiviral vectors in different animal models, a few clinical trials are expected to begin, in which these two classes of vectors are used to transfer the aforementioned neurotrophic factor cDNAs *in vivo*.

Finally, it should be noted that treatment of HD with neurotrophic factors might prevent disease progression, however it is not curative. In contrast, knocking down the mutated *huntingtin* allele might represent a true etiologic therapy. This ambitious goal might be met by the injection of siRNAs, either as naked, synthetic nucleic acids or vectored as shRNAs in the context of AAV or lentiviral vectors – see chapter on 'Therapeutic Nucleic Acids' – specifically targeting the disease allele. Preliminary experiments in murine models of the disease indicate that this approach is feasible, however several technical issues still need to be solved before such a diffuse RNA interference approach might be efficiently applied to the whole brain.

4.7.6
Gene Therapy of Amyotrophic Lateral Sclerosis

Motoneuron disorders are caused by defects in the two-neuron system that regulates voluntary motility. This system consists of an upper motoneuron, the soma of which is localized in layer V of the primary motor cortex, which projects a long axon towards the brain stem and the spinal cord to reach a lower motoneuron, which in turn projects its axon towards the cranial and spinal nerves to reach the muscle fibers. Motor activity only occurs when the upper motoneuron excites the lower motoneuron, which in turn stimulates muscle contraction. The other motor centers of the nervous system (including the cerebellum and the basal ganglia) regulate output of this two-neuron system axis.

The most prevalent motoneuron disorder is amyotrophic lateral sclerosis (ALS), characterized by the progressive degeneration of both upper and lower motoneurons. Other disorders, all transmitted genetically, in contrast with ALS, selectively involve either the upper or the lower motoneurons. Among these, the most relevant one is spinal muscular atrophy (SMA), a degenerative disorder of the α-motoneurons of the anterior horns of the spinal cord, which will be treated in the following section.

Gene therapy currently holds much promise for the development of novel therapeutic approaches for motoneuron disorders. However, this objective is very ambitious, since it requires, in principle, genetic modification, or some kind of treatment, of about 1 million corticospinal neurons and 100,000–200,000 lower motoneurons. Different strategies can be followed to achieve this goal (Figure 4.20). One possibility is to directly transfect or transduce the motoneurons by injecting the therapeutic gene, using an appropriate vector system, into the brain and spinal cord, where the soma of these neurons are located. Alternatively, in the case of lower motoneurons, it is possible to inject viral vectors carrying the gene of interest into the peripheral nerves or the target muscles, thus exploiting the physiological mechanisms of retrograde axonal transport, allowing transfer of biological macromolecules from the axons or the neuromuscular junctions to the cell soma. This gene transfer route is definitely less invasive than intraparenchymal injection into the brain or the spinal cord. Retrograde transport appears to be particularly effective for adenoviral and herpesviral vectors, while it is less efficient (and, anyhow, less investigated at the moment) for AAV and lentiviral vectors. Finally, a third possibility is to exploit intracerebral or intraspinal implantation of *ex vivo* engineered autologous or heterologous cells, separated by a semipermeable membrane allowing release of a potentially therapeutic soluble factor (for example, a neurotrophic factor) while impeding cell diffusion on one hand and contact with cells of the immune system on the other.

ALS, also named motoneuron disease or Lou Gehrig's disease after a famous baseball player affected by the disease, is a neurodegenerative disorder of the motor system that usually affects middle-aged individuals and shows a progressive and inexorable course, causing death in 50% of patients within 3 years of onset. The prevalence of the disease is 6 every 100,000 individuals.

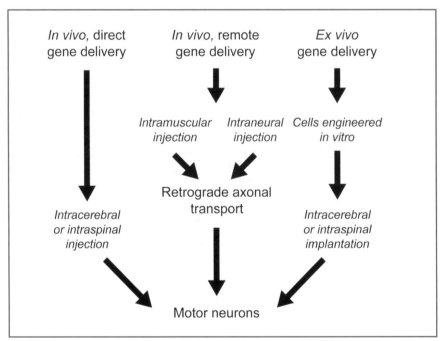

Fig. 4.20 Strategies for gene therapy of motoneuron disorders. Gene transfer procedures can be devised by either direct gene delivery to the motoneuron soma or exploiting retrograde axonal transport. Alternatively, genes can be transferred to *in vitro* cultured cells, which are subsequently implanted *in vivo* to release therapeutic factors

The clinical characteristics of the disease are consequent to the loss of motoneurons at all levels of the motor system, from cortex to spinal cord anterior horns. Depending on the affected regions, the symptoms are mainly characterized by bulbar (dysarthria, dysphagia, dyspnea, emotional instability), cervical (weakness of the upper limb muscles, fasciculations or cramps, spasticity), or lumbar (affection of the lower limbs) disturbances.

The causes of motoneuron degeneration are still unclear, and have been variably related to an excess of intracellular oxidative stress, the presence of excessive levels of excitotoxic neurotransmitters (such as glutamate), the consumption of drugs or other neurotoxic substances, or postulated to be secondary to viral infections. From 5% to 10% of patients have a familial history of disease; in most of these families, the pattern of heredity is autosomal dominant and, in about 20% of these cases, mutations in the gene coding for the Cu/Zn superoxide dismutase (Cu/Sn SOD or SOD1) can be detected; over 100 different mutations are known. Transgenic mice or rats overexpressing the mutated human protein – SOD1(G93A) mice – develop a motoneuron degenerative disorder showing characteristics similar to those of the patients and thus represent the best available animal model for the disease. Mutations in other genes have been identified in other familial cases; in this respect, however, it should be noted that the majority of patients are affected by the sporadic form of the disease, apparently without known mutations.

The only possible drug treatment available for ALS is riluzole, an inhibitor of synaptic release of glutamate. Therapy with this drug, however, increases life expectancy by only a few months.

Different neurotrophic factors have shown efficacy in slowing down degeneration in the murine ALS model. These include neuron-specific factors, including GDNF or CNTF, and also factors such as IGF-1 and VEGF-A or VEGF-B, the action of which has so far been considered prevalent in other areas. However, administration of some of these neurotrophic factors to patients in the form of recombinant proteins injected systemically or into the liquor has generated disappointing results, since the dosage required to provide benefit causes intolerable side effects.

Gene therapy might well overcome this problem by allowing the prolonged local release of these factors in the absence of systemic overdosage. In the murine model, viral vectors based on AAV or EIAV (a lentivirus) delivering the cDNAs for GDNF, IGF-1 or VEGF have shown efficacy by reducing symptoms and prolonging animal survival. In particular, these vectors were injected in the respiratory and lower limb muscles of SOD1(G93A) muscles and, from here, were able to penetrate neuromuscular junctions and be retrogradely transported into the spinal ganglia. Based on these encouraging results, clinical trials in affected patients will certainly follow.

Besides the use of neurotrophic factor, a series of animal studies were based on the transfer of genes preventing neuron apoptosis, including *bcl-2* (which regulates the mitochondrial apoptosis pathway) or *IAP* (inhibitor of apoptosis, which regulates caspase activity). Alternatively, therapeutic efficacy was also observed by decreasing the excitotoxic activity of glutamate, the increase of which is considered one of the most important pathogenetic mechanisms of disease development. One of the genes considered for this purpose is *EEAT2* (excitatory amino acid transporter 2), coding for a membrane protein transporting glutamate outside of the cells.

Finally, treatment of the ALS forms due to SOD1 mutation could also be tackled by using siRNAs targeting the transcript of this gene. A proof-of-principle of the efficacy of this approach was recently obtained using a lentiviral vector expressing a shRNA against the mutated SOD1(G93A) mRNA in the transgenic murine model. The vector was administered intramuscularly to obtain spinal motoneuron transduction through retrograde transport.

4.7.7
Gene Therapy of Spinal Muscular Atrophy

The collective term *spinal muscular atrophies* indicates a series of hereditary and acquired disorders characterized by the selective loss of α-motoneurons of the anterior horns of the spinal cord. SMA is the most common of these disorders, showing autosomal recessive inheritance due to mutations in the SMN (*survival motor neuron*) gene. SMA is the second most common autosomal recessive disorder, following cystic fibrosis, with an incidence of 10 affected children every 100,000 newborns and a frequency of carriers of 1:50. This disease represents the major cause of infant death. Other forms of spinal muscular atrophies include spinal and bulbar muscular atrophy (SBMA) or Kennedy's disease, and the distal spinal muscular atrophies.

The characteristic signs of SMA consist of muscular weakness and atrophy, with a symmetric and proximal pattern of distribution, by which the legs are more affected than the arms, and the arms more than the head and the diaphragm. According to the severity of the affection, the disease is usually classified into three types (I, II, and III). About half of the SMA patients have type I disease (Wednig-Hoffman disease), characterized by important and generalized muscle weakness and hypotonia since birth or within the first 6 months of life. These patients usually die before age 2 due to respiratory insufficiency. In patients with type II SMA, disease onset is between 6 and 18 months of life; in these children, prognosis strictly depends on the extent of respiratory muscle involvement. Finally, patients with type III SMA (Kugelberg-Welander disease) develop symptoms between 6 months and first infancy; they are usually able to stand and walk without assistance during infancy, however end up in a wheelchair when the disease progresses during adolescence or adulthood. In these patients, life expectancy is, however, long. In general, in contrast to ALS, which shows inexorable worsening, patients with SMA experience most neuronal loss at disease onset, followed by subsequent stabilization.

A characteristic common to all SMA patients is the selective loss of lower motoneurons, including those in the anterior horns of the spinal cord and in the brain stem motor nuclei. Cortical motoneurons and the corticospinal tract are, however, unaffected.

The gene responsible for SMA was cloned by linkage analysis in a region of chromosome 5q13. In all individuals, this region contains a duplication of the large sequence, in which 4 different genes are present, each one thus present in two copies. SMA is caused by mutations in the SMN gene located in the duplicated region in telomeric position (gene SMN1); the other SMN gene located in the centromeric duplication (SMN2) differs from SMN1 by only 5 nucleotides, of which one (a C to T substitution) is however crucial, since its presence in the pre-mRNA determines exclusion of an exon (exon 7) from most of the mature mRNAs (Figure 4.21). The patients affected by SMA present homozygous mutations in both SMN1 alleles. The levels of SMN2-encoded protein are insufficient to com-

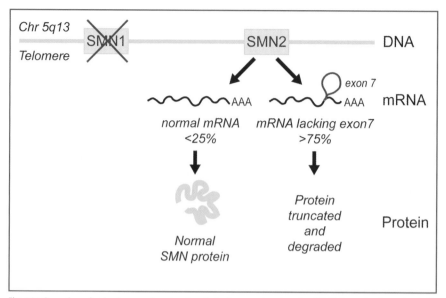

Fig. 4.21 Genetics of spinal muscular atrophy. Spinal muscular atrophy (SMA) results from the deficiency of the SMN protein. In normal individuals, most full-length SMN transcripts and protein arises from the SMN1 gene. Patients with SMA have homozygous mutations in the SMN1 gene but retain at least one copy of the SMN2 gene. The majority of the mature transcripts from this gene, however, lack exon 7 and thus code for a truncated protein that is rapidly degraded. One possible therapeutic approach entails the inhibition of exon 7 exclusion during splicing

pletely compensate the defect, since most of the protein produced does not contain exon 7 and thus is not able to oligomerize and is rapidly degraded.

The SMN protein is ubiquitously expressed and is involved in RNA metabolism. In particular, an SMN-containing protein complex regulates assembly of protein involved in the formation of small nuclear ribonucleoproteins (snRNPs), essential components of the spliceosome. The reason why SMN1 mutation causes a selective defect of motoneurons, while the SMN2 protein is sufficient to compensate its absence in other cell types, is unknown.

Gene therapy for SMA, although not yet at the clinical stage, represents a potentially solid therapeutic promise. The strategies that are currently followed are based on the delivery of a normal copy of SMN into the motoneurons using AAV or lentiviral vectors. Alternatively, it is possible to attempt a modification of the splicing pattern of SMN2 by forcing inclusion of the missing exon 7 in the mature mRNA. This objective can be met by using short antisense oligonucleotides pairing with the splicing regulatory signals present in the pre-mRNA. Obviously, for this approach to be successful, it will be necessary to obtain permanent levels of these oligonucleotides in most motoneurons.

4.8
Gene Therapy of Eye Diseases

The eye is a very interesting organ for gene therapy, since it is easily accessible and has a very compartmentalized anatomical structure. It is thus possible to deliver genetic material locally, without significant systemic diffusion, and the transduced cells are relatively protected from the immune system. Furthermore, the eye contains different non-replicating cell types, which can be efficiently transduced using relatively low amounts of therapeutic genes and vectors. Efficiency of gene therapy can be easily monitored using various non-invasive procedures, such as ophthalmoscopy and electroretinography. Finally, several animal models of disease are currently available that mimic human disease, which can be used for preclinical development of innovative therapeutic strategies.

Figure 4.22 is a schematic representation of the anatomic structure of the eye. Vector or therapeutic genes can be administered by three routes, namely by injection (i) into the anterior chamber; (ii) into the vitreus cavity; and (iii) into the subretinal space.

Inoculation into the anterior chamber can be utilized for the administration of genes having potential therapeutic effect in corneal diseases. In this context, it should also be remembered that corneal gene transfer can also be obtained *ex vivo* in the context of corneal transplant procedures.

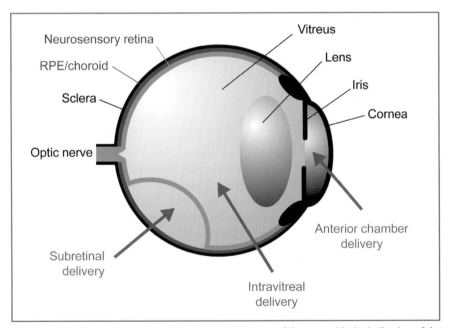

Fig. 4.22 Schematic representation of the anatomic structure of the eye, with the indication of the major routes for gene delivery. Subretinal delivery entails injection between the neurosensory retina and the RPE, with the temporary separation of these layers that rejoin spontaneously

4

Inoculation in the vitreus can, however, reach the retinal ganglion cells. These are specialized neurons that receive visual information from photoreceptors via different types of interneurons and transmit this information from the retina to several regions in the brain. In particular, these cells undergo apoptosis in patients with glaucoma, a disease caused by raised intraocular pressure due to a defect in the efflux of the aqueous humor. The disease is usually controlled by pharmacological therapy aimed at decreasing intraocular pressure. In a few patients, however, neuronal degeneration does not cease with therapy, and in these cases transfer of antiapoptotic genes (for example, *bcl-2*) or genes coding for neurotrophic factors (such as BDNF) might be useful.

Much more appealing due the vast number of applications is the possibility to administer genes and vectors into the subretinal space. This is a virtual cavity between the neurosensory retina and the retinal pigmented epithelium (RPE). Instillation of a solution containing the genetic material determines a temporal dissociation of these two layers, which then reconnect spontaneously. The subretinal route allows access to both RPE cells and photoreceptors, and can thus be utilized for gene therapy of retinal degenerations (including retinitis pigmentosa and other related disorders for which no therapy currently exists), of retinal hyperproliferations (such as retinoblastoma), and of the disorders due to retinal hypervascularization.

As far as the therapeutic genes are concerned, the relatively easy access to the retina and the anatomic compartmentalization of the eye allow utilization of both viral vectors and small regulatory nucleic acids, including chemically modified oligonucleotides, ribozymes, and siRNAs.

Over the last few years, significant progress has been made in gene therapy of the eye in various preclinical models of human disorders, in particular for the treatment of different forms of retinal degenerations, retinoblastoma, and the prevention of neovascularization. Some of these experiments have now reached the clinical stage, and are discussed in detail in the following sections.

4.8.1
Retinal Phototransduction and the Visual Cycle

Phototransduction is a process by which light (photons) is converted into an electric signal, occurring in specialized cells, the photoreceptors (rods and cones). The structure of these cells consists of three portions: an inner segment, containing the nucleus and the cell organelles; an outer segment, characterized by the presence of a series of membrane structures (disks) onto which the photoreactive pigments are localized; and a synaptic termination, allowing transmission of the electric pulse to the neuronal cells (Figure 4.23).

The outer segment of the photoreceptors is in contact with the RPE cells, which form the most external layer of the retina, of neuroectodermic derivation. These cells contain very high levels of melanin, which adsorbs the light that has not been retained by the retina, and thus prevent reflection of this light. Most importantly, these cells have the essential function of regenerating the visual pigments used for phototransduction. Each RPE cell establishes contacts with up to 40 photoreceptors in the region where these cells have their maximum density, that is in the central depressed area of the macula, known as *fovea*.

A schematic representation of visual phototransduction is shown in Figure 4.24.

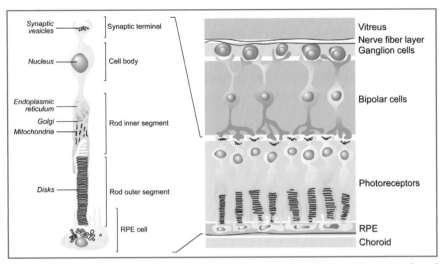

Fig. 4.23 Schematic representation of the retina, showing the major cell layers. Structure of a rod photoreceptor with an underlying RPE cell is shown in the enlargement on the left side

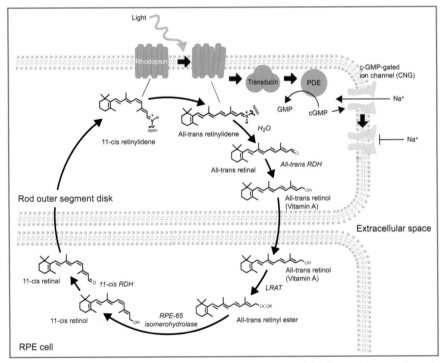

Fig. 4.24 Phototransduction and the visual cycle. The upper part of the figure shows a schematic representation of phototransduction, occurring in photoreceptors. The bottom part shows the biochemical process, known as the visual cycle, that occurs in an adjacent RPE cell and is necessary for the regeneration of retinal. Both processes are described in detail in the text

A characteristic of vertebrate photoreceptors is that they respond to light through hyperpolarization rather than depolarization. In the absence of light, the membrane of the photoreceptor outer segment has a lower potential than the other neurons, determining continuous depolarization of the presynaptic membrane. This membrane potential is generated by the balance between an outward K^+ ion current through selective K^+ channels, which would hyperpolarize the photoreceptors at about -70 mV (the equilibrium potential for K^+), and an inward Na^+ current through cGMP-controlled cyclic nucleotide-gated (CNG) channels, which depolarizes the cells at about -40 mV and thus keeps the voltage-sensitive Ca^{2+} channels open. This current is also known as *dark current*. The high concentration of intracellular Ca^{2+} permits continuous release of the excitatory neurotransmitter glutamate in the synaptic terminations between the photoreceptors and the connected neurons. When light hits the photoreceptors, the levels of cGMP sharply decrease, determining closure of the CNG channels, hyperpolarization of the cells, and consequent decrease of intracellular Ca^{2+}, which eventually results in decreased synaptic glutamate release. The retina efferent pathway originates from the ganglionar neurons of the inner layer of the retina; three classes of interneurons are interposed between the ganglionar cells and the photoreceptors: the bipolar cells, horizontal cells, and amacrine cells.

The essential event in retinal phototransduction is a decrease in cGMP concentration, occurring when light hits the photoreceptors. The membrane of these cells contains a seven-transmembrane, G-protein-coupled receptor (GPCR) protein, named *opsin*, the last segment of which associates with a chromophore, named *retinal*, a derivative of vitamin A that exists in different isomeric conformations. In its inactive form, retinal assumes the 11-cis conformation, which is loosely attached to opsin to form rhodopsin. When a photon hits rhodopsin, the retinal structure is modified by rotation of its terminal chain connected to opsin from the 11-cis to the all-trans form (photoisomerization). Following this conformational change, retinal no longer binds opsin, and thus opsin itself undergoes a conformation modification that activates transducin, a trimeric G protein that in turn activates a cGMP-phosphodiesterase (PDE), which, by hydrolyzing cGMP is eventually responsible for the decrease in the intracellular concentration of this cyclic nucleotide.

Termination of phototransduction occurs by two different mechanisms. On one hand, rhodopsin is phosphorylated by rhodopsin kinase and thus binds arrestin; this process blocks the interaction between rhodopsin and transducin, and thus decreases cGMP-PDE activity. On the other hand, transducin self-inactivates, by hydrolyzing its GTP into GDP.

Parallel to phototransduction termination, to allow further cycles of the process to occur, retinal is reactivated in the so-called *visual cycle* (Figure 4.24). All-trans retinal is first reduced to all-trans retinol (vitamin A), which then leaves the photoreceptor, crosses the subretinal space and enters the RPE cells, which are strictly connected to the outer segment of the photoreceptors. Here, it is first esterified by the enzyme lecithin-retinol acyltransferase (LRAT) and then isomerized to 11-cis retinol by the RPE65 isomerase. Finally, 11-cis retinol can be stored as it is or oxidized to 11-cis retinal, which returns to the photoreceptors to generate new rhodopsin. Therefore, all-trans retinol (vitamin A) is a fundamental compound for the proper function of the visual system. Since this chemical is not synthesized by humans, it must be introduced with the diet to avoid first night blindness, later degeneration of the photoreceptor outer segment and, in the later stages of deprivation, complete blindness.

4.8.2
Gene Therapy of Congenital Retinal Degenerations

The congenital retinal degenerations (or dystrophies), showing hereditary transmission, are the main cause of blindness of non-infectious origin and include a heterogeneous series of molecular defects in the proteins involved in phototransduction, the visual cycle, or in the metabolism of photoreceptor or RPE cells. Over 150 different genes are known, the products of which are important for these processes and that, once mutated, are responsible for retinal degeneration (cf. http://www.sph.uth.tmc.edu/retnet/, provided in the public domain by the University of Texas Houston Health Science Center, Houston, TX). No therapy for these disorders currently exists.

The prototype of retinal degenerations is retinitis pigmentosa (RP), a group of hereditary conditions characterized by pigment deposition at the periphery of the retina, with a worldwide prevalence of approximately 1 in 3500, corresponding to over 1.5 million people worldwide. The disease is characterized by progressive peripheral vision loss and poor night vision (nyctalopia), with the eventual occurrence of central vision loss. Progressive worsening of symptoms is due to early degeneration of rods, followed by degeneration of cones at later times. The disease shows various modes of inheritance, including autosomal dominant, recessive, and X-linked. With advances in molecular research, it is now known that RP consists of many different retinal and RPE dystrophies caused by molecular defects in more than 100 different genes (see: Online Mendelian Inheritance of Man, OMIM: http://www.ncbi.nlm.nih.gov/omim/). The genes most frequently affected are those coding for opsin (the most frequent, about 10–20% of cases), the α and β subunits of cGMP-PDE, and the α subunit of the cGMP-gated channel; other genes involved are those coding for the proteins responsible for maintenance of the outer portion of photoreceptors.

In addition to RP, several other causes of retinal degeneration are known. These include, among others, *Leber's congenital amaurosis* (LCA), characterized by blindness in early infancy, with a prevalence of 1 in 100,000 newborns. This disease results from a defect in at least 11 different genes expressed in photoreceptors or RPE cells. In the former cells, the involved genes code for enzymes of the phototransduction cascade, structural proteins, or photoreceptor-specific transcription factors; in RPE cells, the LCA-associated genes code for proteins involved in vitamin A metabolism or in the recycling of retinoids between photoreceptors and the RPE during the visual cycle (cf. Figure 4.24).

Independent of the underlying causative mutation, therapies are very limited, essentially based on antioxidants, such as vitamin A/beta-carotene, docosahexaenoic acid (DHA), and omega-3 fatty acid, and the antioxidant nutritional supplements lutein and zeaxanthin.

Small (mice and rats) and large (dogs and pigs) animal models exist for a variety of these diseases, either obtained by transgenic technologies or naturally available. These animals offer the immensely valuable possibility to experiment with innovative treatment modalities for these disorders. In particular, various preclinical studies have been conducted over the last ten years by transferring a normal copy of the mutated cDNAs in different autosomal recessive forms of these disorders, or ribozymes or siRNAs targeted against the mutated alleles in the autosomal dominant forms, either as soluble molecules or expressed from viral vectors. Alternatively, various approaches have also exploited the genes coding for neurotrophic (GDNF, CNTF, BDNF, etc.) or antiapoptotic (Bcl-2) factors, similar to gene therapy for neurodegenerative disorders. In particular, CNTF has been shown to slow retinal degeneration in a number of animal models. Based on very encouraging Phase I clinical results, Phase II clinical trials are underway using an encapsulated form of RPE cells producing this factor, surgically placed into the eye. Since the retina is largely composed of non-mitotic, terminally differentiated cells, all these approaches require that treatment is started before retinal cell degeneration has occurred, that is, as early as possible along disease progression.

Attempts made by gene therapy have so far had particular success in a form of LCA due to mutations in the RPE65 gene, which represents about 10% of LCA cases. RPE65 is an essential enzyme in the visual cycle, acting as an isomerase for 11-cis-retinal synthesis (cf. above). Its deficiency leads to complete blindness from infancy; however, in contrast to most other retinal defects, the photoreceptors do not undergo apoptosis and the retina remains anatomically intact for a long time before significant degeneration. For this reason, this disease is an excellent candidate for correction by gene therapy. In addition, a large animal model is available, that is Briard dogs that are affected by a spontaneous mutation of RPE65, which causes a disease similar to the human one. In these dogs, injection of an AAV vector carrying the RPE65 cDNA into the subretinal space resulted in the expression of the functional protein by 20% of RPE cells and rescued retinal function as evaluated by both electrophysiological and, most important, behavioral tests. Restoration of visual function remained stable for at least 3 years after treatment, and will probably last longer.

Based on these exciting results, in 2007 two clinical trials started in the United Kingdom and United States. Studies were initiated in adults for ethical and practical reasons, even though the potential for functional restoration in this group may be limited, since the absence of photoreceptor function since birth leads to eventual retinal degeneration and likely incomplete innervation of the visual cortex. Three subjects from each trial received an AAV vector expressing RPE65, which was found to be safe, with some evidence of modest improvement in the patients' visual function. These results were sufficiently encouraging to consider moving to younger subjects, in whom the potential for more substantial reconstitution of visual function may be possible.

Finally, different forms of RP are inherited as autosomal dominant disorders. Therapy of these forms, therefore, can be attempted using siRNAs or ribozymes specifically targeting the disease allele. In some animal models of disease, this objective was pursued by either instillation of synthetic RNA molecules in the subretinal space or transduction of shRNA-expressing AAV or lentiviral vectors.

4.8.3
Gene Therapy of Retinal Neovascularization

Different common eye disorders are characterized by formation of a hypertrophic, pathological vascular network of the retina and the choroid. These include diabetic retinopathy, age-related macular degeneration (AMD) and retinopathy of prematurity (ROP). In particular, diabetic retinopathy and AMD are among the most common causes of progressive blindness in adult individuals, affecting millions of people worldwide.

In diabetic patients, hyperglycemia induces damage in all blood vessels, especially those with smaller diameter (diabetic microangiopathy). In the retina, this determines formation of microaneurysms, increase of vascular permeability, formation of small areas of swelling that have a yellowish white coloration (the so-called *cotton wools*), and small hemorrhages. This configures the condition of non-proliferative diabetic retinopathy. As the disease progresses, ischemia as a result of microvascular damage becomes progressively more important, and large areas of the retina show hypervascularization (proliferative diabetic retinopathy). The newly formed capillaries progressively develop on the whole surface of the retina and their fragile wall often breaks, generating hemorrhages in the vitreus which, in the long term, lead to retinal fibrosis. The fibrotic tissue, by contracting, can determine retinal detachment. At least 70% of patients with diabetes, after 20–30 years of disease, invariably develop signs of retinopathy. The disease currently represents the main cause of blindness in patients below 50 years of age.

In comparison, AMD is the most common cause of irreversible loss of vision in the elderly, occurring, with differing severity, in more than 30% of patients of 75 years or older; about 7% of individuals in this age group are affected by the most severe forms. The disease is typically manifested by the loss of central vision, thus significantly threatening the quality of life of the affected patients. AMD pathology is characterized by degeneration involving the retinal photoreceptors, RPE cells, and the Bruch's membrane (i.e., the inner layer of the choroid, in immediate contact with the RPE), as well as, in some cases, alterations in choroidal capillaries. AMD is broadly classified as either dry (atrophic, nonexudative) or wet (neovascular, exudative). Individuals with dry AMD typically experience a gradual reduction in central vision due to atrophy of the retina and RPE. In contrast, those with wet AMD suffer a more precipitous and profound loss of vision secondary to the development of choroidal neovascularization (CNV). Although 90% of those with AMD have dry disease, 80–90% of all individuals rendered legally blind from AMD have the wet form of the disease, since the disease develops in proximity to macula, i.e., the region of the retina with the highest concentration of photoreceptors, which facilitates central vision and allows vision at high resolution. Affected patients can no longer read or drive, and do not recognize faces; the disease frequently develops towards complete blindness. Histologically, the primary characteristic of late-stage dry AMD is the appearance of atrophy of the RPE, followed by the gradual degeneration of the nearby photoreceptors, resulting in thinning of the retina. Conversely, CNV is the defining characteristic of the most common wet or neovascular AMD, in which new vessels sprout from the choroidal vessels, penetrate the Bruch's membrane and grow into the subretinal space. No therapies are available for dry AMD. Before the development of molecular therapies – see below – about 5% of patients with the exudative form could benefit, in the early stages of disease, from argon laser photocoagulation or photodynamic therapy. These therapies, however, do not address the primary cause of the disease and are highly destructive of retinal function.

Recent research on the genetic and molecular underpinnings of AMD have brought to light several basic molecular pathways and pathophysiological processes (e.g., oxidative stress, inflammation, alternation of lipid metabolism, pathological angiogenesis) that mediate AMD risk, progression, and/or response to therapy. The retina is highly susceptible to oxidative stress given the elevated oxygen tension, high metabolic activity, exposure to photoradiation, and presence of photosensitizers, as further demonstrated by the increased disease risk with smoking, a potent inducer of oxidizing stress, and the decreased risk with diets rich in antioxidants. Persistent chronic inflammatory damage, including accumulation of macrophages, microglia, and soluble inflammatory components (complement factors and pro-inflammatory cytokines/chemokines), is invariably present in AMD lesions.

In both diabetic retinopathy and AMD, retinal hypervascularization is secondary to the development of hypoxia, which determines local production of pro-angiogenetic factors. This condition can be reproduced in at least three different animal models: (i) by laser-induced breakage of the Bruch's membrane; (ii) by keeping neonatal animals for 7–10 days at a concentration of oxygen higher than 80%, to cause closure of the retinal capillaries and thus determine hypoxia once the animals are exposed again to normal atmosphere (*oxygen-induced retinopathy*, OIR); and (iii) in transgenic mice in which VEGF or IGF-1 are overexpressed in the eye under the control of the opsin promoter. In particular, the last two animal models serve for the development of novel therapeutic strategies for proliferative diabetic retinopathy, since subretinal neovascularization develops from the retinal vasculature and not from the choroid (experimental diabetic animals do not develop retinal pathology similar to humans).

In both proliferative diabetic retinopathy and wet AMD, overproduction of VEGF has an essential pathogenetic role. In the animal models, retinal neovascularization is inhibited by blocking this factor by different strategies, some of which exploit methods of gene therapy, which can be categorized as follows.

(1) Inhibition of VEGF receptor function in endothelial cells. This can be obtained by drugs blocking enzymatic function of the receptors after VEGF activation or with siRNAs reducing their expression.

(2) Blocking of soluble VEGF to prevent receptor binding. This is attained using monoclonal antibodies (bevacizumab, ranibizumab), soluble receptors, or aptamers (pegaptanib). Administration of monoclonal antibodies into the vitreus (one administration per month) currently represents the gold standard of advanced pharmacological therapy for retinal neovascularization.

(3) Inhibition of VEGF production. This objective can be met by antisense oligonucleotides, ribozymes, or, more recently, siRNAs. In particular, the development of siRNAs is currently the focus of interest of different biotechnological companies and has already led to Phase I and II clinical trials. One of the developed products is bevasiranib, an anti-VEGF siRNA, which has already shown safety and efficacy in a multicenter clinical study involving 129 patients. This study was the first siRNA clinical application in humans.

In addition to VEGF, other pro-angiogenic factors sustain retinal neovascularization, including IGF-1, HGF, PDFG, FGF, and TGF-β. These factors can also be the targets of specific molecular therapies. Furthermore, another recently developed treatment is based on siRNA-mediated inhibition of *RTP801*, a gene that is highly induced by hypoxia, independent of VEGF. The application of this siRNA is currently the focus of a Phase I/II trial.

Since the inhibition of angiogenesis has no effect on the primary causes of the disease, the efficacy of these antiangiogenetic drugs is only transitory, and treatment therefore requires their repeated administration, unless some of the therapeutic genes are expressed in the context of viral vectors. In the experimental models, AAV vectors expressing soluble receptors inhibiting VEGF (in particular, soluble VEGR-1) or inhibitory peptides blocking VEGF binding to its receptors were shown to exert protective effects for prolonged periods of time once administered into the vitreus or the subretinal space.

Besides the local overproduction of proangiogenic factors such as VEGF, retinal hyper-vascularization is also due to the diminished release of antiangiogenic factors, which contribute to the homeostasis of blood vessel formation in the retina in normal conditions. One of these antiangiogenic factors, which is selectively produced by RPE cells, is *pigment epithelium-derived factor* (PEDF), which is highly expressed in normal conditions and is responsible for the avascular state of the cornea and the vitreus. The levels of this factor are reduced in patients with diabetic retinopathy or other forms of retinal neovascularization. Based on these observations, a Phase I, multicenter clinical study was recently conducted, entailing administration of a second-generation adenoviral vector (carrying a complete deletion of the E1 and partial deletions of the E3 and E4 regions), expressing PEDF in patients with wet AMD. In most of the 28 treated patients, who had received a single intravitreal vector administration, treatment was well tolerated, with only minor signs of inflammation in a minority of cases and without major adverse events. Based on these results, a larger study is now planned with the purpose of evaluating the clinical efficacy of this approach.

4.8.4
Gene Therapy of Retinoblastoma

Retinoblastoma is the primary intraocular tumor most prevalent in infants and children. The disease is due to the uncontrolled proliferation of cancer cells deriving from the neoplastic transformation of retinal progenitor cells. Transformation is caused by homozygous mutation and inactivation of the *Rb* gene, one of the major controllers of the G1 to S transition of the cell cycle. The affected children are usually treated by enucleation of the affected eye (or of both eyes), followed by radio- or chemotherapy.

In experimental models of the disease, administration of the suicide gene HSV-TK followed by gancyclovir treatment (see section on 'Gene therapy of cancer') was reported to be successful. Following these findings, a Phase I clinical trial was conducted in patients, based on the intravitreal administration of an HSV-TK-expressing adenoviral vector in one of the two affected eyes of 8 patients with bilateral retinoblastoma invasive into the vitreus; vector administration was followed by systemic treatment with gancyclovir. The intravitreal invasion of the tumor was reported to respond well to treatment, although in the presence of a strong inflammatory response. Eventually, however, both eyes were enucleated due to progression of the primary tumor. A positive evaluation of the results of this gene therapy study is thus questionable.

4.9
Gene Therapy of Cardiovascular Disorders

Despite the remarkable progress made over the last several years in early diagnosis and prevention, cardiovascular disorders (CVD) still represent the leading cause of morbidity and mortality in the industrialized world and are a rising concern in most developing countries. Before 1900, infectious diseases and malnutrition were the most common causes of death throughout the world, and CVD were responsible for less than 10% of all deaths. According to the World Health Organization (WHO), today CVD account for ~30% of deaths worldwide, corresponding to 17.5 million deaths in 2005. Of these deaths, 7.6 million (43%) were due to ischemic heart disease (IHD) and 5.7 million (32.5%) to stroke. About 80% of these deaths occurred in low- and middle-income countries (http://www.who.int/cardiovascular_diseases/en/).

The main cause of CVD is atherosclerosis. This is a pathology of medium–large-caliber arteries, characterized by endothelial dysfunction, inflammation, and presence of characteristic plaques, formed by lipids, cholesterol, calcium, and cellular debris, which form inside the artery tunica intima. Several risk factors contribute to plaque formation, some of which are non-modifiable (male gender, age, familiality), while others can be modified with lifestyle (smoking, food, physical activity) or pharmacological therapy (high cholesterol, hypertension, presence of co-morbidity such as diabetes).

Given the social, economic, and sanitary burden of CVD, the identification of novel therapeutic strategies interfering with the molecular mechanisms of disease development is absolutely required. In this context, the appreciation of the potential value of gene therapy has been steadily growing over the last few years. The utilization of nucleic acids as therapeutic tools for CVD is currently envisaged in at least three main areas: (i) treatment of cardiac and peripheral ischemic conditions by the induction of neoangiogenesis; (ii) treatment of heart failure to preserve or improve cardiac function; and (iii) prevention of restenosis after *percutaneous coronary intervention* (PCI). Table 4.14 reports a synopsis of the currently considered strategies for gene therapy in these three conditions, along with a list of the potential therapeutic genes having shown various degrees of success in the animal studies.

4.9.1
Ischemic Heart Disease

IHD is a condition in which there is an inadequate supply of blood and oxygen to a portion of the myocardium. The most common cause of IHD is atherosclerotic disease of epicardial coronary arteries determining inadequate perfusion of the myocardium supplied by the involved arteries.

Patients with IHD fall into one of two large groups: patients with chronic *coronary artery disease* (CAD), who most commonly present with stable *angina pectoris*, characterized by chest or arm discomfort that is reproducibly associated with physical exertion or stress and is relieved by rest and/or sublingual nitroglycerin, and patients with *acute coronary syndromes* (ACSs). The latter group is composed of patients with acute myocardial infarction (MI) with ST-segment elevation on their presenting electrocardiogram (STEMI) and those with unstable angina and non-ST-segment elevation MI (UA/NSTEMI).

Patients with CAD are usually treated with pharmacological therapy to lower risk factors and co-morbidities (for example, statins to lower the levels of LDL-cholesterol and inhibitors of the angiotensin-converting enzyme (ACE) to decrease blood pressure and ventricular dysfunction), reduce the episodes of angina (nitrates), and reduce the myocardial oxygen demand (beta-blockers and calcium-antagonists). Besides pharmacological thera-

Table 4.14 Therapeutic approaches and genes for cardiovascular gene therapy

Purpose	Therapeutic strategy	Therapeutic genes
Induction of therapeutic angiogenesis	Expression of angiogenic growth factor and chemokine genes	VEGF-A and -D
		FGF-1, 2, 4, 5
		PDGF-BB
		HIF-1α/VP16
		HGF
Gene therapy of heart failure	Normalization of the Ca^{2+} cycle	SERCA2a, inhibition of phospholamban (PLB)
	Modulation of β-adrenergic receptors	Inhibition of β-ARK, transfer of adenylate cyclase-6 (AC-6) or β2-AR
	Inhibition of apoptosis	Bcl-2
	Stimulation of compensatory hypertrophy pathways	VEGF-B, IGF-1, AKT, PI3K
Prevention of post-angioplasty restenosis	Expression of proteins inhibiting cell cycle progression	Rb, p53, p16, p21, p27, Gax
	Inhibition of expression of genes essential for cell cycle progression	Antisense oligonucleotides or siRNAs against cdk2, cdc2, c-myb, c-myc, ras, PCNA
	Induction of apoptosis	Antisense oligonucleotides or siRNAs against bcl-2; transfer of FasL
	Induction of cytotoxicity using suicide genes	HSV-TK
	Inhibition of extracellular matrix remodeling	TIMP
	Improvement of endothelial function	VEGF, eNOS, iNOS
	Inhibition of growth factors stimulating smooth muscle cell proliferation or activation	Antisense oligonucleotides or siRNAs against PDGF, TGF-β or their receptors

py, two interventional options are also available to these patients, one based on *coronary artery bypass grafting* (CABG, usually performed by a cardiac surgeon, connecting the internal mammary artery with the coronary artery distal to the obstructive lesion) and the other one, performed by an interventional cardiologist, consisting in PCI, commonly called *balloon angioplasty*. Compared to pharmacological therapy, CABG shows better prognostic efficacy (increase in life expectancy) in only a selected group of patients with high-risk atherosclerotic lesions, localized in specific anatomical segments of the coronary arteries. However, CABG usually decreases symptoms and thus improves quality of life in most patients. In comparison, PCI (defined as "elective" PCI to contrast it with "primary" PCI, performed in patients with acute MI) entails the use of an arterial catheter containing a

4

deflated balloon at its extremity. The catheter is inserted through the femoral artery and the balloon is positioned, under angiographic guidance, at the level of the atherosclerotic plaque in the coronary artery. The balloon is then inflated to mechanically destroy the plaque and restore vessel perviousness. Concomitant with angioplasty, an endovascular tubular prosthesis (*stent*) is usually positioned in correspondence with the artery segment where the balloon was inflated, having the function of keeping the vessel open and preventing restenosis. When compared to pharmacological therapy, PCI does not seem to offer a significant survival advantage. However, it significantly improves the patients' quality of life (episodes of angina, dyspnea, limitations of exercise capacity). PCI thus represents an extremely valuable alternative to CABG in the treatment of angina; as already reported above, CABG is only superior to PCI in a selected group of patients with high-risk lesions.

An acute event, such as rupture of the atherosclerotic plaque, can determine aggregation of platelets and activation of the coagulation cascade, with the formation of an intravascular thrombus, consisting of a solid mass of fibrin, platelets, and other circulating cells. Formation of a thrombus and consequent endothelial constriction determine an abrupt vascular occlusion, with cessation of blood flow. This causes an acute MI, with consequent necrosis of the tissue downstream of the occlusion. About 90% of MIs are due to the sudden formation of a thrombus obstructing an atherosclerotic coronary artery.

The extension of the infarcted area strictly depends on the duration and extent of the perfusion deficit and on the presence of a pre-existing collateral network. About 25–35% of patients with MI die before receiving medical assistance, usually due to ventricular fibrillation. For those who are rapidly hospitalized, prognosis has very significantly improved over recent years thanks to the possibility to revascularize the occluded artery. The sooner revascularization is obtained (possibly within the first 90 min, at most within the first 12 h), the smaller the residual infarcted area is, and the lower the mortality. Revascularization can be obtained by mechanical or pharmacological treatment. The former consists in primary PCI, that is balloon angioplasty performed in emergency conditions, with or without stent placement. Primary PCI usually restores blood flow in more than 90% of patients. If this procedure cannot be performed within 90 min of the initial medical contact, pharmacological thrombolysis can be attempted. This takes advantage of thrombolytic agents, able to digest the fibrin network within the thrombus. These compounds typically consist of plasminogen activators (streptokinase, tissue plasminogen activator (tPA)), which are enzymes transforming plasminogen into plasmin, which in turn enzymatically degrades fibrin and leads to dissolution of the thrombus. Collectively, patients with MI who are rapidly hospitalized and are mechanically or pharmacologically revascularized have a very high probability of survival (90–95%).

4.9.2
Peripheral Artery Disease

Atherosclerosis is a systemic disease that, besides the myocardium, affects medium- and high-caliber arteries in all districts of the body. When this occurs at the levels of the large arteries of the limbs (in particular, the iliac, femoral, popliteal, and tibial arteries) and stenosis significantly impairs perfusion of the muscles supplied by the involved vessels, a condition of *peripheral artery disease* (PAD) ensues. This condition is very frequent, since it affects approximately 12% of individuals above 65 years of age.

Fewer than 50% patients with PAD are asymptomatic and diagnosis is only made after targeted investigation (typically, analysis of blood flow by color Doppler flow imaging). The most common symptom is intermittent claudication: patients, while walking, are forced to stop repeatedly due to leg pain, typically in the calves. Pain is due to the accumulation of lactic acid consequent to muscle activity in anaerobiosis, since the arterial stenosis does not allow proper supply of oxygenated blood – lactic acid stimulates peripheral pain receptors. Patients with more advanced disease also experience pain in the absence of physical exercise, at rest, or at night. Finally, last-stage patients are characterized by the occurrence of ulcerative, necrotic lesions, presence of gangrene, or amputations, sometimes spontaneous, at various levels of the foot or leg; in these patients, perfusion of the most peripheral districts is virtually abolished. Patients with rest pain and/or trophic lesions are commonly associated with the category of "critical ischemia", in which the occlusive plaque, typical of the intermittent claudication stage, further expands to almost completely block the arterial lumen or undergoes rupture or thrombosis. Critical limb ischemia is estimated to develop in 500–1000 individuals per million per year.

The severity and evolution of PAD strictly depend on the associated risk factors, in particular smoking, diabetes, and hypertension. In a relatively low number of cases, ischemia can become so severe as to require amputation of the foot, leg, or entire limb (2–10% of cases). However, in most cases, presence of PAD is a sign of a more general condition of advanced atherosclerosis of all body districts, and thus patients with PAD have a 4–5 times higher risk of MI or stroke than the general population. In particular, patients with critical limb ischemia (corresponding to 5–10% of PAD patients) show a mortality rate at 5, 10, and 15 years of 20, 50, and 70% respectively.

4.9.3
Mechanisms of Blood Vessel Formation

One of the most appealing novel therapeutic strategies, which can find application in the treatment of patients with CAD or PAD and in all conditions in which increased blood perfusion is required such as wound healing or reconstructive surgery, is based on the delivery of factors able to stimulate new blood vessel formation.

Formation of the vasculature is a complex process, which typically follows two pathways: one mainly occurring during embryonic development (vasculogenesis) and the other (angiogenesis) operating in adult life.

The term **vasculogenesis** traditionally refers to the formation of a primitive capillary network starting from pluripotent stem cells, named *angioblasts*, during embryonic development. These cells proliferate to form a primary vascular plexus, which then undergoes pruning and further maturation. Experimental evidence seems to indicate that a similar process might also occur in adult organisms secondary to the mobilization of endothelial progenitor cells (EPCs) from the bone marrow and their recruitment to the sites of new blood vessel formation. Supporting this possibility, antigenic markers common to angioblasts and HSCs have been identified, consistent with the possibility that a common precursor might exist (the hemangioblast), able to originate both hematopoietic and endothelial cells. However, the real existence of EPCs and their relevance in tissue revascularization in adult organisms is now questioned by several investigators.

Traditionally, new blood vessel formation in adult life is believed to occur exclusively through capillary sprouting from pre-existing vessels, a process known as **angiogenesis** (Figure 4.25). The process is initiated by the metabolic activation, proliferation, and migration of endothelial cells, concomitant with vast remodeling of the extracellular matrix. Subsequently, the newly formed capillaries progressively mature by the addition of mural cells (pericytes and smooth muscle cells), allowing proper functionality, and a vascular network formed by larger vessels (arterioles and venules) is eventually formed.

In physiological conditions, the main regulator of angiogenesis is hypoxia. In conditions of low oxygen tension, the cellular transcription factor hypoxia inducible factor 1 (HIF-1) is post-translationally stabilized and activates the expression of a vast series of genes, including VEGF and its receptors, which are required for all phases of angiogenesis (in addition to vasculogenesis in the embryo).

The VEGF family consists of at least five members (VEGF-A, B, C, D, and placental growth factor (PlGF)), with different isoforms arising from alternative splicing. In particular, the VEGF-A gene originates at least four different polypeptides of 189, 165, 149, and 121 amino acids. The action of the VEGF factors is mediated by three specific receptors (VEGFR-1/Flt-1, VEGFR-2/KDR/Flk-1, and VEGFR-3), which interact differently with the various family members. More specifically, VEGF-A binds VEGFR-1 and -2; VEGF-B and PlGF selectively bind VEGFR-1, while VEGF-C and VEGF-D bind VEGFR-3

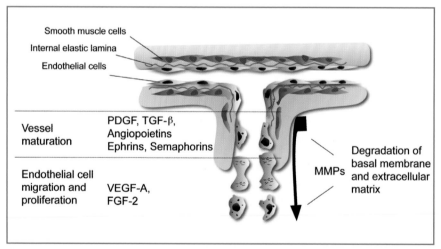

Fig. 4.25 Angiogenesis. The formation of new blood vessels in adult organisms occurs by sprouting from the pre-existing vasculature. Endothelial cells are activated, and proliferate and migrate to form a collateral branch under the control of angiogenic factors such as VEGF-A and FGF-2. Maturation of the newly formed vessel follows, controlled by other soluble factors, including PDGF and TGF-β, which recruit and promote proliferation of smooth muscle cells, and various members of the angiopoietin, ephrin, and semaphorin families. The whole process is accompanied by extensive remodeling of the extracellular matrix by matrix metalloproteases (MMPs)

(Figure 4.26). The last two factors mainly induce a lymphangiogenetic response, leading to the formation of lymphatic rather than hematic vessels. Over the last few years, it has become increasingly apparent that these main receptors act in concert with several other co-receptors, belonging to the integrin, cadherin, neuropilin, and ephrin receptor families; the expression of these co-receptors on endothelial cells potentiates or modulates the proliferative and chemotactic effect of VEGF. In addition, different members of the ephrin and semaphorin (neuropilin ligands) families, originally identified as guides of axonal growth in the developing nervous system, have recently been shown to exert similar roles in the nascent vasculature, where they act in concert with VEGF.

Besides endothelial cell sprouting, functional new blood vessel formation also requires other factors that act at later time points to promote vessel maturation. Among these, an essential role is played by the angiopoietins and, in particular, by angiopoietin-1 (Ang1), which, by interacting with the Tie-2 receptor on the endothelial surface, reduces the permeability of the newly formed vasculature.

In addition to VEGF, several other growth factors are able to activate endothelial cells and promote the angiogenic response. Among these factors, much interest has been engendered by some members of the fibroblast growth factor (FGF) family. By interacting with different ubiquitous receptors, these factors are able to stimulate, on one hand, endothelial cell proliferation, and, on the other hand, the secretion of proteases that are essential for extracellular matrix degradation and cellular migration in the initial phases of the angiogenic process. There are 23 structurally similar forms of FGF. The specific role of these different FGF family members during normal angiogenesis *in vivo* and their relevance in

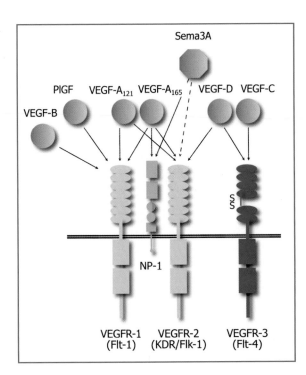

Fig. 4.26 VEGF and VEGFR family members. The five VEGF family members show different specificity for the three canonical VEGF receptors. Neuropilin-1 (NP-1), a co-receptor for the 165 aa-isoform (but not for the 121 aa-isoform) of VEGF-A, is a specific target of semaphorin 3A (Sema3A)

different physiological and pathological conditions still need to be thoroughly addressed. FGF-1, -2, -4, and -5 have been used in angiogenic studies without reporting significant differences in efficacy.

Finally, in addition to vasculogenesis and angiogenesis, a third process of blood vessel formation exists, known as **arteriogenesis**. This consists in the generation of functionally competent medium- and large-caliber arteries, having a tunica media able to properly control patency of the vascular lumen in response to physiological stimuli of vasoconstriction or vasodilatation. A classic example of arteriogenesis in the adult is the formation of a collateral network, visible upon angiography, in response to the progressive occlusion of an arterial vessel by an atherosclerotic plaque.

4.9.4
Gene Therapy for Therapeutic Angiogenesis

Despite the apparent complexity of the molecular and cellular mechanisms leading to the formation of new blood vessels, which require the spatial and temporal coordination of a number of different factors, at least four considerations support the relative feasibility of a gene-therapy-based approach to induce therapeutic angiogenesis in CAD and PAD patients.

First, some of the factors normally involved in angiogenesis (including HIF-1 or VEGF-A) act as master regulatory genes, able alone to trigger activation of the cascade of cytokines required for complete and functional angiogenesis.

Second, basic research over the last several years has provided an impressive list of genes encoding factors that are involved in new blood vessel formation; for several of these genes, preclinical proof of efficacy of their overexpression had already been obtained.

Third, not only the peripheral districts in PAD but also the heart in CAD patients are easily accessible thanks to modern interventional cardiology procedures. The myocardium can be reached by inoculating a therapeutic factor or gene into the coronary arteries through arterial catheterization; or by direct injection through the epicardium from a minithoracotomy or during CABG surgery; or by direct injection through the endocardium, using a catheter positioned inside the left ventricle. Alternatively, it is also possible to insert the therapeutic construct into the coronary sinus by a venous route and, upon increase of venous pressure, retrogradely stimulate gene transfer to the myocardium.

Fourth, and probably most important, gene therapy applications in the cardiovascular field will be successful even if they achieve relatively modest results when compared to the more demanding goals that are required, for example, by gene therapy for hereditary disorders. In the specific context of therapeutic angiogenesis, even a modest increase in cardiac perfusion might significantly improve cardiac function and thus positively impact on the clinical course of the ischemic disease.

After the identification of VEGF and FGF as powerful inducers of angiogenesis, various clinical trials were conducted with the purpose of evaluating the effect of these factors when delivered as recombinant proteins (in particular, TRAFFIC and FIRST, which assessed the effect of recombinant FGF-2 on PAD and CAD respectively, and VIVA, which evaluated recombinant VEGF-A in patients with CAD). Quite unexpectedly, and in contrast to the brilliant results obtained by the same formulations in different animal models of acute or progressive ischemia, the administration of these recombinant proteins produced only modest or no result at all in the treated patients. The negative outcome of these trials might be attributed to a number of different reasons, including the very short half-life of these cytokines *in vivo* and the desensitization of chronic ischemic tissues to growth factor treatment.

4.9.4.1
Trials Based on the Injection of Naked Plasmid DNA

The major gene therapy clinical trials for the induction of therapeutic angiogenesis are summarized in Table 4.15. The first trials were based on a series of animal studies that had shown that the injection of naked DNA plasmids coding for VEGF or FGF-2 was capable of successfully restoring blood flow in various preclinical models of myocardial and limb ischemia, in both small and large animals. The rationale to inject naked DNA is based on the ability of skeletal muscle and cardiac cells to internalize small amounts of DNA from the extracellular environment, as discussed in the Chapter on 'Methods for Gene Delivery'. Clinical experimentation in patients began in the mid-1990s, by identifying patients with critical limb ischemia who had exhausted all conventional options for revascularization. The treatment consisted in the injection of a plasmid coding for the 165-amino acid isoform

Table 4.15 Major gene therapy clinical trials for the induction of therapeutic angiogenesis

Vector	Gene	Disease	Route	Name of trial (year)
Naked DNA plasmid	VEGF-A 165 aa	PAD	Intramuscular	(1998)
		CAD	Intramyocardial through minithoracotomy	(1998–2005)
			Transendocardial catheter-based	(2002) Euroinject 1 (2005)
	FGF-1 (NV1FGF)	PAD	Intramuscular	2002
	HGF	PAD	Intramuscular	2004
	Del-1	PAD	Intramuscular	2004
Plasmid DNA/liposomes	VEGF-A 165 aa	PAD	Intraarterial after PTA	2002
		CAD	Intraarterial after PCI	KAT (2003)
Adenoviral vectors	FGF-4	CAD	Intracoronary	AGENT (2002, 2003)
	VEGF-A 121 aa	PAD	Intramuscular	RAVE (2003)
		CAD	Intramyocardial during CABG or via minithoracotomy	REVASC (1999, 2002)
	VEGF-A 165 aa	PAD	Intraarterial after PTA	(2002)
		CAD	Intraarterial after PCI	KAT (2003)
	HIF1α/VP16	PAD	Intramuscular	(2003)

of VEGF-A. The success of this treatment was evaluated by angiography and nuclear magnetic resonance, revealing formation of new collateral vessels and significant perfusion
improvement in the VEGF-treated group. The only reported major side effect was the
occurrence of a remarkable edema of the leg, which was attributed to the increase in vessel permeability caused by the factor. A series of over 20 clinical trials followed, which
entailed the injection of naked plasmids encoding VEGF to the myocardium through a mini
left anterior thoracotomy or by a transendocardial approach using a dedicated intraventric-
ular catheter, by which injection was performed in the still electrically active however
mechanically silent regions of the myocardium, indicative of a state of myocardial suffer-
ing. Most of these procedures were well tolerated, with few major adverse cardiac events
and without complications directly related to gene expression. The results of these early
studies performed by naked DNA injection should be interpreted with caution, considering
their uncontrolled, open-label design and the significant sham or placebo effect observed
in any intervention in patients with CAD. Indeed, a double-blinded, placebo-controlled trial
performed in Europe a few years later (the Euroinject One trial) failed to show any significant difference between injection of a plasmid encoding VEGF and placebo.

Other ongoing clinical trials, also based on naked plasmid DNA gene transfer, take
advantage of a plasmid encoding FGF-1 (aFGF) fused to a heterologous secretion signal
under the control of a constitutive strong promoter (NV1FGF). A Phase I, multicenter,
open-label clinical study was initially conducted in patients with end-stage PAD by the
intramuscular injection of increasing and repeated doses of this plasmid. Following the
results obtained, a Phase IIb, double-blind, randomized, placebo-controlled clinical trial
took place in 6 European countries in patients with critical limb ischemia at high risk of
amputation. In these trials, no improvement in wound healing was observed, however the
treatment significantly reduced the risk of amputation and might thus lower mortality rates
in the treated patients. This approach is being further evaluated in a larger, Phase III study.

Finally, two additional trials are exploiting naked DNA gene transfer in patients with
PAD. The first is based on the delivery of the hepatocyte growth factor (HGF) cDNA, a
cytokine that, besides mitogenic potential on liver cells, also has angiogenic activity and
plays a role in the regeneration of tissues damaged by various forms of injury. The second
trial is based on the administration of a plasmid encoding the angiomatrix protein Del-1
(developmentally regulated endothelial locus 1), a unique $\alpha v \beta 3$ integrin ligand that is normally produced by endothelial cells and mediates cell attachment, migration, and activation of cytoplasmic signaling molecules in focal contacts.

4.9.4.2
Trials Based on Adenoviral Vectors

Despite the relative ease of production of naked plasmid DNA, the clinical success
obtained by their direct injection is generally poor, and the therapeutic benefit expected
from the ongoing experimentation is thus limited. This is mainly due to the low efficiency of tissue transfection and the relatively short period of transgene expression after naked
DNA injection – see also below. Given these limitations, other clinical trials have exploited first-generation adenoviral vectors to deliver various therapeutic genes.

The angiogenic gene therapy (AGENT) series of studies have addressed the safety and efficacy of the non-surgical, intracoronary injection of an adenoviral vector expressing FGF-4 (Ad5FGF4) in patients with stable angina. These trials established that intracoronary administration of this vector could be performed with reasonable safety in patients with CAD, and that a one-time dose could provide an anti-ischemic effect up to 12 weeks of evaluation. Further appraisal of the efficacy and safety of this vector was then planned in two simultaneous Phase IIb/III multicenter, randomized, double-blind, placebo-controlled pivotal trials in the United States and the European Union, with the designed enrollment of approximately 1000 treated subjects. Unfortunately, however, an interim analysis of the preliminary data from one of these trials has clearly indicated that the trial, as designed, would provide insufficient evidence of efficacy. Therefore, further recruitment was stopped at the beginning of 2004.

Another series of clinical studies was based on the delivery of a defective adenovirus expressing the 121-amino acid isoform of VEGF-A. The RAVE trial was a Phase II study designed to test the efficacy and safety of intramuscular delivery of this vector to the lower extremities of 105 subjects with PAD. This study showed that a single unilateral intramuscular administration of this vector was not associated with improved exercise performance or quality of life, and did not support further experimentation. The same vector was also tested in two different Phase I trials in patients with CAD entailing direct intramyocardial injection during CABG or via a minithoracotomy. Vector administration was well tolerated without any consistent drug-related side effects and there were indications towards improvement in angina symptoms as well as in myocardial perfusion scans. This justified activation of a Phase II randomized, prospective, 'proof-of-concept' trial in 'no-option' patients. This Phase II trial, called REVASC, involved 20 sites in North America; 71 patients with severe CAD were enrolled. The preliminary results showed significant improvements in both cardiac function and quality of life. Another multicenter, randomized, double-blind, placebo-controlled study is being conducted in Denmark, Israel, and the United Kingdom to evaluate the efficacy of Ad-VEGF121 in patients with advanced CAD not amenable to percutaneous coronary revascularization or bypass grafting.

Two trials have assessed the efficiency of an adenovirus expressing the 165-amino-acid isoform of VEGF (VEGF-Adv) or of a plasmid expressing the same gene delivered in a liposome formulation (VEGF-P/L) injected intra-arterially after percutaneous transluminal angioplasty in patients with PAD and CAD (KAT trial). Both the adenovirus and plasmid formulations were reported to improve the vascularity of the treated limbs 3 months after therapy and to enhance myocardial perfusion in the coronary heart disease patients 6 months after therapy.

Finally, a Phase I clinical trial evaluating an adenovirus vector expressing HIF-1α/VP16 (Ad2HIF-1α/VP16) was conducted for the treatment of PAD. HIF-1 is a trimeric master transcription factor controlling the expression of several genes involved in a variety of processes in endothelial cells, including glucose metabolism, erythropoiesis, regulation of vascular tone, and cell proliferation and survival. Among the genes that are activated, several exert powerful angiogenic activity, including VEGF; as neovascularization results from the complex interplay of a variety of factors, an upstream regulatory protein such as HIF-1 could potentially be more effective as a therapeutic agent than any single pro-angiogenic factor. Intracellular oxygen concentration regulates HIF-1 activity by

4

influencing both stability and transcriptional activity of the HIF-1α subunit of HIF-1; under normoxic conditions, HIF-1α is targeted for ubiquitination and proteasomal degradation. A constitutively active version of HIF-1α was generated by replacing the C-terminus of the protein, including its oxygen-dependent and endogenous transactivation domains, with the strong transactivation domain from the herpes virus VP16 protein. After the successful demonstration of the safety of the constructed vector, a Phase II trial was started in patients with PAD. This is a placebo-controlled study that will enroll up to 200 patients at approximately 35 medical centers in the United States and Europe.

What are the main lessons learned from these clinical trials? Despite over 10 years of clinical experimentation, it is very clear that gene therapy for ischemic heart and PAD is still in its infancy. An initial series of at least seven cohort studies, in which patients have been followed up for up to 2 years post-injection, reported highly positive results. These studies, however, essentially suffered from a lack of control groups. In contrast, the overall outcomes of the randomized trials have been much more disappointing. While the definitive results of ongoing experimentation have to be carefully evaluated before a definitive conclusion is drawn, it nevertheless appears that novel vectors and improved delivery methods are needed before definitive clinical success might be met.

Irrespective of the therapeutic gene involved, it appears that the efficiency of gene delivery currently represents one of the major limitations that still hamper clinical success. Plasmid DNA delivery is simple and devoid of major safety concerns; however the efficiency of uptake of naked DNA by muscle and cardiac cells, although surprisingly higher than with most other cell types, is still very poor. Despite preparations of plasmids injected in quantities to the order of hundreds of micrograms or milligrams, the levels of the DNA internalized by the cells remain in orders of magnitude lower than those obtained using viral vectors. In addition, most preclinical investigations have revealed that measurable levels of gene expression are maintained only for the first couple of weeks after injection. This condition might not be sufficient to exert an angiogenic stimulus able to generate a stable neovasculature.

On the other hand, adenoviral vectors are fraught with several problems, mainly due to the strong inflammatory and immune response they elicit. This essentially reduces the vector dose that can be injected, prevents the possibility of vector re-injection, and limits the expression of the transgene to 1–2 weeks after transduction, after which the host's immune system eliminates the transduced cells (see section on 'Viral Vectors'). A completely new generation of adenoviral vectors devoid of viral genes, which are still infectious but less immunogenic, would definitely be desirable. However, after over 10 years of development of gutless adenoviral vectors, the problems intrinsic to the development of these vectors have still not been solved.

Given the current concerns on safety and performance of lentiviral vectors, the only viral vector system that, at present, appears suitable for gene therapy of CVD is that based on AAV. As outlined above, these vectors display a number of appealing features for gene transfer to the heart and the skeletal muscle. Among these features are the lack of relevant immunogenicity, the absence of an inflammatory response at the site of injection, the possibility of obtaining relatively pure vector preparations at high titers, and the capacity to transduce cells at high multiplicity of infection, which allows mixing of different preparations and thus therapy with gene cocktails. In addition, even more relevant characteristics are the specific – and still unexplained – tropism for muscle cells and cardiomyocytes, and the ability of these vectors to drive expression of the therapeutic gene they carry for indefinite periods of time in these post-mitotic tissues. Over the last years, more than 100 AAV variants have been isolated from humans and non-human primates, some of which show even higher transduction of skeletal muscle (AAV1 and AAV6) and heart (AAV8 and AAV9) than the prototype AAV2 serotype.

Besides tropism, another characteristic distinguishing AAV vectors is the prolonged expression of their transgenes, lasting months or years. In this respect, it is worth noting that the induction of therapeutic angiogenesis in patients is a much more ambitious goal than in experimental animals. In the latter, in fact, ischemia is induced acutely (for example, by femoral or coronary artery ligation) or subacutely (by application of an ameroid constrictor around a coronary artery, to induce ischemia in a few weeks' time). In contrast, ischemia in humans ensues chronically as the results of month- or year-long periods of progressive worsening of arterial narrowing due to atherosclerosis. This chronic process usually determines un- or hypo-responsiveness of the ischemic tissue to angiogenic stimulation, and thus sustained and prolonged stimulation with angiogenic factors is likely to be required to induce formation of a neovasculature.

Given all these requirements, it might be reasonably expected that AAV vectors will play a central role in clinical cardiovascular gene therapy experimentation in the near future.

Finally, it is important to note that, in principle, one important concern raised by gene therapy to induce therapeutic angiogenesis is the possible acceleration of tumor growth. Indeed, new blood vessel formation is absolutely required for cancer growth (tumor angiogenesis), and this process might be accelerated by the delivery of angiogenic factors and their release into the circulation. All patients are therefore carefully screened for the possible presence of a hidden tumor before enrollment in a gene therapy clinical trial for therapeutic angiogenesis.

4.9.5
Heart Failure

The term *heart failure* (HF) refers to a pathological condition characterized by an alteration of cardiac function by which the heart is unable to pump sufficient blood to meet the metabolic requirements of the organism. Clinically, the cardinal symptoms of HF are fatigue and dyspnea, associated with the presence of pulmonary and/or peripheral edema.

Approximately one-half of patients who develop HF have a systolic failure, due to a depressed capacity of the left ventricle to contract, while the other half have diastolic failure, when the ventricle wall is incapable of adequately relaxing during diastole. Ventricular function is commonly measured echocardiographically by evaluating the ejection fraction (EF), a parameter measuring the fraction of blood ejected by the ventricle, normally 50–70% relative to its end-diastolic volume. Patients with systolic failure have enlarged ventricles and depressed EF (less than 40%), while EF is relatively preserved in those with diastolic failure (40 and 50%).

The main cause of HF (65–70% of patients) is CAD, which by itself is a progressive disease, eventually determining chronic myocardial dysfunction. In addition, since adult cardiomyocytes are unable to replicate, the acute ischemic events (MI) that are not rapidly or completely revascularized determine substitution of the downstream necrotic myocardium with fibrotic tissue, which is incapable of contracting. Thus, the natural history of IHD inexorably leads towards a condition of HF. Hypertension contributes to HF development in 75% of patients, including most patients with CAD. In 20–30% of HF cases with a depressed EF, the exact etiology is not known, and these patients are referred to as having a primitive disease of the myocardium (idiopathic cardiomyopathy, which, in at least 30% of cases, is genetically determined by mutations of genes encoding cytoskeletal proteins or nuclear membrane proteins). Dilated cardiomyopathy is also associated with Duchenne's, Becker's, and limb-girdle muscular dystrophies (cf. section on 'Gene Therapy of Muscular Dystrophies').

Despite many recent advances in the evaluation and management of HF – especially thanks to the introduction of ACE inhibitors, the antagonists of the β-adrenergic receptors (β-blockers), and the angiotensin II receptor antagonists (sartans), in addition to aldosterone antagonists and diuretics – prognosis of HF remains poor. More than 60% of patients die within 5 years, mainly from worsening HF or as a sudden event (probably because of a ventricular arrhythmia). Up to 16% of patients are readmitted to hospital within 6 months of the first admission. As a consequence, HF is the most frequent (about 20%) cause of hospitalization in the population over 65 years. This percentage has shown a general tendency to increase over the last 30 years, even though the overall prevalence of cardiovascular disease has decreased.

Taken together, these figures underline the absolute need to develop innovative strategies to cure or at least prevent progression of this disease.

4.9.6
Molecular Regulation of Cardiac Contraction in Normal Conditions and During Heart Failure

The possibility to identify novel therapeutic strategies for HF is strictly based on the precise understanding of the molecular correlates of cardiac contraction in normal and pathologic conditions.

Myocardial contractility essentially depends on oscillations in intracellular Ca^{2+} concentration. The mechanism by which an electric signal is converted into mechanical action is known as *excitation–contraction coupling* (ECC) (Figure 4.27). This starts with the depolarization of the plasma membrane (sarcolemma) of cardiomyocytes, which induces the opening of membrane L-type voltage-dependent Ca^{2+} channels, which permit entry of a small quantity of Ca^{2+} into the cytosol; this in turn determines massive release of Ca^{2+} from the sarcoplasmic reticulum stores through tetrameric protein channels formed by the ryanodine receptors (isoform RyR2). Massive entry of Ca^{2+} into the cytosol triggers biochemical coupling between actin and myosin, which is mediated by Ca^{2+} binding to troponin C, and subsequent contraction. In the relaxation phase, RyR2 is inhibited by the FKBP12.6 (FK506 binding protein 12.6 or castalbin-2) protein, which prevents aberrant opening of the channel formed by this protein. The released Ca^{2+} is in part re-conveyed into the sarcoplasmic reticulum by a specific pump, the membrane ATPase SERCA, and in part eliminated from the cell by the Na^+/Ca^{2+} exchanger (NCX). In the myocardium, the main SERCA protein is SERCA2a, while SERCA1 is mainly expressed in the skeletal muscle.

The activity of the SERCA2a pump is normally controlled by association of this protein with phospholamban (PLB). In its non-phosphorylated form, PLB inhibits SERCA2a, while phosphorylation blocks this inhibition. The main kinase phosphorylating PLB in cardiomyocytes (followed by pump activation) is the cAMP-dependent protein-kinase A (PKA), which is typically under the control of β-adrenergic stimulation (cf. below). Besides PLB, SERCA2a activity is regulated by the Ca^{2+}/calmodulin-dependent kinase II, CaMKII, which increases activity of the pump according to intracellular Ca^{2+} concentration.

In addition to the aforementioned regulatory pathways, which are intrinsic to the cardiomyocyte cell, the contractile activity is under extrinsic control by the adrenergic and cholinergic systems, and by the circulating catecholamines. In particular, the adrenergic system (activated by adrenalin and noradrenalin) can significantly increase contractile strength, relaxation, and heart rate. In normal conditions, the cardiomyocyte sarcolemma exposes two types of G-protein-coupled (GPCR) β-adrenergic receptors (β-ARs): β1 (about 75–80%) and β2, associated with a membrane heterotrimeric G protein (with a $G_{\alpha s}$ subunit). β-AR stimulation determines, through this G protein, activation of an adenylate cyclase (AC) located on the cytosolic side of the receptor complex, which catalyzes conversion of ATP to cAMP. This in turn activates PKA, which phosphorylates: (i) the L-type Ca^{2+} channels, thus determining further Ca^{2+} entry during each depolarization cycle; (ii)

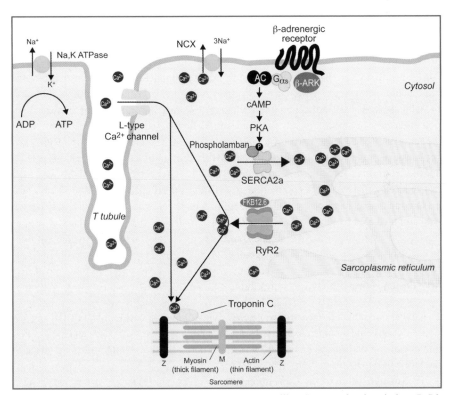

Fig. 4.27 Molecular mechanisms controlling cardiac contractility. See text for description. *RyR2*, ryanodine receptor-2; *β-ARK*, β-adrenergic receptor kinase; *NCX*, Na^+/Ca^{2+} exchanger

4

RyR2, causing dissociation of the inhibitory protein FKB12.6; and (iii) PLB, blocking its inhibitory activity on SERCA2a. These modifications amplify the efficacy of Ca^{2+} release and re-uptake every cardiac cycle, and are part of an evolutionarily conserved mechanism by which the sympathetic system determines rapid increase of cardiac output in response to muscle activity or sudden stress (*fight-or-fly* response).

Multiple experimental evidence over recent decades indicates that, in HF cardiomyocytes, the EEC system is defective. In particular, systolic Ca^{2+} is reduced, diastolic Ca^{2+} is increased, and the Ca^{2+} transient duration is significantly longer. These alterations are likely due to the reduced activity of the SERCA2a pump, of which both the mRNA and protein levels are decreased. In addition, during HF, the PLB/SERCA2a ratio is increased and the phosphorylated fraction of PLB is decreased, following desensitization of the β-adrenergic signaling pathway (cf. below). Besides these modifications in the levels and function of SERCA2a and PLB, the RyR2 receptors, which mediate Ca^{2+} release from the sarcoplasmic reticulum, also appear altered. In particular, they are hyperphosphorylated, and thus dissociated from the FKBP12.6 protein, which contributes to the abnormally high Ca^{2+} levels found in diastole.

One of the peculiar characteristics of HF is the activation of the sympathetic system. β-Adrenergic stimulation initially has an adaptive role, trying to compensate for the relative EEC inefficiency by increased stimulation. However, this prolonged β-adrenergic stimulation determines a progressive desensitization of the receptors: the levels of β1 receptors are diminished, those of the β-adrenergic receptor kinase (β-ARK) are increased (this kinase, also called G-protein coupled receptor kinase 2 (GRK2), phosphorylates the β-ARs, decreasing their activity and determining their degradation), and finally, both β-AR1 and β-AR2 are partially uncoupled from the components of the downstream signal transduction pathway, in particular the $G_{\alpha s}$ and AC proteins. At the same time, the noradrenalin stores in the cardiac nerve terminals are exhausted, thus rendering cardiac function insensitive to physiologically inotropic stimulation, in particular physical exercise. At the intracellular level, the L-type Ca^{2+} channels, the NCX exchanger, and the RyR2 sarcoplasmic pump are all hyperphosphorylated. The net result of these modifications is leakage of Ca^{2+} from its sarcoplasmic stores through the RyR2 channels; this explains why EEC has reduced efficiency and contraction is defective in HF patients. These events also explain the marked therapeutic effect exerted by β-blockers, which inhibit the maladaptive response mediated by the β-ARs and restore the normal Ca^{2+} cycle homeostasis.

4.9.7
Gene Therapy of Heart Failure

The above-outlined pathogenetic mechanisms underlying the development of HF suggest a series of possible therapeutic approaches based on gene therapy. These can be classified into one of two categories, having as a target the normalization of either the Ca^{2+} cycle or the levels of β-AR stimulation.

(i) Normalization of the Ca^{2+} cycle

Experimental evidence in both failing hypertensive rats and larger animals show that overexpression of the SERCA2 gene using adenoviral or AAV vectors significantly improves systolic and diastolic cardiac function. Based on these encouraging results, two Phase I/II clinical trials recently started in HF patients, based on the percutaneous administration of the SERCA2a gene using an AAV6 vector in one case and an AAV1 vector in the other case. These are the first clinical gene therapy trials for HF. The initial results in 9 patients showed safety of the transduction procedure and marginal improvement of HF, thus supporting initiation of Phase II protocols.

An alternative approach to obtain the same functional result consists in the inhibition of PLB function by the knock down of its mRNA using ribozymes or siRNAs, or by using transdominant negative mutants. The mutant that is most used for this purpose contains a phosphomimetic mutation on serine 16 of the protein (PLB S16E), which is the amino acid normally phosphorylated by PKA to release SERCA2a from inhibition.

Additionally, the properties of two Ca^{2+}-binding proteins can be exploited to normalize the Ca^{2+} cycle in the failing cardiomyocytes. The first one is parvalbumin, a protein able to sequester Ca^{2+}, which, in normal conditions, is exclusively expressed in fast-twitch skeletal muscle fibers and neurons. In contrast to SERCA2a overexpression, parvalbumin would allow the energy-independent removal of cytosolic Ca^{2+}, thus preventing the pathological persistence of this ion in diastole without further energy expenditure. The other Ca^{2+}-binding protein is S100A1, which is normally expressed in cardiomyocytes, however at reduced levels during HF. S100A1 stabilizes the RyR2 receptors during diastole and increases Ca^{2+} release during systole. Thus, delivery of the cDNA coding for this protein in a failing heart should restore normal Ca^{2+} levels, on one hand by facilitating their re-uptake into the sarcoplasmic reticulum while, on the other hand, blocking their inappropriate exit through the RyR2 receptors during the relaxation phase.

(ii) Modulation of β-adrenergic receptors

An alternative possibility to exploit gene transfer for therapeutic purposes in HF is to modulate the β-adrenergic response, in particular to combat the receptor desensitization proper of the failing condition. In this respect, however, it is important to remember that this approach has to be followed attentively, since overexpression of β1-AR or the associated $G_{\alpha s}$ protein in transgenic mice has a deleterious effect on cardiac function and prolonged activation of the β-adrenergic system leads to intracellular accumulation of cAMP, a notoriously cardiotoxic and arrhythmogenic molecule.

Despite these caveats, at least two strategies having the β-adrenergic response as a target are currently followed. The first one is based on the overexpression of a β-ARK mutant, corresponding to the C-terminal portion of the protein and thus named β-ARKct. This mutant competes with the endogenous, wild-type kinase and thus alleviates the inhibition that this enzyme imparts on the β-ARs. In the animal models in which this approach was attempted, gene transfer of this mutant appeared to arrest HF evolution.

An alternative approach is instead based on gene transfer of AC. In humans, nine AC isoforms are present, of which AC-5 and AC-6 are mainly expressed in the heart. For still not completely understood reasons, an increase in the levels of AC-6, but not of AC-5, upon gene transfer exerts beneficial effects on a failing heart by increasing β-AR-induced, however not basal, cAMP levels. A Phase I clinical trial based on the intracoronary infusion of an adenoviral vector expressing AC-6 is ongoing.

(iii) Other possible therapeutic genes

Although more distant from clinical experimentation, a series of other genes can be conceived for HF gene therapy, having shown efficacy in animal models mimicking this condition. These include those coding for VEGF-B, a member of the VEGF family of factors that selectively binds VEGFR-1 expressed on cardiomyocytes and elicits a compensatory hypertrophic response; Bcl-2, which counteracts cardiomyocyte apoptosis during HF; IGF-1 and the AKT and PI3K kinases, which are activated upon IGF-1 treatment and also mediate induction of a compensatory hypertrophic response.

4.9.8
Gene Therapy for the Prevention of Post-Angioplasty Restenosis

As discussed above, the possibility of obtaining mechanical dilatation of a stenotic artery or revascularization of a thrombotic vessel has enormous therapeutic impact. These procedures are based on the use of arterial catheters and are collectively known as *percutaneous transluminal angioplasty* (PTA) or *balloon angioplasty* or, in the case of the heart, *percutaneous coronary intervention* (PCI) or *percutaneous transluminal coronary angioplasty* (PTCA).

Despite the great success of this procedure, a relevant percentage of cases, which, in some districts, can be more than 40%, undergo stenosis of the treated vessels over the 6 months following intervention. This is due to the hyperplastic thickening of the tunica intima of the artery, which often leads to the stenosis of the arterial lumen, a process known as *restenosis* (Figure 4.28A). Denudation of the endothelial layer by PCI exposes the underlying tissue, thus determining growth factor secretion and platelet activation, which in turn stimulate the underlying smooth muscle cells (SMCs) to proliferate and migrate into the intima. These cells eventually form a hypercellular, hypo-organized structure, known as *neointima*, which tends to occlude the vascular lumen.

In the 1990s it was first demonstrated that the incidence of restenosis could be significantly reduced to less than 20–25% of cases by the introduction of a *stent*, i.e., an endoluminal metallic device positioned in correspondence to the region of the artery dilated by the balloon. Over the last few years, the efficiency of stents was further improved by the utilization of the so-called *medicated stents* or *drug eluting stents* (DES), which slowly release a drug able to inhibit neointimal hyperplasia. The drugs that are most effective are sirolimus (or rapamycin, a macrolide antibiotic used as an immunosuppressive drugs since it blocks the G1 phase of the cell cycle, eventually resulting in increased levels of p27, one of the cyclin-dependent CDK inhibitors) and paclitaxel (or taxol, a taxane used as an antiblastic drug, which acts by inhibiting microtubuli assembly during cell mitosis). Once released locally from the stent, these drugs act on the flanking segment of the arterial wall by preventing SMC proliferation, thus inhibiting neointima formation. After introduction into clinical practice in 2002–2003, the use of DES was initially hailed as the ultimate solution to the restenosis problem. In reality, however, recent data indicate that DES lead to a significant increase, a few years after placement, of mortality due to cardiovascular events. These would be caused by a process of late in-stent thrombosis, probably due to the inhibition of effective re-endothelization caused by the drugs, which also act by blocking proliferation of endothelial cells in addition to SMCs. Absence of the endothelial layer determines exposure of the sub-endothelial tissue, which is a powerful stimulus to platelet aggregation and activation of the coagulation cascade, eventually triggering formation of a thrombus.

In light of the remaining problems related to the application of stents and the existence of anatomical regions in which these devices cannot be positioned, the use of gene therapy to inhibit restenosis without interfering with the physiological re-endothelization of the vessel wall continues to be appealing.

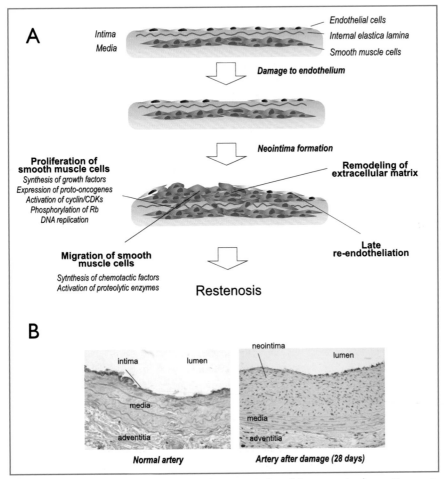

Fig. 4.28 Post-angioplasty restenosis. **A** Schematic representation of the restenosis process. Damage to the endothelial wall determines activation of the tunica media smooth muscle cells, which proliferate and migrate into the intimal layer. This process is stimulated by the release of different growth factors and involves the activation, in the smooth muscle cells, of a vast series of genes involved in the regulation of DNA replication and cell cycle progression. The process is accompanied by extensive remodeling of the extracellular matrix and by a delay in the re-endothelization of the vessel wall. The proliferated cells that invade the intima are responsible for the restenosis of the vessel. **B** Restenosis in the rat carotid model. The component of cellular hyperproliferation proper of human post-PCI restenosis can be mimicked by inducing denudation of the endothelial cell layer in a rat carotid artery. The picture shows the normal histology of a rat carotid artery (*left*) compared to the histology of an artery 28 days after endothelial damage (*right*). Formation of a thick, hypercellular neointima is evident

Restenosis is the consequence of complex pathogenetic mechanisms eventually determining SMC proliferation and migration into the damaged intima, along with extensive remodeling of the extracellular matrix and recruitment of circulating inflammatory and immune cells in response to local release of growth factors and chemokines. All the molecular pathways involved in neointima formation represent potential targets for gene therapy, using either protein coding genes or small regulatory RNAs (Table 4.14).

Efficiency of gene therapy to prevent restenosis has been extensively proven in several animal models in which neointimal hyperplasia can be experimentally induced. For example, in the rat carotid or the pig or rabbit iliac arteries, SMC proliferation is stimulated by the mechanical destruction of the endothelial cell layer using an inflated balloon on the tip of a catheter; this condition exerts a powerful stimulus on SMCs, which in a one-month period generate an important vascular stenosis (Figure 4.28B). It has long been debated whether this experimental setting might truly resemble PCI-induced human restenosis, in which, in addition to cellular hyperplasia, remodeling of the elastic and adventitial layers takes place. Despite this caveat, however, these simple models offer realistic opportunities to at least test the efficacy of treatments aimed at inhibiting SMC proliferation and migration. Analogous to gene therapy of cancer, various experimental studies have reached this objective by lowering expression of various proteins involved in cell cycle progression, including cyclin-CDKs (CDK-4, CDK-6, cyclin D, Cdc2, CDK2), PCNA, activated oncogenes (c-Myb, c-Myc), or, finally, growth factors inducing SMC proliferation (such as FGF or PDGF) using antisense oligonucleotides or siRNAs. Alternatively, inhibition of SMC proliferation was obtained by transferring a constitutively active, mutated form of Rb, or the homeotic *gax* gene (which is downregulated in proliferating SMCs), or the cDNAs coding for the CDK inhibitors p21, p27, and p53, or, finally, a double-stranded DNA, decoy oligonucleotide sequestering transcription factor E2F. In particular, a similar E2F-decoy was also used in a clinical trial entailing prevention of the vascular stenosis occurring in veins when these are used as bypass vessels in CABG; vein stenosis in this case shares similar pathogenetic causes with post-PCI restenosis (see also the section on 'Clinical Applications of Oligonucleotides').

A completely different approach was based on the overexpression of genes improving endothelial cell function, in order to accelerate re-endothelization. In particular, a well known mediator of vascular function is nitric oxide (NO), a molecule synthesized by two enzymes in endothelial cells, one of which is constitutively expressed (eNOS) and the other inducible (iNOS). The vasoprotective action of NO is due, on one hand, to an inhibitory effect on cell adhesion and migration and, on the other hand, to the stimulation of endothelial cell survival and proliferation. In this respect, iNOS appears superior to eNOS in reducing the extent of neointima formation.

Another important mediator of restenosis is extracellular matrix processing by metalloproteases (MP), which render SMC proliferation and migration possible. This process can be efficiently targeted by the local overexpression of the tissue inhibitors of these enzymes (TIMPs). A final, less physiological approach in animals was based on the expression, in proliferating SMCs, of the HSV-TK suicide gene using adenoviral vectors, followed by gancyclovir treatment, similar to gene therapy of cancer (cf. section on 'Gene Therapy of Cancer').

In conclusion, there are several reasons why gene therapy could be considered as an efficient tool to prevent post-PCI restenosis, as an alternative to or in association with stent placement, including: (i) the vast range of therapeutic targets and nucleic acids available, which can also be used in combination; (ii) the efficiency of these approaches in experimental animals; (iii) the possibility to administer therapeutic nucleic acids locally during PCI; (iv) the relatively limited therapeutic goal, when compared to other gene therapy applications, consisting in the treatment of ~1 cm of arterial wall; and (v) the clinical therapeutic benefit still attainable even if only a partial effect is obtained.

In light of these considerations, it appears surprising that none of the gene therapy applications for the prevention of restenosis has reached clinical experimentation so far. In this respect, however, it should be considered that positioning of a drug-eluting stent is a simple and well established procedure, and that the stent market has a current value of over US$5 billion per year and is rapidly expanding. Clinical development of gene therapy for the prevention of restenosis is therefore confronted with practical considerations on one hand and economical issues on the other.

4.10
Gene Therapy of HIV Infection

The pandemic due to HIV infection continues to expand globally at a rate of 7400 new infections every day. The Joint United Nations Program on HIV/AIDS (UNAIDS) estimates that over 30 million people worldwide currently live with HIV/AIDS, of which 2 million are children, and that 25 million people have already died of the disease. In 2007 alone, 2 million people died of HIV-related causes and 2.7 million new infections occurred (UNAIDS/WHO, 2008 Global Report).

Heterosexual transmission remains the dominant mode of transmission (about 85% of cases) of HIV infection worldwide. The epicenter of the pandemic is sub-Saharan Africa: in this geographical region, almost 2 million new infections occur every year. Although the pandemic reached Asia only recently, this continent is becoming a second epicenter, with more than half a million new infections yearly. Outside sub-Saharan Africa, about one third of HIV infections are due to the use of infected syringes, mainly in East Europe and Central and South-East Asia.

Currently available information indicates that the HIV-1 and HIV-2 viruses adapted to humans through multiple zoonotic infections from non-human primates infected with the simian immunodeficiency virus (SIV). This is indicated by the high degree of genetic homology between SIV and HIV-2. In particular, the natural habitat of sooty mangabeys (*Cercocebus atys*), an Old World monkey infected by SIV_{sm}, is the same as the geographical area in which HIV-2 is endemic (West Africa: Guinea Bissau, Gabon, and Ivory Coast). HIV-1, however, is homologous to the SIV_{cpz} strain, originally isolated from the chimpanzee species *Pan troglodytes troglodytes*, an anthropomorphic monkey living in the equatorial zone of Central-Western Africa. Molecular evolutionary studies indicate that passage of SIV to man occurred, from the two monkey species, on multiple occasions approximately 70 years ago.

4

4.10.1
Natural History of HIV-1 Infection

The virus is transmitted by contact with infected biological fluids (blood, sperm, vaginal secretions) from infected individuals. The canonical transmission routes are (i) sexual (homo- and hetero-) intercourse; (ii) use of infected syringes, typically in drug abusers; (iii) vertical transmission from mother to child, with infection occurring during delivery; (iv) infected blood or hemoderivatives; and (v) occasional wounding with infected material, typically in health workers. In the current pandemics in Africa and Asia, transmission is mainly heterosexual, or vertical in children.

After initial contact, the virus is transported into the secondary lymphoid organs, where infection of CD4$^+$ cells (T lymphocytes and monocyte/macrophages) occurs. A phase of rapid viral replication follows, during which viral load (viremia and the number of infected cells) increases exponentially in blood and in lymphoid organs. This phase of primary infection can be asymptomatic; however, 30–70% of infected individuals experience an acute syndrome, characterized by fever, fatigue, lymphoadenomegaly, maculo-papular cutaneous eruption, and, in a few cases, neurological involvement. In most cases, this condition is misdiagnosed or unrecognized.

The rapid viral replication stimulates a robust immune response, with the generation of neutralizing antibodies and CTLs recognizing and destroying the infected cells. This response curtails viral infection and markedly decreases viral load in patients. However, in contrast to other acute viral diseases, the virus is not eradicated from the organism, since it persists in its integrated, proviral DNA form in a reservoir of latently infected cells. These are mainly CD4$^+$ T memory lymphocytes (CD45$^+$ R0$^+$), which do not proliferate and are metabolically inactive. Since these cells do not transcribe the viral genome, no viral protein is expressed, and the cells remained unrecognized by the CTLs. A long period thus begins in which the virus replicates in metabolically active cells (activated T lymphocytes and macrophages) but remains latent in inactive cells. The immune system is unable to eradicate the infection, partly because of this latency phenomenon and partly because the replicating fraction of the virus continuously mutates its sequence and thus generates mutants escaping neutralizing antibodies and CTLs. These mutant variants are continuously selected *in vivo* upon pressure of the immune response. This condition can last several years, in which the patient is asymptomatic or paucisymptomatic, but shows relevant levels of plasma viremia and is thus infective.

This asymptomatic phase of the disease, however, eventually exhausts the immune function, mainly because of the progressive decline in the CD4$^+$ T cells, which are the main targets for infection. Since these cells provide an essential helper function to both antibody production and CTL function, the patients progressively become immunodeficient. This acquired immunodeficiency syndrome (AIDS) phase is characterized by recurrent infections due to normally non-pathogenic microorganisms and by the development of malignant tumors. The infectious diseases of AIDS patients include recurrent pulmonitis by *Pneumocystis carinii*, cerebral toxoplasmosis, cryptosporidiosis, cutaneous infection by herpesviruses, and cryptococcal meningitis. The malignant tumors of these patients include Kaposi's sarcoma and non-Hodgkin's lymphomas. In the absence of therapy, the median life expectancy after HIV-1 infection in about 9.5 years.

The introduction of modern antiretroviral therapy, known as highly active antiretroviral therapy (HAART), has dramatically changed the natural history of the disease. HAART is based on the simultaneous administration of multiple drugs (usually two drugs against the viral reverse transcriptase and one against the viral protease), in order to minimize the probability of occurrence of simultaneous mutations rendering the virus insensitive to treatment. HAART-treated patients live significantly longer in the absence of major signs of immunodeficiency. However, the treatment is unable to eradicate infection and is fraught with major toxicity problems (see below).

4.10.2
Structure of the Genome and Replicative Cycle of HIV-1

The structure of the HIV-1 genome and its replicative cycle are described in the section on 'Viral Vectors'. Briefly, the viral genome contains the three genes common to all retroviruses (*gag*, *pol*, and *env*) and six additional, accessory genes (*tat*, *rev*, *nef*, *vpr*, *vpu*, and *vif*). The *gag* gene codes for the Gag polyprotein, which is then processed by the viral protease to generate MA (p17), CA (p24), NC, and p7. These proteins associate with the viral mRNA or are somehow part of the viral capsid. The *pol* gene, translated into a Gag-Pol polypeptide thanks to a ribosomal frame-shift, codes for the viral enzymes RT (p55/p51), IN (p32), and PR (p11). The *env* gene codes for a precursor Env polyprotein, of 160 kDa, which is cleaved by cellular proteases to generate TM (gp41) and SU (gp120) (see also Figure 3.6). The accessory proteins are essential for viral replication and virion infectivity. Tat is a powerful activator of viral gene expression. The protein binds a structured region present at the 5' end of the transcribed viral RNAs and, from here, recruits different cellular factors to the viral LTR promoter. These factors include the transcriptional co-activator and histone acyltransferase (HAT) p300 and cyclin T1, the cyclin co-factor that, together with the kinase CDK9, forms the P-TEFb complex. On one hand, p300 HAT determines acetylation of histones at the LTR promoter while, on the other hand, P-TEFb phosphorylates the carboxy-terminal tail of RNA polymerase II; these two modifications increase transcriptional initiation and elongation respectively (Figure 3.11). Rev binds a highly structured RNA sequence (Rev-responsive element, RRE), present in correspondence of the *env* gene and thus contained as a potential intron in the fully spliced mRNAs (Figure 3.13). The role of Rev is to mediate transport of these RRE-containing mRNAs outside the nucleus through an interaction with the Crm-1 nuclear pore protein. Nef negatively regulates the expression of CD4, MHC, and the co-stimulatory molecules in the infected cells, thus blocking their recognition by cells of the immune system, besides regulating signal transduction. Vpr blocks progression of the cell cycle in the infected cells by accumulating them in the G2-M phase, is incorporated into virions, and contributes to nuclear transport of the reverse transcribed viral cDNA in the infected cells. Vif is essential for proper virion infectivity since it blocks the function of a cellular enzyme, the RNA deaminase APOBEC, which would otherwise inactivate the viral genome by introducing mutations. Finally, Vpu, which is only present in the HIV-1 and SIVcpz genomes, however not in HIV-2 and SIVmac, increases release of viral particles from the infected cells and contributes to CD4 degradation.

Cell infection with HIV-1 begins with the interaction of virions with the cell surface receptor CD4 and the co-receptors CCR5 (R5 strains, prevalent in the early phases of the disease) or, alternatively, CXCR4 (X4 strains, infecting late-stage patients); the former co-receptor is expressed on macrophages and the latter on T cells. After entry, the capsid is removed (the process involves the action of the cellular protein TRIM-5α) and the viral genome is reverse transcribed by RT. The pre-integration complex is then transported to the nucleus, where IN catalyzes integration of the viral cDNA into the host cell genome to generate a provirus. Transcription of the provirus is controlled by the 5' LTR and is highly sensitive to the metabolic activation of the infected cells. In resting T lymphocytes, such as memory T cells, the promoter is silent and its chromatin structure is deacetylated and compacted. When the cell is stimulated, as occurs after antigen-induced activation, the promoter becomes active and proviral transcription starts. A single RNA is generated by cellular RNA polymerase II, which corresponds to the viral genome that will eventually become packaged into the virions. The majority of this primary transcript is however used to synthesize the viral proteins: the mRNA undergoes multiple, alternative splicing events to generate more than 35 shorter mRNAs, each one coding for a single protein (Figure 3.12). Assembly of the virion occurs close to the plasma membrane and is stimulated by the Gag polypeptide in association with some cellular proteins. The Env protein is concentrated on the plasma membrane in correspondence to the regions of budding and is incorporated into the virions thanks to its interaction with the N-terminal portion of Gag. Once outside the cells, the virion undergoes maturation, by which the Gag and Gag-Pol polyproteins generate their final polypeptides upon cleavage by the viral protease present in the virions.

4.10.3
Gene Therapy of HIV-1 Infection: General Considerations

The beginning of the gene therapy era (late 1980s) coincided with a period in which the molecular mechanisms that control HIV-1 replication were sufficiently deciphered to allow the design of gene-based strategies that might suppress viral infection. In those years, patients and clinicians were witnessing the poor performance of anti-HIV monotherapy and the need for alternative approaches to treat HIV disease was urgent.

From the mid-1990s, a variety of approaches started to be described for the suppression of HIV-1 infection or replication in cultured cells. These have included the expression of (i) intracellular antibodies to viral proteins; (ii) antisense RNAs that inhibit reverse transcription, processing, and translation of HIV-1 RNAs; (iii) mutant HIV structural or regulatory genes with dominant repressor activity (including Rev, Gag, and Tat); (iv) RNA decoys that inhibit HIV-1 transcription (multimeric TAR) and processing (multimeric RRE); and (v) ribozymes to catalytically cleave and thus inactivate the various HIV-1 RNA species. Taken together, this work performed in cell culture systems has indicated overall that the goal of rendering cells resistant to HIV-1 infection by gene transfer was attainable.

Based on these *in vitro* results, during the first half of the 1990s, a dozen gene therapy clinical trials were initiated, aimed at achieving resolution of HIV infection by a variety of approaches. These included the induction of resistance in peripheral blood T cells or in hematopoietic progenitors, the *ex vivo* expansion of cytotoxic T cells recognizing

HIV-infected lymphocytes, and the active immunization of patients against HIV antigens. The overall outcome of these trials has been largely disappointing. In particular, their results highlighted the paramount difference existing between the relative ease of inducing resistance to HIV infection in cultured cells and the difficulty of achieving such a goal in infected patients.

The subsequent advent of the HAART era did then revolutionize the natural history of HIV infection and partially obscured the need for alternative therapies. Indeed, HAART has proven markedly successful in restoring or maintaining immune function and in reducing the risk of opportunistic disease and mortality. In addition, the drugs available for HAART have progressively improved in recent years, allowing a significant reduction in the burden of pill taking and the emergence of undesired side effects.

At the end of the first decade since its introduction, however, the initial enthusiasm for HAART is now giving way to several concerns of note.

First, HAART is still fraught with important long-term toxicity, including an increased cardiovascular risk. This is mainly due to the use of HIV protease inhibitors, which determine hyperlipidemia and thus increase coronary risk.

Second, according to a recent survey involving the analysis of 12 European and North American prospective cohort studies, improvements in virologic response after starting HAART do not translate into a decrease in mortality. This study, which analyzed more than 20,000 antiretroviral-naïve individuals starting HAART in 1995–2003, indicated that despite improved initial HIV virological control in patients taking HAART, in more recent years there were no significant improvements in early immunological response, no reduction in all-cause mortality, and a significant increase in combined AIDS/AIDS-related death.

Third, the emergence of drug-resistant HIV variants remains a major barrier to the successful use of HAART in HIV-infected patients, with resistance to one drug often resulting in cross-resistance to many, if not all, others in the same class. In addition, the rise in the incidence of drug-resistant variants among newly infected patients also represents a formidable challenge for clinicians.

Fourth, and perhaps most important, whereas HAART significantly reduces the levels of viral RNA in plasma and lymphoid tissues, cessation of even prolonged HAART regimens results in viral load rebound to pre-therapy levels, indicating the inability of therapy to eradicate HIV-1 infection. This failure has been attributed to the presence of a long-lived, stable population of latently infected cells that are not eliminated by the antiviral treatment, since this treatment only targets the replicating fraction of the virus. Several of these cells are long-lived memory T cells that have an integrated proviral DNA that is kept in a transcriptionally silent state. The persistence of latent HIV-1 reservoirs is the principal barrier to the complete eradication of HIV-1 infection in patients by HAART at present.

In light of these considerations, and given the need for alternative therapies that might control HIV infection, it is not surprising that gene therapy has of late regained much popularity.

Gene therapy can contribute to the treatment of HIV-1 infection by at least three different modalities, namely: (1) by rendering the target cells resistant to infection or viral replication ("intracellular immunization") by targeting either the viral proteins or the viral RNAs; (2) by inducing the selective activation of suicide or antiviral genes upon HIV-1 infection; and (3) by activating the immune system to recognize and destroy the infected cells (Table 4.16).

Table 4.16 Strategies for gene therapy of HIV-1 infection

"Intracellular immunization" by targeting viral proteins	Intracellular antibodies
	RNA or DNA decoys
	Transdominant negative mutants
"Intracellular immunization" by targeting viral RNAs	Antisense RNAs or DNAs
	Ribozymes
	siRNAs
"Intracellular immunization" by targeting cellular factors essential for HIV infection or HIV replication	Expression of cellular factors that restrict HIV replication (e.g., rhesus monkey TRIM5α)
	Inhibition of cellular factors essential for HIV infection (e.g., CCR5)
Selective activation of suicide or antiviral genes upon HIV-1 infection	e.g., LTR-TK, LTR-IFN
Activation of the host immune system	Vaccination against HIV proteins
	Genetic modification of CD8$^+$ T cells

4.10.4
Gene Therapy of HIV Infection by "Intracellular Immunization"

As early as 1988, D. Baltimore (Nobel Price in Physiology or Medicine for the discovery of the retroviral enzyme reverse transcriptase in 1975) put forward the concept of "intracellular immunization" as a strategy to inhibit the replication of HIV-1 at the cellular level. This consists in the introduction, into the virus target cells, of a therapeutic nuclei acid interfering with one or more phases of the viral life cycle, with the ultimate purpose of rendering the cells resistant to viral replication. The available therapeutic genes include both protein-coding cDNAs and small regulatory RNAs. The former category includes cDNAs coding for dominant negative mutants of the HIV-1 structural or regulatory proteins, able to block the function of the corresponding wild-type proteins, or intracellular antibodies against the viral proteins. The category of the regulatory RNAs instead includes antisense RNAs pairing with various regions of the viral genome (or its transcripts), ribozymes or siRNAs targeted against different species of viral mRNAs, or RNA decoys blocking transcription or transport of viral RNAs.

In principle, a cell resistant to HIV-1 infection, once reinfused into an infected patient, possesses a selective advantage over the other non-resistant cells. Infection itself, therefore, should exert a selective pressure able to selectively expand the genetically modified cells *in vivo*.

The cells that are targets for gene therapy are mainly CD4$^+$ T-lymphocytes, representing the main HIV-1 target *in vivo* and the most affected population along the course of HIV disease. Therapeutic gene expression in CD4$^+$ T lymphocytes can be obtained by directly transferring the therapeutic genes into these cells upon their culture *ex vivo* or by gene transfer into CD34$^+$ HSCs, from which CD4$^+$ T cells derive. The former possibility offers the advantage of obtaining high levels of transduction with gammaretroviral or lentiviral vectors and significantly expanding the target cells *ex vivo*. In contrast, gene transfer into the HSCs is much less efficient and effective transplantation requires partial bone marrow ablation (cf. section on 'Gene Therapy of Hematopoietic Stem Cells'). However, HSCs are, in principle, capable of completely repopulating the whole hematopoietic system, thus also offering protection, in addition to lymphocytes, to macrophages and APCs. In this respect, however, it should be considered that the selective pressure exerted by infection on the resistant cells, which can potentially lead to their amplification *in vivo*, does not apply to CD34$^+$ HSCs, since these cells are not a target for HIV-1.

Among the vectors having interesting properties for HSC transduction are the HIV-1-based lentiviral vectors. Multiple viral proteins, including MA, Vpr, and IN, confer the wild-type HIV-1 pre-integration complex the capacity to enter the nucleus of non-replicating cells. While Vpr is not included in the third-generation packaging systems for lentiviral vectors, both MA and IN are essential components of these vectors and are sufficient to mediate their nuclear import; these vectors are therefore capable of transducing non-replicating cells (cf. section on 'Viral Vectors'). However, several issues concerning both the efficiency and the safety of these vectors remain to be elucidated. These include the definition of the optimal vector design, the exact understanding of their mutagenic potential, the characterization of their propensity of being silenced over time in different tissues, and safety concerns regarding their clinical utilization. These safety concerns mainly relate to the possibility of recombination of a lentiviral vector into a replication-competent lentivirus (RCL) that might represent a novel pathogen, and the possibility that, in an HIV-1-infected patient, the lentiviral vector construct might be mobilized by the wild-type virus (cf. section on 'Viral Vectors'). Indeed, the latter event was already observed in patients in a clinical trial using a lentiviral vector expressing an antisense *env* gene – see below. Vector mobilization, however, which *per se* is a beneficial event since it extends vector spread, cannot occur in SIN vectors, in which the vector U3 region is modified or deleted.

One important question that needs to be thoroughly addressed is whether genetically modified HSCs can reconstitute the immune system in adults. Although studies have demonstrated that the adult uninfected thymus maintains the ability to support T-lymphopoiesis, after transplantation of HIV-negative patients functional recovery of lymphoid and immune effectors cells occurs gradually, and reconstitution of normal humoral and cellular immunity may take a year or more. In uninfected individuals, T-cell reconstitution takes place by either peripheral expansion of the already existing T-cell pool or by renewal of thymopoiesis. In HIV-infected individuals transplanted with HSCs carrying HIV-resistant cells, thymopoiesis is absolutely required to generate a non-skewed repertoire of HIV-resistant, naïve T-cells. Indeed, in HIV-infected patients, thymic function was found to be depressed but still present, supporting the conclusion that immune reconstitution after transplantation of genetically modified hematopoietic progenitors is an attainable objective in these patients.

Most of the HIV gene therapy clinical trials aimed at inducing intracellular resistance to infection were initiated in the mid-1990s, before the introduction of HAART, while experimentation almost stopped afterwards since the suppression of viral replication imposed by this therapy abolishes the selective pressure that is at the basis of the intracellular immunization approach. However, the incapacity of HAART to eradicate the disease, the emergence of resistant strains, and the poor compliance of patients to therapy due to its significant side effects are now revitalizing this field, and new trials have been initiated over the last few years.

The main clinical trials so far conducted are summarized in Table 4.17 according to therapeutic gene, vector and target cells used, and are discussed in the following paragraphs.

Table 4.17 Major gene therapy clinical trials aimed at rendering CD4$^+$ T cells resistant to HIV-1 infection

Therapeutic gene	Vector	Target cells
RevM10 (dominant negative Rev mutant)	Plasmid transfection using gold microparticles	CD4$^+$ T lymphocytes
	Retrovirus	CD4$^+$ T lymphocytes
		CD34$^+$ bone marrow cells
		Mobilized allogenic CD34$^+$ cells from peripheral blood
RevM10 and/or RevM10 plus antisense TAR	Retrovirus	CD4$^+$ T lymphocytes from HIV-negative identical twins
RRE decoy	Retrovirus	CD34$^+$ bone marrow cells
Antisense RNAs against TAR and Tat/Rev (HGTV43)	Retrovirus	CD34$^+$ bone marrow cells
Long antisense against env, driven by the HIV-1 LTR (VRX496)	Lentivirus	CD4$^+$ T lymphocytes
Hairpin ribozyme against HIV-1 U5 leader sequence	Retrovirus	CD4$^+$ T lymphocytes
Hammerhead ribozyme targeted to tat and rev	Retrovirus	CD34$^+$ bone marrow cells
		CD34$^+$ bone marrow cells after myeloablation
Hammerhead ribozyme against the translation initiation region of tat (Rz2, OZ1)	Retrovirus	CD4$^+$ T lymphocytes from HIV-negative identical twins
		Mobilized CD34$^+$ cells from peripheral blood
Gp41-derived peptide blocking fusion (M87o)	Retrovirus	CD4$^+$ T lymphocytes
TAR decoy RNA, siRNA against Tat and Rev, ribozyme against cellular CCR5	Lentivirus	CD34$^+$ bone marrow cells

4.10.4.1
Dominant Negative Forms of the HIV-1 Rev Protein

Perhaps the most investigated gene for gene therapy of HIV-1 is that coding for a dominant negative mutant of Rev. Rev is a 116-amino-acid viral protein essential for viral replication, which is translated from a fully spliced mRNA expressed early in viral infection. The protein shuttles between the nucleus and the cytoplasm and, as reported above, acts post-transcriptionally to mediate the cytoplasmic export of unspliced and singly spliced viral RNAs. This function requires the direct and highly specific binding of Rev to the RRE present in these unspliced RNAs. A mutant protein, RevM10, which blocks Rev function, has been described. This mutant, which bears two amino acid substitutions in the highly conserved leucine-rich region of the protein, is still able to bind RRE RNA, but no longer binds the Crm-1 nuclear export factor. Since it still multimerizes with other Rev monomers, RevM10 acts as a dominant negative inhibitor of Rev and blocks transport of the incompletely spliced HIV-1 transcripts into the cytoplasm. When delivered by retroviral vectors to T-cell lines and primary bone marrow cells, RevM10 acts as a powerful suppressor of viral replication.

An initial series of clinical studies assessed the feasibility of gene therapy of HIV-1 infection first by delivering RevM10 to CD4$^+$ T cells from three HIV-infected individuals by plasmid transfection using gold microparticles and later using gammaretroviral vectors in CD34$^+$ HSCs. The most encouraging outcome of these trials was the observation that T lymphocytes expressing RevM10 effectively possess a selective advantage *in vivo* over normal cells and thus expand *in vivo* upon the selective pressure imposed by HIV replication. Additional clinical trials exploiting the dominant negative properties of Rev mutants were performed by infusing CD4$^+$ T lymphocytes transduced with a retroviral vector containing negative dominant Rev in combination with an antisense molecule that inhibits viral replication by binding the transactivation response element (TAR). The protocol involved isolating CD4$^+$ T lymphocytes from HIV-negative identical twins of HIV-positive patients, *ex vivo* transduction followed by infusion into an HIV-positive sibling. Also in this trial, the engineered cells could be detected for prolonged periods after infusion. Of interest, the ratio of therapeutic to control vector-containing cells markedly increased in one patient who discontinued HAART treatment, further supporting the conclusion that HIV-1 infection imposes a selective pressure on resistant cells.

A potential obstacle to the persistence of cells expressing viral proteins *in vivo* is the development of host immune responses against the cells expressing these proteins. However, though RevM10 proved to be immunogenic in mice, no apparent cellular immune response against the transduced cells was detected in these patients.

In all these early clinical trials, however, the number of HIV-1-resistant cells was too low to confer a true therapeutic benefit to patients, most likely as a consequence of the low efficiency of gammaretroviral vectors to transduce HSCs. More recently, due to the problems caused by HAART and the poor compliance to therapy of a large number of patients, experimentation with RevM10 was resumed, leading to a trial in which the gene is transferred into HSCs using a lentiviral vector and the transduced cells are reinfused after partial myeloablation, in light of the success of this procedure in gene therapy of ADA deficiency and CGD (see section on 'Gene Therapy of Hematopoietic Stem Cells').

4.10.4.2
RRE Decoy

The potency of RevM10 to suppress HIV-1 replication clearly indicates that the Rev function is essential for HIV-1 replication. Consistent with this conclusion, the overexpression of the RRE RNA sequences as part of the transcript from the viral LTR of a retroviral vector was found to inhibit HIV-1 replication in both T lymphocytes and the progeny of transduced CD34+ HSCs. In contrast to RevM10, this RRE decoy has the additional advantage of not being potentially immunogenic. Following these considerations, four HIV-1-infected children and adolescents underwent bone marrow harvest from which CD34+ cells were isolated and transduced by a retroviral vector carrying this RRE decoy gene. The cells were reinfused into the subjects, without complications. However, also in this case, gene-containing leukocytes in the peripheral blood were seen only at a low level (1–3 cells for every 1×10^5 peripheral blood cells) and only in the first months following cell infusion.

4.10.4.3
Intracellular Antibodies

A class of molecules that shows promise for anti-HIV intracellular immunization is that of intracellular antibodies (or intrabodies), which represent the antibody single-chain variable fragments (scFvs) selected against various HIV proteins – see section on 'Antibodies and Intracellular Antibodies'. For the past 15 years, virtually every HIV-1 protein has been targeted by intrabodies, including structural proteins (matrix, nucleocapsid, and envelope), enzymes (integrase and reverse transcriptase), and regulatory proteins (Tat, Rev, and Nef). So far, progress in the clinics has been slow, possibly because of the relatively low anti-viral effect shown by these molecules when compared to other therapeutics. However, intrabodies remain an attractive option when protein half-life is long and, most importantly, when a protein has more than one protein-interaction domain, because it is possible to develop a reagent that prevents particular associations but spares others. This might turn out to be an especially interesting tool to selectively target some of the cellular proteins that are essential for HIV-1 replication, by preserving their cellular function while inhibiting their pro-viral activity.

4.10.4.4
Ribozymes

Another class of molecules that has been widely utilized over the last several years to inhibit HIV-1 replication by cleaving the viral genome and transcripts is that of ribozymes. These RNA enzymes have the potential to act at several stages in the HIV infectious cycle, including upon the initial entry of genomic viral RNA into the target cell, during the transcription of genomic RNA molecules, prior to and during translation of mRNA to viral proteins, and prior to packaging of the genomic RNA. The cleavage of HIV RNA by ribozymes at any of these stages can significantly decrease or block intracellular viral replication. When choosing a ribozyme, due to the sequence variation among HIV-1 isolates and the rapid mutation rate in response to anti-retroviral treatment, it is imperative to select target sites that are critical for viral replication and highly conserved in sequence between clades. Over the last several years, a number of hammerhead and hairpin ribozymes have been described that fulfill these criteria, and have been proven to reduce HIV-1 replication in cell culture.

In the early 1990s it was first demonstrated that a gene encoding a hairpin ribozyme targeted to the HIV-1 U5 leader sequence conferred resistance to HIV-1 infection when delivered to T-cell lines and primary lymphocytes using a retroviral vector. Based on these results, a Phase I clinical trial was initially conducted on six HIV-infected individuals, by transducing their peripheral blood lymphocytes with a retrovirus expressing this ribozyme. The initial results of this trial showed a modest survival advantage for cells expressing the ribozyme. Other initial attempts at exploiting anti-HIV-1 ribozymes for gene therapy of HSCs have also been quite disappointing. Five healthy HIV-positive subjects who received autologous $CD34^+$ cells transduced with a retroviral vector expressing ribozymes targeted to the Tat and Rev mRNAs showed minimal transient engraftment of the transduced cells. Slightly better results were obtained in five other AIDS lymphoma patients, who were reinfused with $CD34^+$ cells modified using the same retrovirus after myeloablative treatment. These patients showed a significant increase in gene-marked cells post-transplant. However, the durability of this engraftment was short-lived, as indicated by the loss of observable gene marking 6 months post-transplant, most likely indicating transduction of an already committed progenitor cell population.

More encouraging results were obtained by using another gammaretroviral vector expressing a hammerhead ribozyme, named Rz2 or OZ-1, targeting the conserved translation initiation region of the HIV-1 *tat* gene. Phase I clinical trials involving the delivery of this vector were conducted by transducing $CD4^+$ T lymphocytes in identical twins discordant for infection or autologous mobilized peripheral blood $CD34^+$ HSCs, which were reinfused without myelosuppression. In both trials, separate populations of cells were transduced with either a retroviral vector containing the ribozyme or with the vector alone. Equal numbers of the two transduced cell types were then reintroduced into the recipient patients, in order to monitor the survival of the anti-HIV-1 ribozyme-expressing cells. The clinical trial conducted by gene transfer in CD34+ cells indicated that the transgene could be detected for up to 30 months in multiple hematopoietic lineages, with a frequency of $1{:}1{\times}10^4$ to $1{:}1{\times}10^5$ of hematopoietic cells analyzed, including naïve T cells. However, since the $CD4^+$ cells expressing the ribozyme had no selective advantage, the patients being on HAART, the number of transduced cells remained low and no significant enrichment of the Rz2-expressing cells could be detected over the control cells.

Based on these relatively encouraging results, the same ribozyme-expressing vector was later used for $CD34^+$ HSC transduction in a multicenter, Phase II trial in 74 patients in the US and Australia, in which selective pressure was applied to the transduced cells by two interruptions in the patients' antiretroviral therapy. The recently published results of this trial showed modest effects on viral load and number of $CD4^+$ T cells in the treated patients.

4

4.10.4.5
Antisense RNAs

Over the last several years, different investigators have reported the efficacy of intracellularly expressed antisense RNA molecules targeted against different HIV-1 genes in preventing HIV-1 infection in cell culture. In addition, retrovirus-mediated delivery of antisense genes targeting *tat* and *rev* to rhesus macaque CD4+ T lymphocytes using gammaretroviral vectors followed by the infusion of the transduced cells was found to significantly reduce viral load after challenge with SIV. These antisense RNAs pair with the viral genomic mRNA or with some of the spliced transcripts generated from this mRNA and block their function, by both preventing translation and inducing their inactivation, probably as a consequence of the extensive deamination of adenosines occurring on double-stranded RNA.

As discussed in the Chapter on 'Methods for Gene Delivery', one efficient way to express such antisense RNAs is as fusions to the U1 snRNA transcript, an essential component of the cellular splicing machinery that is abundantly and constitutively expressed in all cells. A retroviral vector expressing three antisense RNA sequences targeting TAR and two sequences in the *tat/rev* genes as fusions to U1 was used to deliver these genes into CD34+ peripheral blood HSCs of five HIV-infected individuals. These patients were subsequently engrafted with the modified cells without bone marrow ablation. Similar to the previously discussed studies, persistence of the resistant cells could be detected for several years after transduction, albeit their frequency has remained below the threshold required to provide significant therapeutic benefit.

These approaches are now exploiting the possibility to improve gene transfer efficiency by delivering the therapeutic RNAs using either gammaretroviral vectors after bone marrow conditioning, or lentiviral vectors. In particular, the first lentiviral vector trial was initiated for gene therapy of HIV infection in 2003. The vector expressed a long (937 nt) antisense RNA against the HIV envelope gene from the HIV LTR and was used to transduce autologous CD4+ T cells. This Phase I trial was conducted in five subjects with chronic HIV infection who had failed to respond to at least two antiviral regimens by a single infusion of gene-modified cells. The results obtained showed prolonged engraftment with lentivirus-modified T cells in three of the patients for at least one year after infusion, albeit at low levels. No statistically significant anti-HIV effects were observed, however one patient developed a sustained decrease in viral load. Follow-on studies will evaluate the potential of this approach when given in repeated doses and in the context of structured interruption in HAART therapy, in order to provide selective advantage to the modified cells.

4.10.4.6
siRNAs

The process of double-stranded RNA-mediated RNA interference (RNAi) was originally discovered by Fire and Mello in the worm *C. elegans* as a powerful mechanism of suppression of gene expression – see section on 'Small Regulatory RNAs'. The observation that RNAi also occurs in mammalian cells as part of a larger network of RNA-silencing mechanisms that share common pathways has rapidly prompted its possible utilization as a tool to combat viral infections by targeting the destruction of viral RNAs. In the early 2000s, it was demonstrated that synthetic siRNAs against the cellular CD4 or the viral Gag gene inhibit HIV-1 infection and, most notably, that siRNAs against different regions of the HIV-1 genome could be generated inside the cells after transfection of plasmids expressing short hairpin RNAs (shRNAs). In subsequent years, a number of studies have extended these observations, by targeting various other regions of the HIV genome, by inserting shRNA-expression cassettes into viral vectors for improved gene delivery and by expanding the repertoire of target sequences to cellular genes known to be required for efficient HIV-1 infection.

More generally, it now appears that the attempt to inhibit HIV infection by exogenous therapeutic siRNAs should deal with the complex interplay existing between HIV nucleic aids and the host cell machinery that regulates microRNA (miRNA) production and activity. Accumulating evidence in fact indicates that the cellular miRNA-silencing machinery restricts HIV-1 replication on one hand, while, on the other hand, the virus has evolved ways to cope with this inhibition. First, upon HIV-1 infection, the expression levels of several cellular miRNAs are significantly altered; some of these miRNAs are putatively involved in the regulation of cellular factors that are essential for HIV replication. Second, the miRNA machinery directly generates miRNAs from the HIV-1 RNA itself, which regulate HIV-1 infection. Third, inhibition of Drosha (required for primary miRNA processing in the nucleus to generate 60–70-nt-long miRNA precursors) or Dicer (which, in the cytoplasm, activates miRNAs by generating miRNA:miRNA* duplexes and promotes their incorporation into effective miRNA-containing ribonucleoprotein complexes) significantly increases HIV-1 replication, further indicating that the miRNA pathway contributes to the suppression of HIV replication. Fourth, overexpression of Tat attenuates silencing of reporter genes when this is induced by shRNAs but not by siRNAs, since the protein directly inhibits Dicer. Any possible gene therapy treatment aimed at inhibiting HIV-1 infection by RNAi should evidently deal with this intricate pathway of reciprocal regulation as well as evade the evolutionary mechanisms that the virus has evolved to escape inhibition by cellular miRNAs.

One crucial concern in developing a therapeutic strategy based on RNAi is that the therapeutic siRNA or shRNA does not exert off-target effects on other important cellular genes or that their overexpression (in the case of shRNAs) does not saturate the whole RNAi machinery, thus leading to unspecific cell death. Another concern, more specific to HIV-1 infection, is the possibility of the emergence of escape mutants. Indeed, HIV-1 has been shown to easily evade RNAi by the selection of mutants encoding the same viral proteins but with silent mutations impairing siRNA recognition or by evolving alternative structures in its RNA genome that occlude the siRNA binding site. This problem might be overcome by the simultaneous expression of several individual siRNAs against multiple targets. For example, in cultured T cells, HIV-1 can escape from shRNA inhibition by mutating after just 25 days, but infection is controlled for at least several months by using a combination of four shRNAs. An even more effective strategy to avoid the selection of mutants resistant to RNA interference is to target a cellular gene, such as the CCR5 co-receptor (see also below).

A clinical trial based on the utilization of a lentiviral vector transferring multiple therapeutic genes, including an siRNA against the Tat and Rev transcripts, a ribozyme against CCR5, and a decoy sequence for Tat, consisting in a multimerized TAR sequence, is currently ongoing.

4

4.10.4.7
Targeting HIV-1 Internalization

Most of the above-described therapeutic genes (antisense, ribozymes, siRNAs, intracellular antibodies, decoys, and others) block HIV replication after the cell has internalized the virus and, most likely, after integration of the provirus into the host cell genome. One attractive possibility would instead be to render the cells refractory to HIV infection by directly impeding its actual infection.

One drug that has recently entered the market is enfuvirtide, commonly known as T20. T20 derives from a 26-amino-acid peptide from the C-terminus of HIV-1$_{HXB2}$ gp41 (C36 peptide), which blocks HIV entry by inhibiting the conformational changes needed for fusion of the viral envelope with the cellular membrane. For a gene therapy approach, this peptide was modified with an anchor protein for cell surface expression, and further optimized for reduced immunogenicity and improved expression and stability; the final version of the construct (called M87o) was expressed using a gammaretroviral vector. Transduced cells express this peptide on their surface, a strategy that allows the attainment of a sufficiently high local concentration to inhibit fusion of the viral envelope to the cell membrane, exactly as T20 does. A pilot clinical trial in 10 patients with late-stage HIV disease was performed by infusion of CD4$^+$ T-cells transduced with the retroviral vector. Initial results from this trial indicated that the approach was safe and that enrichment for the transduced cells was detectable in the peripheral blood of some of the patients, although no changes in viral load were observed. One major concern about the clinical utilization of M87o stems from the observation that T20 elicits the rapid emergence of resistant viruses, and therefore it needs to be used in combination with other antiretroviral drugs. It is therefore likely that the success of M87o will depend on its utilization in the context of multi-strategy gene therapy.

Further, with respect to viral entry, one of the most striking discoveries in the HIV research field has been the observation that individuals with homozygous deletions in the CCR5 chemokine receptor gene are genetically resistant to HIV-1 infection, independent of the route of transmission. Most importantly, these individuals do not appear to be associated with clinical conditions, suggesting that the biologic function of CCR5, in contrast to that of CXCR4, is compensated by other chemokine receptors, probably due to the redundancy of the chemokine family. These observations demonstrated both the critical importance of CCR5 for HIV-1 infection and highlighted the dispensable nature of its function, thus suggesting that inactivation of CCR5 in lymphocytes or stem cells might be of therapeutic value.

Inactivation of CCR5 expression or function has been attempted by a variety of means, which include peptides derived from the natural chemokine ligands RANTES and MIP-1α, chemical drugs, antisense peptide nucleic acids, as well as by different genetic approaches for the phenotypic knock down of the protein, which can be exploited by gene therapy. These include the delivery of both proteins and nucleic acids. The former category includes the utilization of anti-CCR5 intracellular antibodies, dominant negative mutants, and intrakines (modified RANTES and MIP-1α CCR5 ligands targeted to the endoplasmic reticulum, which block the surface expression of newly synthesized CCR5). Among the therapeutic nucleic acids, several groups have described resistance to HIV-1 infection of cell lines treated with anti-CCR5 ribozymes, antisense, or, more recently, siRNAs.

As reported above, an anti-CCR5 ribozyme is one of the three therapeutic genes delivered by a lentiviral vector in a recent clinical trial.

4.10.5
Gene Therapy of HIV-1 Infection by Inducing the Selective Activation of Suicide or Antiviral Genes upon Viral Infection

A potential therapeutic strategy, alternative to the induction of resistance in CD4$^+$ T cells, is to modify these cells in order to induce their selective death or to obtain the production of antiviral factors after infection with HIV-1. These objectives can be met by the *ex vivo* transduction of expanded T cells with suicide gene constructs (for example, HSV-TK or diphtheria toxin) or antiviral genes (for example, the α-interferon gene) under the control of the viral LTR promoter. In non-infected cells, the promoter is transcriptionally silent and thus the therapeutic gene is not expressed. Upon HIV-1 infection, the promoter becomes active and the cells either die or somehow produce the desired antiviral factor. This approach, which was initially proposed in the early 1990s, has now been abandoned as its success would in principle require gene transfer into the vast majority of target cells. In addition, if a gene inducing cell death is used, the cells expressing these genes are negatively selected by the infection rather than having a selective advantage, as in the case of intracellular immunization.

4.10.6
Gene Therapy for Anti-HIV-1 Immunotherapy

An alternative therapeutic strategy for HIV/AIDS is to exploit gene transfer to increase antiviral immune response. This objective can be reached following two routes: by active vaccination or adoptive immunotherapy, i.e., by modifying the specificity of CD8$^+$ CTLs to direct them against viral antigens.

In the case of active immunotherapy (**vaccination**), the development of anti-HIV vaccine has turned out to be much more challenging than originally anticipated in the early days when the virus was first isolated. HIV sequences are very variable from patient to patient – variability can occur for as much as 35% of the *env* gene amino acids; a population of latently infected cells is established very soon after infection; the virus continuously evolves variants escaping neutralizing antibodies and CTLs; no attenuated strains are available for vaccination; finally, no simple animal model exists for the disease: these are some of the major difficulties that have so far hampered the development of an effective vaccine. Even more relevant is the observation that, despite over 60 million individuals having been infected with HIV over the last 25 years, no convincing proof exists that any of these patients has been cured. This observation clearly highlights the incapacity of our immune system to eradicate this pathogen. Thus, not only we do not have a vaccine, today we do not even understand which correlates of immune protection we should look for.

From the experimental point of view, the strategies followed to obtain an anti-HIV vaccine are similar to those exploited for other viruses and can be categorized into traditional and innovative approaches. The traditional approaches consist in the use of attenuated or inactivated viruses or, alternatively, purified or recombinant viral proteins. In the case of HIV-1, the use of attenuated replicative viruses, despite their efficacy in preventing SIV infection in monkeys, cannot be reasonably considered in humans for safety reasons. In contrast, inactivated viruses appear unable to sufficiently stimulate broad neutralizing activity and are weak in stimulating a CTL response. As far as recombinant proteins are concerned, most early trials focused on the products of the *env* gene (uncleaved gp160 and mature gp120) as immunogens, since antibodies neutralizing infection should be directed against this protein. Unfortunately, in two large Phase III trials the injection of monomeric gp120 protein failed to induce broadly reactive neutralizing antibodies and only elicited strictly type-specific responses.

The innovative approaches for vaccination consist of genetic vaccination using plasmids or biological vectors, which, in the case of HIV, include both viruses (mostly adenovirus and poxvirus) and bacteria (*Salmonella* and *Listeria*). So far studies based on the administration of a canarypox vector expressing gp120, followed by a boost using recombinant gp120, have been unsatisfactory. As discussed in the section on 'Gene Therapy of Cancer', genetic vaccination using plasmid DNA is simple and versatile, however usually requires multiple injections of large quantities of DNA, and its success depends on the identification of efficient adjuvants able to increase the strength of the immune response.

Alternative to focusing on Env as a target for neutralizing antibody production, other studies have exploited an adenoviral vector serotype 5 (Ad5) to express the HIV-1 Gag, Pol, and Nef proteins, with the purpose of eliciting a CTL-based response. A Phase IIb efficacy study on 6000 individuals, in collaboration between a pharmaceutical company and the National Institute of Health (NIH) of the United States (the STEP study), was terminated at the end of 2007 when the first planned interim analysis showed that this vaccine not only failed to protect against infection or to reduce viral load after infection, but also that vaccinees with pre-existing Ad5-specific neutralizing antibodies exhibited an enhanced rate of HIV-1 infection. There is as yet no explanation for these finding, however it is important to note that 30–40% of individuals in Western countries and 80–90% of those in sub-Saharan Africa have anti-Ad5 antibodies, which would anyhow render the use of this vector inefficacious or potentially harmful.

In conclusion, development of an HIV-1 vaccine is still in its infancy. Major unsolved problems remain, and a renewed commitment to basic discovery research and preclinical studies is now absolutely required to move the field forward.

Adoptive immunotherapy instead consists in the *ex vivo* expansion and activation of CTLs directed against viral antigens, followed by their *in vivo* re-administration to patients with the purpose of recognizing and destroying virus-infected cells. In HIV-1 patients, several CTL clones can be isolated that naturally recognize different viral proteins. However, clinical experimentation has so far been unable to show long-term therapeutic efficacy of these clones, probably because of the difficulty in obtaining a sufficient number of antigen-specific CTLs.

Much more interesting is, therefore, the possibility to engineer CD8$^+$ lymphocytes to retarget these cells toward HIV-1 antigens. Similar to adoptive immunotherapy for cancer therapy (cf. section on 'Gene Therapy of Cancer'), retargeting of CD8$^+$ T cells can be obtained in at least two manners. The first is based on transfer, into CD8$^+$ cells recovered from the patients and expanded *ex vivo*, of the α and β chains of the TCR obtained from antigen-specific CTLs, using gammaretroviral or lentiviral vectors. This strategy, which is however still experimental as far as HIV-1 infection is concerned, allows the generation of CTL clones expressing a defined TCR, able to recognize the desired antigen at high affinity and specificity. Alternatively, antigen recognition specificity can be conferred by using a T-body, consisting of a genetic fusion between a single-chain antibody recognizing the antigen of interest and the TCR CD3ζ chain, able to activate the signal transduction cascade leading to cell activation. This latter modality renders the engineered CTL independent of MHC restriction and prevents mispairing between the exogenous and endogenous TCR α and β chains. A particular T-body is composed of the extracellular portion of CD4 fused to the CD3ζ chain (CD4ζ); since the CD4 molecule recognizes HIV-1 gp120 exposed on the infected cell surface, it directs CTL activity against these cells. This approach was tested in a few clinical trials, which showed that the cells expressing CD4ζ persist for very long periods in the patients and show antiviral activity *in vivo*. Based on these encouraging results, further Phase II trials are currently ongoing.

Finally, it is important to observe that the clinical efficacy of all CTL-based approaches strictly depends on the presence of functional CD4$^+$ T lymphocytes providing helper function. However, the number of these cells is significantly reduced in the course of advanced HIV disease. Expansion of anti-HIV-1 CTLs, therefore, needs to occur concomitantly with expansion of CD4$^+$ T cells rendered resistant to HIV-1 infection by gene therapy using one of the intracellular immunization approaches described above.

4.11
Gene Therapy of Liver Diseases

The liver is an organ showing a series of uniquely attractive characteristics for *in vivo* and *ex vivo* gene therapy. It is the largest organ of the human body and the only one that has two circulatory systems, one connected to the systemic circulation by which the hepatic artery transports oxygenated blood from the aorta (30% of blood) and one connected to the portal vein, which brings nutrients from the intestine. Furthermore, the liver possesses a ductal system transporting a series of metabolized compounds from liver to intestine, solubilized in the bile.

The liver can be divided into different segments according to the anatomy of its vascularization; each of these segments can be individually removed, until resecting up to 70% of the total hepatic mass. Hepatocytes represent about 70% of the liver cells, which can be easily separated from the other cell types by differential centrifugation ad can be cultured *ex vivo* and modified by gene transfer.

Less than 1% of liver cells divide in normal conditions. However, when the organ is damaged by different insults, hepatocytes have the capacity to reenter the cell cycle within a few hours of stimulation, also followed by the other cell types. In response to a partial hepatectomy removing about 70% of the liver parenchyma, over 95% of the hepatocytes continue to replicate until full liver regeneration is complete. This extraordinary regenerative capacity has been known since ancient times and has generated the myth of Prometheus who, having offended the gods, was condemned every day to see his liver devoured by an eagle, while the organ regenerated during the night. In this respect, it is important to remark that liver regeneration is not sustained by a reservoir of stem cells, as occurs, for example, in the bone marrow, but occurs thanks to the replication of the mature surviving hepatocytes. However, liver stem cells also exist, known as *oval cells*, which are mobilized and differentiate into hepatocytes only when proliferation of mature hepatocytes is blocked and when hepatocytes are progressively and chronically destroyed.

Currently, the only curative treatment for last-stage liver disorders is orthotopic liver transplantation. However, several patients cannot find matched donors or succumb to hepatic insufficiency while waiting for a donor (about 15% of patients unfortunately fall into the latter category). Therefore the development of innovative approaches to treat liver disease, including gene therapy, appears most desirable, and justifies the great efforts made in this field over the last 20 years.

4.11.1
Gene Transfer to the Liver

Plasmid DNA or viral vectors can be administered to the liver *in vivo*, using either the arterial (hepatic artery) or the venous (portal vein) routes, or *ex vivo*, by isolating primary hepatocytes from the liver and exploiting the regenerative capacity of these cells after their reinfusion.

As far as *in vivo* administration is concerned, this is very efficient in the liver, since the hepatic sinusoid epithelium presents fenestrations with a diameter of about 100 nm, large enough to allow passage of large multimolecular complexes or viral particles, which thus have direct access to the hepatocytes. In addition, hepatocyte blood perfusion corresponds to about 1/5 of all cardiac output; therefore, any compound or vector present in the circulation very easily reaches the liver.

In vivo gene transfer into the liver has been the subject of several studies having as an objective the development of technologies for cell targeting (for example, use of lipid complexes containing asialoglycoproteins, which specifically bind a receptor expressed on the hepatocyte surface) or naked DNA delivery (for example, hydrodynamic gene transfer) – these approaches are detailed in the Chapter on 'Methods for Gene Delivery'. In the case of viral vectors, hepatocytes are very efficiently transduced with adenoviral, lentiviral and AAV vectors. In animal models, gammaretroviral vectors are also effective in transducing hepatocytes *in vivo*, however only after hepatic regeneration is induced by partial hepatectomy or using compounds inducing liver damage. Alternatively, these vectors can be considered for neonatal liver transduction, when hepatocytes are still in active replication. The latter approach was successfully used in a dog model of type VII mucopolysaccharidosis, an inherited metabolic disease due to β-glucuronidase deficiency, in which a gammaretroviral vector expressing the correct cDNA was injected in the liver of neonatal dogs genetically defective for the production of this enzyme.

The possibility to transduce isolated hepatocytes *ex vivo* followed by their reintroduction into the liver is an equally attractive possibility. These studies are based on the experience gained from the clinical trials based on allogenic transplantation of liver cells, performed for the treatment of various metabolic diseases, including Crigler-Najjar disease type I, deficit of OTC, and type I glycogenosis. In these studies, it was shown that transplantation of a number of normal hepatocytes from a donor equal to 1% of the total number of liver hepatocytes is sufficient to provide therapeutic benefit to the patients. Although this number appears low, it must be remembered that, in an adult weighting 70 kg, the liver contains about 2×10^{11} hepatocytes. If the objective is to correct at least 1% of these cells, it means that it is necessary to inoculate at least 3–5 billion cells, assuming that the efficiency of repopulation of each of these cells is higher than 50%. In case autologous transduced hepatocytes are considered, these evaluations demand that gene transfer efficiency is close to 100%. It is thus clear that autologous transplantation of *ex vivo* genetically modified hepatocytes poses formidable technological problems.

In the *ex vivo* hepatocyte transduction attempts, when the goal is to correct a hereditary disorder, retroviral vectors are commonly considered, due to their property of permanently modifying the genome of transduced cells. Gammaretroviral vectors, however, are unable to transduce quiescent cells, and gene transfer must therefore be performed after *ex vivo* culture and expansion of these cells. This represents a difficult task, given the vast number of cells required to attain clinical efficacy. In contrast to gammaretroviral vectors, lentiviral vectors appear capable of transducing isolated primary hepatocytes in the absence of stimulation, without requiring passage in cell culture.

The liver disorders that are candidates for a gene therapy approach can be schematically divided into one of three different categories: (i) primary disorders in which accumulation of toxic products in hepatocytes leads to extensive hepatotoxicity; (ii) metabolic disorders of the liver with mainly extrahepatic manifestations; and (iii) acquired liver disorders, such as viral hepatitis and hepatocarcinoma. Since gene therapy of hepatocellular carcinoma is not conceptually different from gene therapy of other solid cancers (cf. section on 'Gene Therapy of Cancer') and clinical gene therapy of hepatitis is still in its infancy (the first Phase I clinical trial for hepatitis B using RNA interference being initiated in 2008), the following two sections will be devoted to the first two categories of disorders.

4.11.2
Gene Therapy of Hereditary Metabolic Disorders of the Liver Impairing Liver Function

The hereditary metabolic disorders of the liver include a series of variegated disorders in which accumulation of toxic products in the liver leads to progressive death of hepatocytes and prevents normal liver function, eventually determining general hepatotoxicity. These pathologic conditions include $\alpha 1$-antitrypsin deficiency, type I tyrosinemia, familial progressive intrahepatic cholestasis type III, and Wilson's disease. In these disorders, normal or *ex vivo* genetically corrected hepatocytes possess a selective advantage over endogenous mutated hepatocytes.

4.11.2.1
AAT Deficiency

Of particular relevance because of its prevalence is the deficiency of α1-antitrypsin (AAT), an inherited disorder leading to the development of chronic obstructive pulmonary disease (COPD), often accompanied by hepatic cirrhosis and panniculitis. AAT is a 52-kDa acute-phase glycoprotein, synthesized by the liver and subsequently secreted in plasma. The protein is also produced, in lower amounts, by alveolar macrophages, circulating monocytes, and, probably, lung epithelial cells. The main function of AAT is to protect tissues from elastase, an enzyme produced by neutrophils, and, to a lesser extent, from other serine-proteases, including cathepsin G and proteinase 3. These enzymes, produced during inflammation, digest elastin and other components of the extracellular matrix, thus damaging epithelial cells. In experimental models, their experimental activation induces pulmonary damage leading to emphysema. The role of AAT, which, structurally, belongs to the serpin (inhibitors of serine-proteases) family of enzymes, is to counteract the action of these enzymes and block their inappropriate diffusion. Over 30 human serpins are known to date; other members of the family are antithrombin and antiplasmin.

AAT is encoded by the Pi (protease inhibitor) locus on chromosome 14q, a highly polymorphic region in which approximately 125 single-nucleotide polymorphisms (SNPs) are known to influence protein levels or function. Some of these SNPs are *de facto* mutations, since they generate a non-functional protein or block protein production at all. This is the case of the amino acid substitution PiZ (Glu342Lys), which determines protein multimerization in the liver and lack of secretion in plasma. On one hand, this causes hepatic damage, while, on the other hand, lack of the enzyme exposes the lung to uncontrolled elastase activity. The ensuing disease has a broad phenotypic spectrum, also related to the presence of environmental co-factors. The PiZ allele has a frequency of 0.5–4% in Europe.

Several preclinical gene therapy studies in animal AAT models have shown the feasibility of functional correction of the enzymatic defect by gene transfer using retroviral, adenoviral, and AAV vectors, or using non-viral methods, by which the therapeutic genes are administered to the liver through the portal vein, or to the respiratory tree by aerosol. In principle, direct AAT expression in the lung would offer an advantage, since 90% of patients with the defect show pulmonary symptoms. A first Phase I clinical trial was conducted to assess efficacy of AAT cDNA transfection to the nasal epithelium using a plasmid complexed with cationic liposomes. However, this approach suffered from low efficiency of gene transfer hampering non-viral transfection, a problem similarly encountered by the clinical trials of cystic fibrosis.

An alternative modality of treatment is to use AAV vectors to transfer the AAT cDNA into the skeletal muscle, to attain secretion of the enzyme in the circulation. A first clinical trial used an AAV2 vector for this purpose, however the levels of circulating protein were still modest. An additional trial was recently completed in 9 patients who received different doses of the same vector but packaged in the AAV1 serotype, which shows higher transduction efficiency in skeletal muscle. The results of this trials showed that, in the patients receiving the highest doses of vectors, the enzyme was stably produced in serum for at least 1 year after transduction, albeit at levels that were 0.1% of normal. Although these levels are still not considered therapeutic, this approach definitely warrants further experimentation. In this respect, it is important to observe that AAT production from the skeletal muscle is potentially able to prevent damage of the lung, the protein being secreted in plasma, but not damage of the liver, which is caused by the accumulation of the endogenous mutated protein in the hepatocytes.

4.11.3
Gene Therapy of Hereditary Metabolic Disorders of the Liver Causing Extrahepatic Damage

Different hereditary metabolic disorders due to altered activity of the liver often manifest themselves by causing damage in other organs. This category includes metabolic defects such as OTC deficiency (see below), Crigler-Najjar syndrome type I (a defect of bilirubin conjugation due to absence of the enzyme bilirubin-UDP-glucuronosyltransferase), familial hypercholesterolemia type IIa, and the coagulation defects causing hemophilia A (Factor VIII), hemophilia B (Factor IX), and afibrinogenemia. Gene therapy of the hemophilias is presented in the dedicated section.

4.11.3.1
OTC Deficiency

OTC deficiency is a rare metabolic disorder, occurring in one out of every 80,000 births. OTC is an essential enzyme of the urea cycle, in the absence of which nitrogen deriving from protein catabolism accumulates in blood in the form of ammonia, which is highly toxic to the brain. Children born with complete OTC deficiency show early mortality due to hyperammonemia untreatable with any pharmacological or dietary therapy. Since allogenic hepatocyte transplantation is successful in completely eliminating the disease, OTC deficiency could be definitely cured by transfer of the cDNA coding for the enzyme in autologous hepatocytes.

Gene therapy of OTC represents one of the most dramatic moments of the whole history of gene therapy. In 1999, a first Phase I clinical trial for this disease was conducted at the University of Pennsylvania at Philadelphia. This was a safety trial aimed at assessing safety of the infusion, through the hepatic artery, of a second-generation adenoviral vector, deleted in the E1 and E4 genes (cf. section on 'Vectors Based on Adenoviruses'), expressing the OTC cDNA. One of the enrolled patients was a young man of 18 years, with a mild form of the disease. This patient, a few hours after injection of a relatively high dose of vector, started experiencing serious signs of systemic toxicity, including massive hepatic damage and disseminated intravascular coagulation. After 4 days, the patient eventually died due to multiorgan failure. Death of this patient was later attributed to a massive, systemic inflammatory response to the injected adenoviral vector, probably due to a cytokine storm triggered by the viral capsid. Since other patients of the same trial had received the same dose of vector without experiencing important adverse events, it was concluded that the death of this patient had to be attributed to an unpredictable, exceptionally vigorous, individual response to the vector. Despite its exceptionality, this tragic event highlighted the potential danger of first- and second-generation adenoviral vectors when administered in high doses and demanded extreme caution in all other trials employing these vectors.

As far as gene therapy for OTC deficiency is concerned, other trials are now considering gene transfer of the enzyme cDNA using other vector systems, in particular AAVs.

4.11.3.2
Familial Hypercholesterolemia

Familial hypercholesterolemia (FH) is an autosomal co-dominant disorder characterized by elevated plasma levels of low-density lipoprotein (LDL) cholesterol with normal triglycerides, tendon xanthomas, and premature coronary atherosclerosis. The disease is caused by a mutation of the LDL receptor gene; over 900 different mutations are known. Heterozygous FH, in which only one mutant LDL receptor allele is inherited, is very frequent since it affects 1 in 500 individuals, making it one of the most common diseases with Mendelian inheritance worldwide. The age of onset of CAD is highly variable in these patients and depends on the type of molecular defect of the LDL receptor gene and on the presence of coexisting risk factors. Collectively, heterozygous FH patients have a ~50% chance of having a MI before age 60.

Individuals homozygous for a defective LDL receptor gene are instead relatively rare (1:1 million persons worldwide) and show accelerated atherosclerosis, with can result in disability and death in childhood. These patients can be classified into one of two groups, based on the residual amount of LDL receptor activity. Those patients with <2% of normal LDL receptor activity (receptor-negative patients) show devastating complications of accelerated atherosclerosis and rarely survive beyond the second decade. Patients with receptor-defective LDL receptor defects instead have a better prognosis, although invariably develop vascular disease by age 30 or earlier.

Given the net distinction in prognosis between receptor-negative and receptor-defective homozygous FH, a reasonable goal of gene therapy is to restore LDL receptor activity at a level at least 2% or of normal. The animal model for this disease is the hypercholesterolemic Watanabe rabbit, which, analogous to human disease, develops atherosclerosis. In these rabbits, it was demonstrated that the implantation of 200 million hepatocytes (corresponding to about 2% of total liver mass), transduced *ex vivo* with a gammaretroviral vector expressing the LDL receptor gene, reconstituted LDL receptor activity at a level of 2–4% of normal and significantly reduced the amount of circulating cholesterol.

Based on these encouraging results, a first gene therapy clinical trial was conducted in the early 1990s. This consisted in the isolation of hepatocytes from the resected liver left lateral lobe of five FH patients, *ex vivo* transduction of these cells with a gammaretroviral vector expressing the LDL receptor gene, and transplantation back to the patients. The results of this trial were encouraging, however modest in terms of overall efficacy. This was mainly due to the poor gene transfer efficiency attained, most likely consequent to the use of gammaretroviral vectors and the poor replicative activity of *ex vivo* cultured primary hepatocytes. In addition, this trial further emphasized the difficulty of isolating, culturing, transducing, and reimplanting a very high number of primary hepatocytes (the study was conducted by transplanting over 5 billion cells per patient, corresponding to 800 culture dishes!).

4.1 Clinical Applications of Gene Therapy: General Considerations

Further Reading

Alexander BL, Ali RR, Alton EW et al (2007) Progress and prospects: gene therapy clinical trials (part 1). Gene Ther 14:1439–1447

Alton E, Ferrari S, Griesenbach U (2007) Progress and prospects: gene therapy clinical trials (part 2). Gene Ther 14:1555–1563

Edelstein ML, Abedi MR, Wixon J (2007) Gene therapy clinical trials worldwide to 2007: an update. J Gene Med 9:833–842

Edelstein ML, Abedi MR, Wixon J, Edelstein RM (2004) Gene therapy clinical trials worldwide 1989–2004: an overview. J Gene Med 6:597–602

Fischer A, Cavazzana-Calvo M (2008) Gene therapy of inherited diseases. Lancet 371:2044–2047

Porteus MH, Connelly JP, Pruett SM (2006) A look to future directions in gene therapy research for monogenic diseases. PLoS Genet 2:e133

Relph K, Harrington K, Pandha H (2004) Recent developments and current status of gene therapy using viral vectors in the United Kingdom. BMJ 329:839–842

Schenk-Braat EA, van Mierlo MM, Wagemaker G et al (2007) An inventory of shedding data from clinical gene therapy trials. J Gene Med 9:910–921

4.2 Gene Therapy of Hematopoietic Stem Cells

Further Reading

Ambudkar SV, Kimchi-Sarfaty C, Sauna ZE, Gottesman MM (2003) P-glycoprotein: from genomics to mechanism. Oncogene 22:7468–7485

Baker SJ, Rane SG, Reddy EP (2007) Hematopoietic cytokine receptor signaling. Oncogene 26:6724–6737

Cavazzana-Calvo M, Fischer A (2007) Gene therapy for severe combined immunodeficiency: are we there yet? J Clin Invest 117:1456–1465

Cavazzana-Calvo M, Lagresle C, Hacein-Bey-Abina S, Fischer A (2005) Gene therapy for severe combined immunodeficiency. Annu Rev Med 56:585–602

Greenberger JS (2008) Gene therapy approaches for stem cell protection. Gene Ther 15:100–108

Hawley RG, Sobieski DA (2002) Of mice and men: the tale of two therapies. Stem Cells 20:275–278

Hossle JP, Seger RA, Steinhoff D (2002) Gene therapy of hematopoietic stem cells: strategies for improvement. News Physiol Sci 17:87–92

Licht T, Herrmann F, Gottesman MM, Pastan I (1997) In vivo drug-selectable genes: a new concept in gene therapy. Stem Cells 15:104–111

Nienhuis AW (2008) Development of gene therapy for blood disorders. Blood 111:4431–4444

Sands MS, Davidson BL (2006) Gene therapy for lysosomal storage diseases. Mol Ther 13:839–849

Tey SK, Brenner MK (2007) The continuing contribution of gene marking to cell and gene therapy. Mol Ther 15:666–676

Thrasher AJ, Gaspar HB, Baum C et al (2006) Gene therapy: X-SCID transgene leukaemogenicity. Nature 443:E5–6; discussion E6–7

Zielske SP, Braun SE (2004) Cytokines: value-added products in hematopoietic stem cell gene therapy. Mol Ther 10:211–219

4

Selected Bibliography

Aiuti A, Cattaneo F, Galimberti S et al (2009) Gene therapy for immunodeficiency due to adenosine deaminase deficiency. N Engl J Med 360:447–458

Aiuti A, Slavin S, Aker M et al (2002) Correction of ADA-SCID by stem cell gene therapy combined with nonmyeloablative conditioning. Science 296:2410–2413

Aiuti A, Vai S, Mortellaro A et al (2002) Immune reconstitution in ADA-SCID after PBL gene therapy and discontinuation of enzyme replacement. Nat Med 8:423–425

Alexander IE, Cunningham SC, Logan GJ, Christodoulou J (2008) Potential of AAV vectors in the treatment of metabolic disease. Gene Ther 15:831–839

Baum C, von Kalle C, Staal FJ et al (2004) Chance or necessity? Insertional mutagenesis in gene therapy and its consequences. Mol Ther 9:5–13

Beck M (2007) New therapeutic options for lysosomal storage disorders: enzyme replacement, small molecules and gene therapy. Hum Genet 121:1–22

Biffi A, Naldini L (2005) Gene therapy of storage disorders by retroviral and lentiviral vectors. Hum Gene Ther 16:1133–1142

Bonini C, Bondanza A, Perna SK et al (2007) The suicide gene therapy challenge: how to improve a successful gene therapy approach. Mol Ther 15:1248–1252

Brenner MK (1996) Gene marking. Gene Ther 3:278–279

Cattoglio C, Facchini G, Sartori D et al (2007) Hot spots of retroviral integration in human CD34+ hematopoietic cells. Blood 110:1770–1778

Cavazzana-Calvo M, Hacein-Bey S, de Saint Basile G et al (2000) Gene therapy of human severe combined immunodeficiency (SCID)-X1 disease. Science 288:669–672

Cheng SH, Smith AE (2003) Gene therapy progress and prospects: gene therapy of lysosomal storage disorders. Gene Ther 10:1275–1281

Dave UP, Jenkins NA, Copeland NG (2004) Gene therapy insertional mutagenesis insights. Science 303:333

Deisseroth AB, Zu Z, Claxton D et al (1994) Genetic marking shows that Ph+ cells present in autologous transplants of chronic myelogenous leukemia (CML) contribute to relapse after autologous bone marrow in CML. Blood 83:3068–3076

Dinauer MC, Orkin SH (1992) Chronic granulomatous disease. Annu Rev Med 43:117–124

Gaspar HB, Bjorkegren E, Parsley K et al (2006) Successful reconstitution of immunity in ADA-SCID by stem cell gene therapy following cessation of PEG-ADA and use of mild preconditioning. Mol Ther 14:505–513

Gaspar HB, Parsley KL, Howe S et al (2004) Gene therapy of X-linked severe combined immunodeficiency by use of a pseudotyped gammaretroviral vector. Lancet 364:2181–2187

Hacein-Bey-Abina S, von Kalle C, Schmidt M et al (2003) A serious adverse event after successful gene therapy for X-linked severe combined immunodeficiency. N Engl J Med 348:255–256

Hacein-Bey-Abina S, Von Kalle C, Schmidt M et al (2003) LMO2-associated clonal T cell proliferation in two patients after gene therapy for SCID-X1. Science 302:415–419

Hodges BL, Cheng SH (2006) Cell and gene-based therapies for the lysosomal storage diseases. Curr Gene Ther 6:227–241

Kohn DB, Sadelain M, Glorioso JC (2003) Occurrence of leukaemia following gene therapy of X-linked SCID. Nat Rev Cancer 3:477–488

Muul LM, Tuschong LM, Soenen SL et al (2003) Persistence and expression of the adenosine deaminase gene for 12 years and immune reaction to gene transfer components: long-term results of the first clinical gene therapy trial. Blood 101:2563–2569

Ott MG, Schmidt M, Schwarzwaelder K et al (2006) Correction of X-linked chronic granulomatous disease by gene therapy, augmented by insertional activation of MDS1-EVI1, PRDM16 or SETBP1. Nat Med 12:401–409

Schwarzwaelder K, Howe SJ, Schmidt M et al (2007) Gammaretrovirus-mediated correction of SCID-X1 is associated with skewed vector integration site distribution in vivo. J Clin Invest 117:2241–2249

Shou Y, Ma Z, Lu T, Sorrentino BP (2006) Unique risk factors for insertional mutagenesis in a mouse model of XSCID gene therapy. Proc Natl Acad Sci U S A 103:11730–11735

Woods NB, Bottero V, Schmidt M et al (2006) Gene therapy: therapeutic gene causing lymphoma. Nature 440:1123

4.3 Gene Therapy of Cystic Fibrosis

Further Reading

Anson DS, Smith GJ, Parsons DW (2006) Gene therapy for cystic fibrosis airway disease: is clinical success imminent? Curr Gene Ther 6:161–179

Flotte TR, Ng P, Dylla DE et al (2007) Viral vector-mediated and cell-based therapies for treatment of cystic fibrosis. Mol Ther 15:229–241

Griesenbach U, Alton EW (2009) Gene transfer to the lung: lessons learned from more than 2 decades of CF gene therapy. Adv Drug Deliv Rev 61:128–139

Griesenbach U, Geddes DM, Alton EW (2006) Gene therapy progress and prospects: cystic fibrosis. Gene Ther 13:1061–1067

O'Sullivan BP, Freedman SD (2009) Cystic fibrosis. Lancet 373:1891–1904

Riordan JR (2008) CFTR function and prospects for therapy. Annu Rev Biochem 77:701–726

Selected Bibliography

Crystal RG, McElvaney NG, Rosenfeld MA et al (1994) Administration of an adenovirus containing the human CFTR cDNA to the respiratory tract of individuals with cystic fibrosis. Nat Genet 8:42–51

Kinsey BM, Densmore CL, Orson FM (2005) Non-viral gene delivery to the lungs. Curr Gene Ther 5:181–194

Kremer KL, Dunning KR, Parsons DW, Anson DS (2007) Gene delivery to airway epithelial cells in vivo: a direct comparison of apical and basolateral transduction strategies using pseudo-typed lentivirus vectors. J Gene Med 9:362–368

Li W, Zhang L, Johnson JS et al (2009) Generation of novel AAV variants by directed evolution for improved CFTR delivery to human ciliated airway epithelium. Mol Ther 17:2067–2077

Tagalakis AD, McAnulty RJ, Devaney J et al (2008) A receptor-targeted nanocomplex vector system optimized for respiratory gene transfer. Mol Ther 16:907–915

4.4 Gene Therapy of Muscular Dystrophies

Further Reading

Athanasopoulos T, Graham IR, Foster H, Dickson G (2004) Recombinant adeno-associated viral (rAAV) vectors as therapeutic tools for Duchenne muscular dystrophy (DMD). Gene Ther 11[Suppl 1]:S109–121

Chakkalakal JV, Thompson J, Parks RJ, Jasmin BJ (2005) Molecular, cellular, and pharmacological therapies for Duchenne/Becker muscular dystrophies. FASEB J 19:880–891

Foster K, Foster H, Dickson JG (2006) Gene therapy progress and prospects: Duchenne muscular dystrophy. Gene Ther 13:1677–1685

Selected Bibliography

Alter J, Lou F, Rabinowitz A et al (2006) Systemic delivery of morpholino oligonucleotide restores dystrophin expression bodywide and improves dystrophic pathology. Nat Med 12:175–177

Cerletti M, Negri T, Cozzi F et al (2003) Dystrophic phenotype of canine X-linked muscular dystrophy is mitigated by adenovirus-mediated utrophin gene transfer. Gene Ther 10:750–757

Duan D (2006) Challenges and opportunities in dystrophin-deficient cardiomyopathy gene therapy. Hum Mol Genet 15[Spec No 2]:R253–261

Goyenvalle A, Vulin A, Fougerousse F et al (2004) Rescue of dystrophic muscle through U7 snRNA-mediated exon skipping. Science 306:1796–1799

Gregorevic P, Blankinship MJ, Allen JM et al (2004) Systemic delivery of genes to striated muscles using adeno-associated viral vectors. Nat Med 10:828–834

Heemskerk H, de Winter CL, van Ommen GJ et al (2009) Development of antisense-mediated exon skipping as a treatment for duchenne muscular dystrophy. Ann N Y Acad Sci 1175:71–79

Lu QL, Rabinowitz A, Chen YC et al (2005) Systemic delivery of antisense oligoribonucleotide restores dystrophin expression in body-wide skeletal muscles. Proc Natl Acad Sci U S A 102:198–203

Mendell JR, Rodino-Klapac LR, Rosales-Quintero X et al (2009) Limb-girdle muscular dystrophy type 2D gene therapy restores alpha-sarcoglycan and associated proteins. Ann Neurol 66:290–297

Nelson R (2004) Utrophin therapy for Duchenne muscular dystrophy? Lancet Neurol 3: 637

Romero NB, Braun S, Benveniste O et al (2004) Phase I study of dystrophin plasmid-based gene therapy in Duchenne/Becker muscular dystrophy. Hum Gene Ther 15:1065–1076

Scott JM, Li S, Harper SQ et al (2002) Viral vectors for gene transfer of micro-, mini-, or full-length dystrophin. Neuromuscul Disord 12[Suppl 1]:S23–29

van Deutekom JC, Janson AA, Ginjaar IB et al (2007) Local dystrophin restoration with antisense oligonucleotide PRO051. N Engl J Med 357:2677–2686

Wang Z, Zhu T, Qiao C et al (2005) Adeno-associated virus serotype 8 efficiently delivers genes to muscle and heart. Nat Biotechnol 23:321–328

4.5 Gene Therapy of Hemophilia

Further Reading

Bolton-Maggs PH, Pasi KJ (2003) Haemophilias A and B. Lancet 361:1801–1809

Foster K, Foster H, Dickson JG (2006) Gene therapy progress and prospects: Duchenne muscular dystrophy. Gene Ther 13:1677–1685

Graw J, Brackmann HH, Oldenburg J et al (2005) Haemophilia A: from mutation analysis to new therapies. Nat Rev Genet 6:488–501

Hasbrouck NC, High KA (2008) AAV-mediated gene transfer for the treatment of hemophilia B: problems and prospects. Gene Ther 15:870–875

Mingozzi F, High KA (2007) Immune responses to AAV in clinical trials. Curr Gene Ther 7:316–324

Murphy SL, High KA (2008) Gene therapy for haemophilia. Br J Haematol 140:479–487

Selected Bibliography

Jiang H, Pierce GF, Ozelo MC et al (2006) Evidence of multiyear factor IX expression by AAV-mediated gene transfer to skeletal muscle in an individual with severe hemophilia B. Mol Ther 14:452–455

Kay MA, Manno CS, Ragni MV et al (2000) Evidence for gene transfer and expression of factor IX in haemophilia B patients treated with an AAV vector. Nat Genet 24:257–261

Manno CS, Pierce GF, Arruda VR et al (2006) Successful transduction of liver in hemophilia by AAV-Factor IX and limitations imposed by the host immune response. Nat Med 12:342–347

Margaritis P, Roy E, Aljamali MN et al (2009) Successful treatment of canine hemophilia by continuous expression of canine FVIIa. Blood 113:3682–3689

Wang L, Nichols TC, Read MS et al (2000) Sustained expression of therapeutic level of factor IX in hemophilia B dogs by AAV-mediated gene therapy in liver. Mol Ther 1:154–158

4.6 Gene Therapy of Cancer

Further Reading

Aghi M, Hochberg F, Breakefield XO (2000) Prodrug activation enzymes in cancer gene therapy. J Gene Med 2:148–164

Anderson RJ, Schneider J (2007) Plasmid DNA and viral vector-based vaccines for the treatment of cancer. Vaccine 25[Suppl 2]:B24–34

Cattaneo R, Miest T, Shashkova EV, Barry MA (2008) Reprogrammed viruses as cancer therapeutics: targeted, armed and shielded. Nat Rev Micro 6:529–540

Heath WR, Belz GT, Behrens GMN et al (2004) Cross-presentation, dendritic cell subsets, and the generation of immunity to cellular antigens. Immunol Rev 199:9–26

Hermiston TW, Kirn DH (2005) Genetically based therapeutics for cancer: similarities and contrasts with traditional drug discovery and development. Mol Ther 11:496–507

June CH (2007) Adoptive T cell therapy for cancer in the clinic. J Clin Invest 117:1466–1476

June CH (2007) Principles of adoptive T cell cancer therapy. J Clin Invest 117:1204–1212

Larin SS, Georgiev GP, Kiselev SL (2004) Gene transfer approaches in cancer immunotherapy. Gene Ther 11[Suppl 1]:S18–25

Liu TC, Kirn D (2008) Gene therapy progress and prospects cancer: oncolytic viruses. Gene Ther 15:877–884

McNeish IA, Bell SJ, Lemoine NR (2004) Gene therapy progress and prospects: cancer gene therapy using tumour suppressor genes. Gene Ther 11:497–503

Offringa R (2006) Cancer. Cancer immunotherapy is more than a numbers game. Science 314:68–69

Palmer DH, Young LS, Mautner V (2006) Cancer gene-therapy: clinical trials. Trends Biotechnol 24:76–82

Rice J, Ottensmeier CH, Stevenson FK (2008) DNA vaccines: precision tools for activating effective immunity against cancer. Nat Rev Cancer 8:108–120

Rossig C, Brenner MK (2004) Genetic modification of T lymphocytes for adoptive immunotherapy. Mol Ther 10:5–18

Selected Bibliography

Bonini C, Ferrari G, Verzeletti S et al (1997) HSV-TK gene transfer into donor lymphocytes for control of allogeneic graft-versus-leukemia. Science 276:1719–1724

Breckpot K, Aerts JL, Thielemans K (2007) Lentiviral vectors for cancer immunotherapy: transforming infectious particles into therapeutics. Gene Ther 14:847–862

Cesco-Gaspere M, Zentilin L, Giacca M, Burrone OR (2008) Boosting anti-idiotype immune response with recombinant AAV enhances tumour protection induced by gene gun vaccination. Scand J Immunol 68:58–66

Finke LH, Wentworth K, Blumenstein B et al (2007) Lessons from randomized phase III studies with active cancer immunotherapies: outcomes from the 2006 meeting of the Cancer Vaccine Consortium (CVC). Vaccine 25[Suppl 2]:B97–B109

Folkman J, Watson K, Ingber D, Hanahan D (1989) Induction of angiogenesis during the transition from hyperplasia to neoplasia. Nature 339:58–61

Johnson LA, Morgan RA, Dudley ME et al (2009) Gene therapy with human and mouse T-cell receptors mediates cancer regression and targets normal tissues expressing cognate antigen. Blood 114:535–546

McNeish IA, Bell SJ, Lemoine NR (2004) Gene therapy progress and prospects: cancer gene therapy using tumour suppressor genes. Gene Ther 11:497–503

Morgan RA, Dudley ME, Wunderlich JR et al (2006) Cancer regression in patients after transfer of genetically engineered lymphocytes. Science 314:126–129

Nestle FO, Farkas A, Conrad C (2005) Dendritic-cell-based therapeutic vaccination against cancer. Curr Opin Immunol 17:163–169

Parmigiani RB, Bettoni F, Vibranovski MD et al (2006) Characterization of a cancer/testis (CT) antigen gene family capable of eliciting humoral response in cancer patients. Proc Natl Acad Sci U S A 103:18066–18071

Pulkkanen KJ, Yla-Herttuala S (2005) Gene therapy for malignant glioma: current clinical status. Mol Ther 12:585–598

Rapoport AP, Stadtmauer EA, Aqui N et al (2005) Restoration of immunity in lymphopenic individuals with cancer by vaccination and adoptive T-cell transfer. Nat Med 11:1230–1237

Ringdén O, Karlsson H, Olsson R et al (2009) The allogeneic graft-versus-cancer effect. Br J Haematol 147:614–633

Rosenberg SA, Aebersold P, Cornetta K et al (1990) Gene transfer into humans: immunotherapy of patients with advanced melanoma, using tumor-infiltrating lymphocytes modified by retroviral gene transduction. N Engl J Med 323:570–578

Rosenthal R, Viehl CT, Guller U et al (2008) Active specific immunotherapy phase III trials for malignant melanoma: systematic analysis and critical appraisal. J Am Coll Surg 207:95–105

Tasciotti E, Zoppe M, Giacca M (2003) Transcellular transfer of active HSV-1 thymidine kinase mediated by an 11-amino-acid peptide from HIV-1 Tat. Cancer Gene Ther 10:64–74

Terando AM, Faries MB, Morton DL (2007) Vaccine therapy for melanoma: current status and future directions. Vaccine 25[Suppl 2]:B4–16

4.7 Gene Therapy of Neurodegenerative Disorders

Further Reading

Azzouz M (2006) Gene therapy for ALS: progress and prospects. Biochim Biophys Acta 1762:1122–1127

Baker D, Hankey DJ (2003) Gene therapy in autoimmune, demyelinating disease of the central nervous system. Gene Ther 10:844–853

Bradbury J (2005) Hope for AD with NGF gene-therapy trial. Lancet Neurol 4:335

Burton EA, Glorioso JC, Fink DJ (2003) Gene therapy progress and prospects: Parkinson's disease. Gene Ther 10:1721–1727

Choudry RB, Cudkowicz ME (2005) Clinical trials in amyotrophic lateral sclerosis: the tenuous past and the promising future. J Clin Pharmacol 45:1334–1344

Federici T, Boulis N (2007) Gene therapy for peripheral nervous system diseases. Curr Gene Ther 7:239–248

Fiandaca M, Forsayeth J, Bankiewicz K (2008) Current status of gene therapy trials for Parkinson's disease. Exp Neurol 209:51–57

Kaspar BK, Llado J, Sherkat N et al (2003) Retrograde viral delivery of IGF-1 prolongs survival in a mouse ALS model. Science 301:839–842

Mitchell JD, Borasio GD (2007) Amyotrophic lateral sclerosis. Lancet 369:2031–2041

Palfi S (2008) Towards gene therapy for Parkinson's disease. Lancet Neurol 7:375–376

Sheridan C (2007) Positive clinical data in Parkinson's and ischemia buoy gene therapy. Nat Biotechnol 25:823–824

Sumner CJ (2006) Therapeutics development for spinal muscular atrophy. NeuroRx 3:235–245

Tuszynski MH (2002) Growth-factor gene therapy for neurodegenerative disorders. Lancet Neurol 1:51–57

Zacchigna S, Giacca M (2009) Chapter 20: Gene therapy perspectives for nerve repair. Int Rev Neurobiol 87:381–392

Selected Bibliography

Azzouz M, Le T, Ralph GS et al (2004) Lentivector-mediated SMN replacement in a mouse model of spinal muscular atrophy. J Clin Invest 114:1726–1731

Azzouz M, Ralph GS, Storkebaum E et al (2004) VEGF delivery with retrogradely transported lentivector prolongs survival in a mouse ALS model. Nature 429:413–417

Bloch J, Bachoud-Levi AC, Deglon N et al (2004) Neuroprotective gene therapy for Huntington's disease, using polymer-encapsulated cells engineered to secrete human ciliary neurotrophic factor: results of a phase I study. Hum Gene Ther 15:968–975

Christine CW, Starr PA, Larson PS et al (2009) Safety and tolerability of putaminal AADC gene therapy for Parkinson disease. Neurology 73:1662–1669

Eberling JL, Jagust WJ, Christine CW et al (2008) Results from a phase I safety trial of hAADC gene therapy for Parkinson disease. Neurology 70:1980–1983

Hua Y, Vickers TA, Okunola HL et al (2008) Antisense masking of an hnRNP A1/A2 intronic splicing silencer corrects SMN2 splicing in transgenic mice. Am J Hum Genet 82:834–848

Kaplitt MG, Feigin A, Tang C et al (2007) Safety and tolerability of gene therapy with an adeno-associated virus (AAV) borne GAD gene for Parkinson's disease: an open label, phase I trial. Lancet 369:2097–2105

Kennington E (2009) Gene therapy delivers an alternative approach to Alzheimer's disease. Nat Rev Drug Discov 8:275

Lewis TB, Standaert DG (2008) Design of clinical trials of gene therapy in Parkinson disease. Exp Neurol 209:41–47

Mandel RJ, Burger C (2004) Clinical trials in neurological disorders using AAV vectors: promises and challenges. Curr Opin Mol Ther 6:482–490

Mandel RJ, Burger C, Snyder RO (2008) Viral vectors for in vivo gene transfer in Parkinson's disease: properties and clinical grade production. Exp Neurol 209:58–71

Nagahara AH, Merrill DA, Coppola G et al (2009) Neuroprotective effects of brain-derived neurotrophic factor in rodent and primate models of Alzheimer's disease. Nat Med 15:331–337

Ralph GS, Radcliffe PA, Day DM et al (2005) Silencing mutant SOD1 using RNAi protects against neurodegeneration and extends survival in an ALS model. Nat Med 11:429–433

Storkebaum E, Lambrechts D, Carmeliet P (2004) VEGF: once regarded as a specific angiogenic factor, now implicated in neuroprotection. Bioessays 26:943–954

Towne C, Schneider BL, Kieran D et al (2009) Efficient transduction of non-human primate motor neurons after intramuscular delivery of recombinant AAV serotype 6. Gene Ther 17:141–146

Tuszynski MH, Thal L, Pay M et al (2005) A phase 1 clinical trial of nerve growth factor gene therapy for Alzheimer disease. Nat Med 11:551–555

4.8 Gene Therapy of Eye Diseases

Further Reading

Bainbridge JW, Ali RR (2008) Success in sight: the eyes have it! Ocular gene therapy trials for LCA look promising. Gene Ther 15:1191–1192

Bainbridge JW, Tan MH, Ali RR (2006) Gene therapy progress and prospects: the eye. Gene Ther 13:1191–1197

Bennett J, Maguire AM (2000) Gene therapy for ocular disease. Mol Ther 1:501–505

Buch PK, Bainbridge JW, Ali RR (2008) AAV-mediated gene therapy for retinal disorders: from mouse to man. Gene Ther 15:849–857

Jager RD, Mieler WF, Miller JW (2008) Age-related macular degeneration. N Engl J Med 358:2606–2617

Kaiser J (2008) Gene therapy. Two teams report progress in reversing loss of sight. Science 320:606–607

Smith AJ, Bainbridge JW, Ali RR (2009) Prospects for retinal gene replacement therapy. Trends Genet 25:156–165

Selected Bibliography

Acland GM, Aguirre GD, Ray J et al (2001) Gene therapy restores vision in a canine model of childhood blindness. Nat Genet 28:92–95

Bainbridge JW, Smith AJ, Barker SS et al (2008) Effect of gene therapy on visual function in Leber's congenital amaurosis. N Engl J Med 358:2231–2239

Dinculescu A, Glushakova L, Min SH, Hauswirth WW (2005) Adeno-associated virus-vectored gene therapy for retinal disease. Hum Gene Ther 16:649–663

Gehrs KM, Anderson DH, Johnson LV, Hageman GS (2006) Age-related macular degeneration: emerging pathogenetic and therapeutic concepts. Ann Med 38:450–471

Le Meur G, Stieger K, Smith AJ et al (2007) Restoration of vision in RPE65-deficient Briard dogs using an AAV serotype 4 vector that specifically targets the retinal pigmented epithelium. Gene Ther 14:292–303

Maguire AM, Simonelli F, Pierce EA et al (2008) Safety and efficacy of gene transfer for Leber's congenital amaurosis. N Engl J Med 358:2240–2248

4.9 Gene Therapy of Cardiovascular Disorders

Further Reading

Adams RH, Alitalo K (2007) Molecular regulation of angiogenesis and lymphangiogenesis. Nat Rev Mol Cell Biol 8:464–478

Augustin HG, Koh GY, Thurston G, Alitalo K (2009) Control of vascular morphogenesis and homeostasis through the angiopoietin-Tie system. Nat Rev Mol Cell Biol 10:165–177

Bhargava B, Karthikeyan G, Abizaid AS, Mehran R (2003) New approaches to preventing restenosis. BMJ 327:274–279

Carmeliet P (2005) Angiogenesis in life, disease and medicine. Nature 438:932–936

Crook MF, Akyurek LM (2003) Gene transfer strategies to inhibit neointima formation. Trends Cardiovasc Med 13:102–106

Giacca M (2007) Virus-mediated gene transfer to induce therapeutic angiogenesis: where do we stand? Int J Nanomed 2:527–540

Olsson AK, Dimberg A, Kreuger J, Claesson-Welsh L (2006) VEGF receptor signalling: in control of vascular function. Nat Rev Mol Cell Biol 7:359–371

Rissanen TT, Yla-Herttuala S (2007) Current status of cardiovascular gene therapy. Mol Ther 15:1233–1247

Stewart S, MacIntyre K, Hole DJ et al (2001) More 'malignant' than cancer? Five-year survival following a first admission for heart failure. Eur J Heart Fail 3:315–322

Vincent KA, Jiang C, Boltje I, Kelly RA (2007) Gene therapy progress and prospects: therapeutic angiogenesis for ischemic cardiovascular disease. Gene Ther 14:781–789

Vinge LE, Raake PW, Koch WJ (2008) Gene therapy in heart failure. Circ Res 102:1458–1470

Yancopoulos GD, Davis S, Gale NW et al (2000) Vascular-specific growth factors and blood vessel formation. Nature 407:242–248

Yla-Herttuala S, Alitalo K (2003) Gene transfer as a tool to induce therapeutic vascular growth. Nat Med 9:694–701

Yla-Herttuala S, Markkanen JE, Rissanen TT (2004) Gene therapy for ischemic cardiovascular diseases: some lessons learned from the first clinical trials. Trends Cardiovasc Med 14:295–300

Yla-Herttuala S, Martin JF (2000) Cardiovascular gene therapy. Lancet 355:213–222

Selected Bibliography

Arsic N, Zacchigna S, Zentilin L et al (2004) Vascular endothelial growth factor stimulates skeletal muscle regeneration in vivo. Mol Ther 10:844–854

Arsic N, Zentilin L, Zacchigna S et al (2003) The biology of VEGF and its receptors. Nat Med 9:669–676

Baumgartner I, Pieczek A, Manor O et al (1998) Constitutive expression of phVEGF165 after intramuscular gene transfer promotes collateral vessel development in patients with critical limb ischemia. Circulation 97:1114–1123

del Monte F, Harding SE, Schmidt U et al (1999) Restoration of contractile function in isolated cardiomyocytes from failing human hearts by gene transfer of SERCA2a. Circulation 100:2308–2311

Ferrarini M, Arsic N, Recchia FA et al (2006) Adeno-associated virus-mediated transduction of VEGF165 improves cardiac tissue viability and functional recovery after permanent coronary occlusion in conscious dogs. Circ Res 98:954–961

Grines CL, Watkins MW, Helmer G et al (2002) Angiogenic Gene Therapy (AGENT) trial in patients with stable angina pectoris. Circulation 105:1291–1297

Inagaki K, Fuess S, Storm TA et al (2006) Robust systemic transduction with AAV9 vectors in mice: efficient global cardiac gene transfer superior to that of AAV8. Mol Ther 14:45–53

Lafont A, Guerot C, Lemarchand P (1995) Which gene for which restenosis? Lancet 346:1442–1443

Melo LG, Pachori AS, Kong D et al (2004) Gene and cell-based therapies for heart disease. FASEB J 18:648–663

Simons M, Bonow RO, Chronos NA et al (2000) Clinical trials in coronary angiogenesis: issues, problems, consensus: an expert panel summary. Circulation 102:E73–E86

Sinagra G, Giacca M (2003) Induction of functional neovascularization by combined VEGF and angiopoietin-1 gene transfer using AAV vectors. Mol Ther 7:450–459

Tafuro S, Ayuso E, Zacchigna S et al (2009) Inducible adeno-associated virus vectors promote functional angiogenesis in adult organisms via regulated vascular endothelial growth factor expression. Cardiovasc Res 83:663–671

Vale PR, Losordo DW, Milliken CE et al (2000) Left ventricular electromechanical mapping to assess efficacy of phVEGF(165) gene transfer for therapeutic angiogenesis in chronic myocardial ischemia. Circulation 102:965–974

Zacchigna S, Tasciotti E, Kusmic C et al (2007) In vivo imaging shows abnormal function of vascular endothelial growth factor-induced vasculature. Hum Gene Ther 18:515–524

4.10 Gene Therapy of HIV Infection

Further Reading

Baltimore D (1988) Gene therapy. Intracellular immunization. Nature 335:395–396

Dropulic B, June CH (2006) Gene-based immunotherapy for human immunodeficiency virus infection and acquired immunodeficiency syndrome. Hum Gene Ther 17:577–588

Fillat C, Carrio M, Cascante A, Sangro B (2003) Suicide gene therapy mediated by the Herpes Simplex virus thymidine kinase gene/Ganciclovir system: fifteen years of application. Curr Gene Ther 3:13–26

Giacca M (2008) Gene therapy to induce cellular resistance to HIV-1 infection: lessons from clinical trials. Adv Pharmacol 56:297–325

Haasnoot J, Westerhout EM, Berkhout B (2007) RNA interference against viruses: strike and counterstrike. Nat Biotechnol 25:1435–1443

Manilla P, Rebello T, Afable C et al (2005) Regulatory considerations for novel gene therapy products: a review of the process leading to the first clinical lentiviral vector. Hum Gene Ther 16:17–25

Morris K.V, Rossi JJ (2006) Lentiviral-mediated delivery of siRNAs for antiviral therapy. Gene Ther 13:553–558

Rossi JJ (2006) RNAi as a treatment for HIV-1 infection. Biotechniques [Suppl]:25–29

Rossi JJ, June CH, Kohn DB (2007) Genetic therapies against HIV. Nat Biotechnol 25:1444–1454

Strayer DS, Akkina R, Bunnell BA et al (2005) Current status of gene therapy strategies to treat HIV/AIDS. Mol Ther 11:823–842

Wolkowicz R, Nolan GP (2005) Gene therapy progress and prospects: novel gene therapy approaches for AIDS. Gene Ther 12:467–476

Selected Bibliography

Buchbinder SP, Mehrotra DV, Duerr A et al (2008) Efficacy assessment of a cell-mediated immunity HIV-1 vaccine (the Step Study): a double-blind, randomised, placebo-controlled, test-of-concept trial. Lancet 372:1881–1893

Das AT, Brummelkamp TR, Westerhout EM et al (2004) Human immunodeficiency virus type 1 escapes from RNA interference-mediated inhibition. J Virol 78:2601–2605

Levine, BL, Humeau, LM, Boyer J et al (2006) Gene transfer in humans using a conditionally replicating lentiviral vector. Proc Natl Acad Sci U S A 103:17372–17377

Li MJ, Kim J, Li S et al (2005) Long-term inhibition of HIV-1 infection in primary hematopoietic cells by lentiviral vector delivery of a triple combination of anti-HIV shRNA, anti-CCR5 ribozyme, and a nucleolar-localizing TAR decoy. Mol Ther 12:900–909

Mitsuyasu RT, Merigan TC, Carr A et al (2009) Phase 2 gene therapy trial of an anti-HIV ribozyme in autologous CD34+ cells. Nat Med 15:285–292

Morgan RA, Walker R (1996) Gene therapy for AIDS using retroviral mediated gene transfer to deliver HIV-1 antisense TAR and transdominant Rev protein genes to syngeneic lymphocytes in HIV-1 infected identical twins. Hum Gene Ther 7:1281–1306

Nisole S, Stoye JP, Saib A (2005) TRIM family proteins: retroviral restriction and antiviral defence. Nat Rev Microbiol 3:799–808

Novina CD, Murray MF, Dykxhoorn DM et al (2002) siRNA-directed inhibition of HIV-1 infection. Nat Med 8:681–686

Poeschla E, Corbeau P, Flossie W-S (1996) Development of HIV vectors for anti-HIV gene therapy. Proc Natl Acad Sci U S A 93:11395–11399

Rondon IJ, Marasco WA (1997) Intracellular antibodies (intrabodies) for gene therapy of infectious diseases. Annu Rev Microbiol 51:257–283

Sarver N, Cantin EM, Chang PS et al (1990) Ribozymes as potential anti-HIV-1 therapeutic agents. Science 247:1222–1225

Schambach A, Schiedlmeier B, Kuhlcke K et al (2006) Towards hematopoietic stem cell-mediated protection against infection with human immunodeficiency virus. Gene Ther 13:1037–1047

ter Brake O, Konstantinova P, Ceylan M, Berkhout B (2006) Silencing of HIV-1 with RNA interference: a multiple shRNA approach. Mol Ther 14:883–892

Westerhout EM, Ooms M, Vink M et al (2005) HIV-1 can escape from RNA interference by evolving an alternative structure in its RNA genome. Nucleic Acids Res 33:796–804

4.11 Gene Therapy of Liver Diseases

Further Reading

Broedl UC, Rader DJ (2005) Gene therapy for lipoprotein disorders. Exp Opin Biol Ther 5:1029–1038

Nguyen TH, Ferry N (2004) Liver gene therapy: advances and hurdles. Gene Ther 11[Suppl 1]:S76–84

Nguyen TH, Mainot S, Lainas P et al (2009) Ex vivo liver-directed gene therapy for the treatment of metabolic diseases: advances in hepatocyte transplantation and retroviral vectors. Curr Gene Ther 9:136–149

Stecenko AA, Brigham KL (2003) Gene therapy progress and prospects: alpha-1 antitrypsin. Gene Ther 10:95–99

Wood AM, Stockley RA (2007) Alpha one antitrypsin deficiency: from gene to treatment. Respiration 74:481–492

Selected Bibliography

Grossman M, Rader DJ, Muller DW et al (1995) A pilot study of ex vivo gene therapy for homozygous familial hypercholesterolaemia. Nat Med 1:1148–1154

Grossman M, Raper SE, Kozarsky K et al (1994) Successful ex vivo gene therapy directed to liver in a patient with familial hypercholesterolaemia. Nat Genet 6:335–341

Kozarsky KF, Jooss K, Donahee M et al (1996) Effective treatment of familial hypercholesterolaemia in the mouse model using adenovirus-mediated transfer of the VLDL receptor gene. Nat Genet 13:54–62

Miranda PS, Bosma PJ (2009) Towards liver-directed gene therapy for Crigler-Najjar syndrome. Curr Gene Ther 9:72–82

Stoller JK, Aboussouan LS (2005) Alpha1-antitrypsin deficiency. Lancet 365:2225–2236

Ethical and Social Problems of Gene Therapy

5

Since the early days of gene therapy at the end of the 1980s, both the scientific community and the public have perceived the ethical and social problems intrinsic to this discipline.
On one hand, the technologies for gene transfer are still largely experimental and thus pose
important safety issues. On the other hand, the objective of several gene therapy applications is the stable modification of the genetic characteristics of an individual. Whether such
modification is ethically acceptable and might also be applied to the embryo or fetus before
birth, or to the germinal cells, has always been the subject of intense debate. Finally, while
gene therapy is largely accepted when its application is to allow survival or improvement
in the health of an individual, the gene transfer technologies of this discipline can also be
exploited for the improvement of aesthetic appearance or physical and intellectual performance. These, and other major themes currently at the center of the ethical and social debate
accompanying the development of gene therapy, are discussed in this chapter.

5.1
Safety of Clinical Experimentation

As outlined in Chapter 4 ('Clinical Applications of Gene Therapy'), a gene therapy clinical experimentation must fulfill the same requirements as any other pharmacological trial,
including the assessment of safety in Phase I studies. In principle, there are a few reasons
why gene transfer might be potentially dangerous.

First, nucleic acids themselves can be toxic, given their chemical and biological properties. For example, siRNAs or oligonucleotides, when administered systemically as
naked nucleic acids, can interfere with the coagulation system or determine thrombocy-
topenia; once internalized by the cells, they can cause off-target effects leading to cell
death or interference with cell function.

Second, and more pertinent, the delivery systems that are used for gene transfer might
cause damage both at the systemic and at the cellular levels. The non-viral systems based
on cationic polymers or dendrimers, once administered systemically, can affect the coag-

ulation system or the complement cascade or, if the polymer/DNA complexes are large, be nephrotoxic. On the other hand, viral vectors can exert toxicity as a consequence of the biological properties of the wild-type viruses from which they derive. Adenoviral vectors are inflammatory and immunogenic, gammaretroviral vectors can lead to cell transformation, lentiviral vectors pose the concern that novel viruses might arise as the consequence of unforeseen recombination events, and herpesviruses might be neurovirulent once injected *in vivo*. These and other safety issues have been extensively treated in the sections dealing with the properties of these vectors.

Third, gene delivery procedures themselves can be toxic. For example, gene transfer to the arterial wall during angioplasty demands the use of catheters partially obstructing blood flow to allow injection of the therapeutic nucleic acids, gene therapy of hematopoietic stem cells is only effective after myelosuppression using the cytotoxic drug busulfan, and so on.

Despite these various sources of potential toxicity, from the results of the over 1500 clinical trials conducted so far, it can be convincingly concluded that, in the vast majority of cases, gene therapy is safe. Thus, the major current concern is probably efficacy rather than safety. There are, however, a few exceptions to this general conclusion, since some major adverse events have been observed in a few clinical trials.

In 1999, a patient with ornithine transcarbamylase (OTC) deficiency, enrolled in a gene therapy clinical trial in Philadelphia, died after administration of a high dose of a second-generation adenoviral vector in the hepatic artery (see section on 'Gene Therapy of Liver Diseases'); death was attributed to massive cytokine production triggered by the viral capsid. The decision to include this patient in the trial was later strongly criticized, since he was a young adult in whom the disease had been relatively well controlled by support therapies. It was also criticized that a hereditary disorder was treated using an adenoviral vector, since these vectors do not integrate into the host genome and are known to drive short-term transgene expression. The important inflammatory response elicited by first- and second-generation adenoviruses has also been observed in other trials based on these vectors, particularly in the treatment of cystic fibrosis by aerosol administration. These episodes of toxicity have now significantly narrowed the possibility of using adenoviruses clinically; ongoing trials are based on the injection of very low viral doses and aim at treating diseases in which prolonged transgene expression is not required (e.g., for vaccination).

A second series of major adverse events was observed in two gene therapy clinical trials for a severe immunodeficiency due to a defect in the common γ-chain, a protein required for the function of different interleukin receptors (SCID-X1; see section on 'Gene Therapy of Hematopoietic Stem Cells'). In Paris, and later in London, at least 5 patients treated by *ex vivo* gene transfer into hematopoietic stem cells with gammaretroviral vectors coding for the normal common γ-chain developed T-cell leukemia a few years after treatment. Development of uncontrolled cell proliferation was, at least in part, due to the insertion of the provirus in correspondence with the cellular LMO2 oncogene, leading to its transcriptional activation. These results dramatically brought to general attention the very well known problem of insertional mutagenesis due to gammaretroviral vectors: the insertion of the proviral DNA form of these vectors into the host cell genome occurs in close proximity to the gene promoters, thus potentially causing their activation. If the gene is a protooncogene, its inappropriate expression might continuously trigger cell proliferation, eventually leading to the accumulation of additional genetic defects contributing to full neoplastic transformation.

Molecular studies in several cultured or primary cell types later showed that a number of cellular genes become transcriptionally activated upon transduction with gammaretroviral vectors, further underlying the mutagenic and oncogenic potential of these vectors. It is still unclear why the vast majority of other clinical trials based on gene transfer into hematopoietic stem cells using these vectors did not experience their leukemogenic potential. The unfortunate case of SCID-X1 is most likely the consequence of the cooperative oncogenic effect of the unregulated expression of the common γ-chain and the activated LMO2 oncogene, since the simultaneous presence of high levels of both proteins might result in continuous cell proliferation and render the probability of additional mutagenic events more likely. In other words, the same reason that gene therapy was successful in these patients, namely the proliferative advantage of the transduced clones, was also the reason leukemia eventually occurred in some of the treated patients.

The tumorigenic potential of gammaretroviruses would instinctively suggest a halt to all trials based on the use of these vectors. However, this conclusion needs to be opportunely balanced with proper evaluation of the benefits that may, nonetheless, be derived. For example, in the case of SCID-X1, no therapeutic alternative to allogenic bone marrow transplantation exits. Furthermore, most of the patients treated with gene therapy were effectively cured of their disease and now live a normal life. Finally, from a scientific point of view, the results of these trials provide a definitive indication that gene therapy is effective for this disease and that, quite simply, improved formulation for gene delivery needs to be identified, possibly based on the use of SIN gammaretroviral or lentiviral vectors, or on the expression of the therapeutic gene from regulated promoters.

5.2
Gene Therapy of Germ Cells

Any permanent gene therapy modification of somatic cells will eventually vanish with the death of the patient; in contrast, should this modification occur in the patient's germ cells (i.e., the spermatozoa and egg progenitor cells), this will be transferred to future generations. Over the last several years, the ethical acceptability of such a possibility has been vastly debated.

Strong arguments exist both in favor of and against the ethical acceptability of germ cell gene transfer. Among the positive arguments, it should be considered that it is a moral obligation to provide the best available treatment to any patient. Should a gene therapy-based approach cure an otherwise lethal or invalidating hereditary disorder, the treated patient will have the right to reproduce. However, should the correction of the genetic defect occur in germ cells, the progeny will be affected by the disorder, thus rendering necessary an additional gene therapy intervention. In contrast, gene therapy of germ cells would cure the defect in a permanent manner.

An evolutionary argument against gene therapy of germ cells is the observation that mutated alleles are normally counterselected by evolution, while, thanks to gene therapy, they would be artificially maintained in the population, although compensated for by the presence of the therapeutic gene added by gene transfer. Even more intriguing is the observation that the border between a frank pathologic condition and an unfavorable physical or mental characteristic is often very indistinct (for example, dwarfism compared to short stature). Thus, gene therapy intervention aimed at curing a pathologic characteristic might well extend to improving the "normal" somatic characteristics of the population. Should the possibility to modify height, strength, intelligence or memory by gene therapy be legally allowed, this would greatly interfere with the normal variability that is the basis for evolution and, once adopted in medical practice, would be extremely difficult to control. Finally, a pragmatic additional argument against germ cell gene therapy is that current gene transfer technologies are primitive and highly inaccurate and the possibility of causing unpredictable effects by the random insertion of a gene into the genome is unacceptable at this time.

In light of these considerations, a vast consensus exists that gene therapy, at least at present, must be limited to somatic cells and not be applied to germ cells.

5.3
In Utero Gene Therapy

One relevant therapeutic option for some hereditary conditions is fetal gene therapy, or *in utero* gene therapy. This kind of treatment might represent a suitable therapeutic possibility for pregnant women in whose fetus a hereditary disorder has been diagnosed and who do not wish to undergo therapeutic abortion.

In comparison to post-natal gene therapy, there are several potential benefits of early *in utero* treatment. Correction of a genetic defect during development allows the fetus (i) to avoid the early manifestations of the genetic defect, which can influence development (malformations) or brain function (mental retardation) and can thus be irreversible at birth; (ii) to obtain the permanent correction of all, or at least most, cells of an organ, since the therapeutic gene might be transferred in the organ progenitor cells during development; (iii) to avoid the problem of the immune response against the transgene, since the organism will consider the therapeutic factor as a self protein, this being expressed before the development of the immune system.

Advanced medical technologies for diagnosis and early treatment during pregnancy now allow the injection of therapeutic nucleic acids, under echographic guidance, directly into the fetus during intrauterine life. Over the last few years, proof-of-feasibility of such gene therapy procedures has been obtained in several small animal models of human disease, including those of Crigler-Najjar disease (the Gunn rat, with a defect in UDP-glucuronosyltransferase enzyme), Leber's congenital amaurosis (the RPE65 knock-out mouse), Pompe's disease (α-glucosidase knock-out mouse), and hemophilia B (Factor IX knock-out mouse), in addition to fetuses of larger animals, mostly sheep. Since *in utero* gene therapy requires a stable cell modification, gammaretroviral or lentiviral vectors have been used in most of these experiments.

In utero gene therapy is, however, not without risks. These include the possibility to cause developmental abnormalities or increase the probability of cell transformation, especially when retroviral vectors are used. In addition, the chance of gene delivery into germ cells is much higher during *in utero* gene therapy compared to adult treatment. In this respect, however, the risk should not be overestimated, since, at the gestational time when a vector can be systemically administered to the fetus, the germ cells are already completely compartmentalized in their definitive organs (testis and ovary).

5.4
Gene Therapy of the Embryo

Even more controversial is the possibility to exploit gene transfer technologies for the genetic manipulation of embryos or embryo-derived cells. Indeed, these procedures are more pertinent to the field dealing with the generation of genetically modified organisms, rather than to gene therapy. In several animal species, it is possible to stably modify the genetic information by the injection of DNA into the pronucleus of a fertilized egg (transgenesis) or by gene transfer into embryonic stem cells (ES). ES cells are stable cell lines of totipotent cells, originally generated in the 1980s in the mouse by culturing cells from the inner mass of a blastocyst. ES cells can be manipulated *in vitro*, for example by transfecting a gene of interest or a mutant construct recombining with an endogenous gene to knock this down, followed by *in vitro* selection of a clone with the desired properties. Once the modified ES cells are introduced into a new blastocyst, a chimeric animal is generated, in which several organs are derived from the modified ES cells; should these also contribute to the formation of the germ cells, the genetic modification is transmitted permanently to the progeny. This technology has been extensively used over the last several years in biomedical research for the generation of genetically modified organisms, in which a gene is added, knocked out, or modified. At the end of the 1990s, it was originally demonstrated that ES cells can also be obtained from human blastocysts, and there is no reason to doubt that these cells have the potential to contribute to the generation of a new individual, similar to the mouse.

The debate on the acceptability of genetic research on human ES cells was further ignited in recent years by the observation that cells with properties virtually identical to those of ES cells can also be generated by cell cloning or gene transfection. Cell cloning, which originally led to the generation of Dolly the sheep in the late 1990s, is based on the transfer of a somatic cell nucleus into an oocyte in which the endogenous nucleus has been previously removed. By still, largely, incompletely understood molecular mechanisms entailing the reset of epigenetic information, the transferred nucleus starts a full developmental program of a true ES cell. In a conceptually similar manner, it was recently shown that reprogramming an adult somatic cell nucleus to generate an ES cell can also be obtained by transferring the genes coding for four transcription factors (c-myc, Oct4, Klf4, and Sox2); the cells generated by this procedure are known as induced pluripotent stem (iPS) cells.

Blastocyst-derived ES cells, ES cells obtained by nuclear transfer, and iPS cells can all potentially become recipients of therapeutic genes, for example to correct a hereditary defect. These cells can be cultured and expanded in the laboratory, and thus offer the additional possibility to select for specific clones displaying a desired phenotype. This property is of great interest for gene transfer, for example for the selection of clones in which homologous recombination has erased an inherited mutation, or in which a retroviral vector has become integrated into a specific region of the genome.

All these procedures are of paramount interest to advance our knowledge of the molecular and cellular mechanisms of development and differentiation, as well as to generate genetically modified animals useful for biomedical research. However, conceived as a tool for genetic modification of human cells, they inevitably elicit a vibrant moral, sociological, political, and religious debate. Without entering in depth into these themes here, it is certainly worth remembering that the possibility of exploiting ES cell manipulation still appears too rudimentary and fraught with technical problems to be reasonably considered for therapeutic purposes in humans at this time. Should the ultimate purpose be the prevention of a hereditary disease, a much more realistic alternative to genetic manipulation of ES cells is preimplantation diagnosis in the embryo, by which, in the context of *in vitro* fertilization (IVF) and embryo transfer procedures, single cells from a few-cell embryo can be analyzed for the presence of a given mutation, followed by implantation of the embryo only if this has not been detected.

5.5
Gene Transfer for Cosmetic Appearance and Gene Doping

Although conceived with the essential aim of addressing the cure of otherwise untreatable diseases, over the years gene therapy has developed a series of technologies that, in principle, could also be used to improve appearance or performance, for example by increasing muscular mass, memory, learning, body weight, height, resistance to stress, or prevention of hair loss. Western society confers great importance on appearance and performance, and commonly accepts a series of medical or para-medical practices not directly addressing a medical need. Among these, the use of type A toxin from *Clostridium botulinum* now represents the most common aesthetic treatment in the United States and Europe; several drugs, together with hair follicle transplantation, are extensively applied to combat hair loss; finally, a mounting number of people undergo plastic surgery to improve their appearance or reduce body weight. Therefore, an obvious question is why should we not also use gene transfer for cosmetic purposes?

Table 5.1 reports a list of genes that can be considered to improve aesthetic appearance or physical and intellectual performance.

As far as **body weight** is concerned, the balance between factors increasing the sense of hunger and those suppressing it is essential to determine the extent of food intake. Leptin, a hormone produced by the adipose tissue and the hypothalamus, plays a fundamental role in this control, by inhibiting the action of ghrelin, a hormone produced by the stomach and the hypothalamus to stimulate hunger. One mediator of the effects of leptin is the melanocortin-derived pre-hormone, known as pro-opiomelanocortin (POMC). In experimental animals, gene transfer in the liquor of the leptin or POMC genes using AAV vectors inhibits fat accumulation and food intake.

In mice, fur loss consequent to chemotherapy is inhibited by gene transfer of Sonic hedgehog (Shh), a morphogen that, during development, is essential in controlling fur growth. The same gene could be considered to prevent **hair loss** or stimulate hair re-growth in humans.

When **cognitive capacities** are considered, these could also be targeted by increasing expression of various genes. For example, in the rat, the intracerebral administration of a gene coding for a chimeric estrogen-progesterone receptor, using a herpesviral vector, was shown to inhibit the negative effect of glucocorticoids on memory and learning – these hormones are secreted from the adrenal glands in response to stress.

The possibility to exploit gene therapy to improve **athletic performance** deserves a more detailed description, since it currently represents one of the major concerns of the World Anti-Doping Agency (WADA), the international organization based in Montreal (Canada) that coordinates anti-doping activity in sports. Known as *gene doping*, since 2003 the use of gene and cell transfer to improve performance has been included in the list of medicines and practices forbidden in sport.

One of the genes considered in gene doping is the one coding for erythropoietin (EPO), an acid glycoprotein normally secreted by the kidney and promoting erythrocyte

Table 5.1 Possible gene therapy applications to improve physical or intellectual performance or cosmetic appearance. The examples are taken from current studies in animal models

Aim	Gene	Route of administration	Mechanism of action
Improvement of muscle performance	Erythropoietin (EPO)	Intramuscular, subcutaneous, engineered cells	Increase of hematocrit
	IGF-1	Intramuscular	Induction of muscle hypertrophy, increase of muscle strength
	Myostatin inhibitors (siRNAs, dominant negative mutants)	Intramuscular, systemic	Increase of muscle hypertrophy
	VEGF	Intramuscular	Increase of muscle vascularization
	Growth hormone (GH)	Intramuscular, subcutaneous	Increase of muscle trophism
	PPAR-δ	Intramuscular	Increase of muscle resistance to prolonged exercise
Improvement of intellectual performance	Chimeric estrogen or glucocorticoid receptor	Intracranial, intracerebroventricular	Improvement of spatial memory and intellectual performance
	Constitutively active protein kinase C (PKC)	Intracranial, intracerebroventricular	Increase of learning ability
Improvement of cosmetic appearance	Leptin	Intracerebroventricular	Decrease of body weight
	Propriomelanocortin (POMC)	Intracranial	Decrease of body weight, improvement of glucose and lipid metabolism
	Sonic hedgehog (Shh)	Subcutaneous	Increase of hair (fur) growth, prevention of hair loss

differentiation and stimulating production of hemoglobin. The administration of recombinant EPO increases the hematocrit of an athlete and thus, by improving blood supply to the muscles, illegally improves athletic performance in endurance sports (typically, cycling and Nordic skiing). A striking example of the effects exerted by EPO stimulation on athletic performance is provided by the Finn, Eero Mäntyranta, a Nordic skier who won two gold medals at the 1964 Olympic games and was among the best skiers ever. The genome of this athlete and of several of his relatives had a rare mutation in the EPO receptors, which rendered this receptor particularly responsive to its ligand, with a consequent para-physiological increase of red blood cells.

Another obvious objective of gene doping is to increase muscle mass and strength. One of the known proteins exerting such activities is insulin-like growth factor-1 (IGF-1), a growth factor able to determine important muscle fiber hypertrophy. The effects of IGF-1 are exemplified by the phenotype of the transgenic mouse overexpressing this protein, which shows remarkable muscle hypertrophy and is thus known as the "Schwarzenegger mouse", named after Arnold Schwarzenegger, the former body-builder/actor, now governor of California. Several of the effects of IGF-1 on muscle function and trophism are also exerted by the growth hormone (GH), which can also be used for the same purpose.

Other possible genetic treatments for gene doping target myostatin, a member of the TGF-β family of growth factors that acts as a muscle hypertrophy inhibitor. Animals naturally deficient in this protein, such as Belgian blue cattle, show an impressively hypertrophic muscular mass. The expression of myostatin can be inhibited using siRNAs targeting the protein mRNA or expressing a truncated form of the factor with a transdominant negative activity, or using an antibody against the protein.

Finally, an additional group of genes known to improve muscle performance belong to the peroxisome-proliferator-activated receptor (PPAR) family. In particular, PPAR-δ mediates an adaptive response of muscle to environmental stimuli, by stimulating promoting the generation of slow-twitch, Type I/IIa muscle fibers, which have a mainly oxidative metabolism and are resistant to prolonged work. The presence of this type of fiber is essential in endurance athletes, for example marathon runners, cyclists, or long-distance swimmers. Not surprisingly, the mouse that is transgenic for PPAR-δ is extremely resistant to prolonged fatigue, albeit in the absence of significant increase of muscular mass, and has thus been named the "marathon mouse". This mouse has enough physical endurance to run twice as far as normal mice and shows an innate resistance to weight gain, even when fed a high-fat diet that causes normal mice to become obese.

Besides fraud in competitive sports, the health risk of gene doping is enormous. This is mainly due to the cardiovascular effects of most doping genes and to the intrinsic risk of permanent gene transfer, especially when using AAV vectors for gene delivery, since transgene expression from these vectors – which appear the most indicated for muscle gene transfer – is prolonged or permanent.

To identify an athlete who has been administered gene doping is not simple: siRNAs are short-lived and doping genes are transcribed and translated by the muscle cells, thus generating proteins identical to the endogenous ones, except when the natural protein is produced by a different organ (for example, endogenous EPO has a different post-translational glycosylation pattern from EPO produced by muscle fibers). For this reason, WADA currently collaborates with several research centers worldwide to develop techniques suitable for the identification of gene doping signatures in blood or urine, based on the recognition of metabolic or proteomic modifications induced by transgene overexpression, or on the detection of traces of transgene DNA released by muscle cells undergoing lysis upon intense physical exercise.

The boundary between therapeutic and non-therapeutic use of gene therapy cannot always be defined in a precise manner. Several of the genes considered for cosmetic or gene doping purposes have a therapeutic indication for specific diseases. For example, recombinant GH is used to treat growth retardation and EPO is used to treat anemia in patients with chronic kidney failure, thalassemia, or after treatment with antineoplastic chemotherapy. The myostatin inhibitors, in the form of siRNAs or peptides binding the myostatin receptor and blocking its function, could find application in the treatment of muscle dystrophies. Finally, transfer of the Ssh gene in hair follicle cells was proposed as a tool to combat hair loss following chemotherapy, clearly an ethically acceptable suggestion.

In conclusion, it is not always easy to establish a correct boundary between the therapeutic applications of gene therapy and those that are unacceptable on ethical grounds. Considering the rapid pace of advancement of the identification of genes controlling our physical, temperamental, and intellectual characteristics, it is likely that these problems will become more and more compelling in the near future. In this respect, a particularly worrisome field is the study of the genetic and molecular mechanisms of aging: once these mechanisms are identified, it appears inevitable that pharmacological or genetic treatments will be developed, aimed at increasing the life span of individuals.

Further Reading

Caplan AL (2008) If it's broken, shouldn't it be fixed? Informed consent and initial clinical trials of gene therapy. Hum Gene Ther 19:5–6

Chan S, Harris J (2006) Cognitive regeneration or enhancement: the ethical issues. Regen Med 1:361–366

Chan S, Harris J (2006) The ethics of gene therapy. Curr Opin Mol Ther 8:377–383

Coutelle C, Themis M, Waddington SN et al (2005) Gene therapy progress and prospects: fetal gene therapy – first proofs of concept – some adverse effects. Gene Ther 12:1601–1607

Deakin CT, Alexander IE, Kerridge I (2009) Accepting risk in clinical research: is the gene therapy field becoming too risk-averse? Mol Ther 17:1842–1848

Harris J, Chan S (2008) Enhancement is good for you!: understanding the ethics of genetic enhancement. Gene Ther 15:338–339

Kahn J (2008) Informed consent in human gene transfer clinical trials. Hum Gene Ther 19:7–8

Kiuru M, Crystal RG (2008) Progress and prospects: gene therapy for performance and appearance enhancement. Gene Ther 15:329–337

Spink J, Geddes D (2004) Gene therapy progress and prospects: bringing gene therapy into medical practice: the evolution of international ethics and the regulatory environment. Gene Ther 11:1611–1616

Wells DJ (2008) Gene doping: the hype and the reality. Br J Pharmacol 154:623–631

Selected Bibliography

Batshaw ML, Wilson JM, Raper S et al (1999) Recombinant adenovirus gene transfer in adults with partial ornithine transcarbamylase deficiency (OTCD). Hum Gene Ther 10:2419–2437

Frank KM, Hogarth DK, Miller JL et al (2009) Investigation of the cause of death in a gene-therapy trial. N Engl J Med 361:161–169

Grimm D, Streetz KL, Jopling CL et al (2006) Fatality in mice due to oversaturation of cellular microRNA/short hairpin RNA pathways. Nature 441:537–541

Kaiser J (2007) Clinical research. Death prompts a review of gene therapy vector. Science 317:580

Reay DP, Bilbao R, Koppanati BM et al (2008) Full-length dystrophin gene transfer to the mdx mouse in utero. Gene Ther 15:531–536

Subject Index

Printed in March 2010